AF841851

VISCOELASTICITY – FROM THEORY TO BIOLOGICAL APPLICATIONS

Edited by **Juan de Vicente**

Viscoelasticity – From Theory to Biological Applications
http://dx.doi.org/10.5772/3188
Edited by Juan de Vicente

Contributors

L. A. Dávalos-Orozco, Takahiro Tsukahara, Yasuo Kawaguchi, B.N. Narahari Achar, John W. Hanneken, Kejian Wang, Naoki Sasaki, Supriya Bhat, Dong Jun, Biplab C. Paul, Tanya E. S Dahms, Tetsuya Nemoto, Ryo Kubota, Yusuke Murasawa, Zenzo Isogai, Jun Xi, Lynn S. Penn, Ning Xi, Jennifer Y. Chen, Ruiguo Yang, Tomoki Kitawaki, Ioanna G. Mandala, Luis Carlos Platt-Lucero, Benjamín Ramírez-Wong, Patricia Isabel Torres-Chávez, Ignacio Morales-Rosas, Elisa Magaña-Barajas, Benjamín Ramírez-Wong, Patricia I. Torres-Chávez, I. Morales-Rosas, Youhong Tang, Ping Gao, Takaya Kobayashi, Masami Sato, Yasuko Mihara, B.S. K. K. Ibrahim, M.S. Huq, M.O. Tokhi, S.C. Gharooni, Hayssam El Ghoche

Published by InTech

Janeza Trdine 9, 51000 Rijeka, Croatia

Copyright © 2012 InTech

All chapters are distributed under the Creative Commons Attribution 3.0 license, which allows users to download, copy and build upon published articles even for commercial purposes, as long as the author and publisher are properly credited, which ensures maximum dissemination and a wider impact of our publications. However, users who aim to disseminate and distribute copies of this book as a whole must not seek monetary compensation for such service (excluded InTech representatives and agreed collaborations). After this work has been published by InTech, authors have the right to republish it, in whole or part, in any publication of which they are the author, and to make other personal use of the work. Any republication, referencing or personal use of the work must explicitly identify the original source.

Notice

Statements and opinions expressed in the chapters are these of the individual contributors and not necessarily those of the editors or publisher. No responsibility is accepted for the accuracy of information contained in the published chapters. The publisher assumes no responsibility for any damage or injury to persons or property arising out of the use of any materials, instructions, methods or ideas contained in the book.

Publishing Process Manager Marina Jozipovic
Technical Editor InTech Prepress, Novi Sad
Cover InTech Design team

Additional hard copies can be obtained from orders@intechopen.com

Viscoelasticity – From Theory to Biological Applications, Edited by Juan de Vicente
p. cm.
ISBN 978-953-51-0841-2

We are IntechOpen, the world's leading publisher of Open Access books

Built by scientists, for scientists

4,100+
Open access books available

116,000+
International authors and editors

120M+
Downloads

151
Countries delivered to

Our authors are among the
Top 1%
most cited scientists

12.2%
Contributors from top 500 universities

CLARIVATE ANALYTICS BOOK CITATION INDEX INDEXED

WEB OF SCIENCE™

Selection of our books indexed in the Book Citation Index
in Web of Science™ Core Collection (BKCI)

Interested in publishing with us?
Contact book.department@intechopen.com

Numbers displayed above are based on latest data collected.
For more information visit www.intechopen.com

Meet the editor

Juan de Vicente received Ph.D. degrees in Physics from the University of Granada (Spain) and the University of Nice-Sophia Antipolis (France) with European Mention and Extraordinary Doctorate Awards. He has carried out research stays at the LPMC - University of Nice, RRC - University of Wisconsin-Madison, and Vakgroep Reologie - Universiteit Twente, under a FPU Predoctoral Fellowship. He has been Marie Curie Postdoc and Marie Curie ERG Fellow at Unilever Corporate Research and Imperial College London. He is recipient of the "Young Investigator Award" from the Social Council and the "Physics Research Award" from the Academy of Sciences. He is presently Associate Professor of Applied Physics at the University of Granada. His research interests include magnetorheology and soft-elasto-ferrohydrodynamic lubrication.

Contents

Preface

The word "viscoelastic" means the simultaneous existence of viscous and elastic responses of a material. Hence, neither Newton's law (for linear viscous fluids) nor Hooke's law (for pure elastic solids) suffice to explain the mechanical behavior of viscoelastic materials. Strictly speaking all materials are viscoelastic and their particular response depends on the Deborah number, that is to say the ratio between the natural time of the material (relaxation time) and the time scale of the experiment (essay time). Thus, for a given material, if the experiment is slow, the material will appear to be viscous, whereas if the experiment is fast it will appear to be elastic. Many materials exhibit a viscolastic behavior at the observation times and the area is relevant in many fields of study from industrial to technological applications such as concrete technology, geology, polymers and composites, plastics processing, paint flow, hemorheology, cosmetics, adhesives, etc.

In this book, 16 chapters on various viscoelasticity related aspects are compiled. A number of current research projects are outlined as the book is intended to give the readers a wide picture of current research in viscoelasticity balancing between fundamentals and applied knowledge. For this purpose, the chapters are written by experts from the Industry and Academia.

The first part of the book is dedicated to theory and simulation. The first chapter, by Dávalos-Orozco is a review of the theory of linear and nonlinear natural convection of fluid layers between two horizontal walls under an imposed vertical temperature gradient. Chapter 2 by Tsukahara and Kawaguchi deals with the turbulent flow of viscoelastic fluids through complicated geometries such as orifice flows. Next, in chapter 3, Narahari and Hanneken describe a microscopic formulation of fractional theory of viscoelasticity. Finally, in chapter 4, Kejian revisits the die swell problem of viscoelastic polymeric systems.

The second part of the book covers important aspects of viscoelasticity in biological systems. The first chapter by Sasaki highlights the importance of viscoelasticity in the mechanical properties of biological materials. Next, Dahms and coworkers summarize the current techniques used to probe viscoelasticity with special emphasis on the application of Atomic Force Microscopy to microbial cell mechanics. In chapters 7 and 8 Zenzo and Xi and coworkers focus on the viscoelastic properties of human dermis and epithelial cells. Last chapter in this section cover aspects related to the blood flow, where Kitawaki proposes a numerical model for the viscoelasticity of arterial walls.

The third part of the book is devoted to the study of the viscoelastic properties of food colloids. Chapter 10 is an attempt to clarify the relationship between the viscoelastic properties of starches, and their mixtures, and texture in real foods. In chapter 11 Ramirez-Wong and coworkers determine the effect of xantham gum on viscoelastic and textural characteristics of masa and tortilla from extruded nixtamalized corn flour. Finally, in chapter 12, stress-relaxation and dynamic tests are performed to evaluate the viscoelastic properties of dough from soft wheat cultivars.

The last part of the book deals with other miscellaneous applications. Tang and Gao perform a micro-rheological study of fully exfoliated organoclay modified thermotropic liquid crystalline polymers (TLCP). Chapter 14 is an attempt to estimate the thermal deformation in laminated printed circuit boards by the application of a layered plate theory that includes energy transport. In the next chapter, chapter 15, Ibrahim and coworkers describe an approach for the dynamic characterization of passive viscoelasticity of a paraplegic's knee joint. This last section finishes with chapter 16, by Hayssam, and describes a nonlinear viscoelastic model to be applied on compressed plastic films for light-weight embankment.

The format of this book is chosen to enable fast dissemination of new research, and to give easy access to readers. The chapters can be read individually.

I would like to express my gratitude to all the contributing authors that have made a reality this book. I wish to thank also InTech staff and their team members for the opportunity to publish this work, in particular, Ana Pantar, Dimitri Jelovcan, Romana Vukelic and Marina Jozipovic for their support which has made my job as editor an easy and satisfying one.

Finally, I gratefully acknowledge financial support by the Ministerio de Ciencia e Innovación (MICINN MAT 2010-15101 project, Spain), by the European Regional Development Fund (ERDF), and by the projects P10-RNM-6630 and P11-FQM-7074 from Junta de Andalucía (Spain).

Juan de Vicente
University of Granada
Spain

Section 1

Theory and Simulations

Chapter 1

Viscoelastic Natural Convection

L. A. Dávalos-Orozco

Additional information is available at the end of the chapter

http://dx.doi.org/10.5772/49981

1. Introduction

Heat convection occurs in natural and industrial processes due to the presence of temperature gradients which may appear in any direction with respect to the vertical, which is determined by the direction of gravity. In this case, natural convection is the fluid motion that occurs due to the buoyancy of liquid particles when they have a density difference with respect the surrounding fluid. Here, it is of interest the particular problem of natural convection between two horizontal parallel flat walls. This simple geometry brings about the possibility to understand the fundamental physics of convection. The results obtained from the research of this system may be used as basis to understand others which include, for example, a more complex geometry and a more complex fluid internal structure. Even though it is part of our every day life (it is observed in the atmosphere, in the kitchen, etc.), the theoretical description of natural convection was not done before 1916 when Rayleigh [53] made calculations under the approximation of frictionless walls. Jeffreys [27] was the first to calculate the case including friction in the walls. The linear theory can be found in the monograph by Chandrasekhar [7]. It was believed that the patterns (hexagons) observed in the Bénard convection (see Fig. 1, in Chapter 2 of [7] and the references at the end of the chapter) were the same as those of natural convection between two horizontal walls. However, it has been shown theoretically and experimentally that the preferred patterns are different. It was shown for the first time theoretically by Pearson [45] that convection may occur in the absence of gravity assuming thermocapillary effects at the free surface of a liquid layer subjected to a perpendicular temperature gradient. The patterns seen in the experiments done by Bénard in the year 1900, are in fact only the result of thermocapillarity. The reason why gravity effects were not important is that the thickness of the liquid layer was so small in those experiments that the buoyancy effects can be neglected. As will be shown presently, the Rayleigh number, representative of the buoyancy force in natural convection, depends on the forth power of the thickness of the liquid layer and the Marangoni number, representing thermocapillary effects, depends on the second power of the thickness. This was not realized for more than fifty years, even after the publication of the paper by Pearson (as seen in the monograph by Chandrasekhar). Natural convection may present hexagonal patterns only when non

INTECH
open science | open minds

© 2012 The Author(s). Licensee InTech. This chapter is distributed under the terms of the Creative Commons Attribution License http://creativecommons.org/licenses/by/3.0), which permits unrestricted use, distribution, and reproduction in any medium, provided the original work is properly cited.

Boussinesq effects [52] occur, like temperature dependent viscosity [57] which is important when temperature gradients are very large. The Boussinesq approximation strictly assumes that all the physical parameters are constant in the balance of mass, momentum and energy equations, except in the buoyancy term in which the density may change with respect to the temperature. Any change from this assumption is called non Boussinesq approximation. When the thickness of the layer increases, gravity and thermocapillary effects can be included at the same time [40]. This will not be the subject of the present review. Here, the thickness of the layer is assumed large enough so that thermocapillary effects can be neglected.

The effects of non linearity in Newtonian fluids convection were taken into account by Malkus and Veronis [33] and Veronis [65] using the so called weakly non linear approximation, that is, the Rayleigh number is above but near to the critical Rayleigh number. The small difference between them, divided by the critical one, is used as an expansion parameter of the variables. The patterns which may appear in non linear convection were investigated by Segel and Stuart [57] and Stuart [61]. The method presented in these papers is still used in the literature. That is, to make an expansion of the variables in powers of the small parameter, including normal modes (separation of horizontal space variables in complex exponential form) of the solutions of the non linear equations. With this method, an ordinary non linear differential equation (or set of equations), the Landau equation, is obtained for the time dependent evolution of the amplitude of the convection cells. Landau used this equation to explain the transition to turbulent flow [31], but never explained how to calculate it. For a scaled complex A(t), the equation is:

$$\frac{dA}{dt} = rA - |A|^2 A. \tag{1}$$

In some cases, the walls are considered friction free (free-free case, if both walls have no friction). One reason to make this assumption is that the nonlinear problem simplifies considerably. Another one is that the results may have relevance in convection phenomena in planetary and stellar atmospheres. In any way, it is possible that the qualitative results are similar to those of convection between walls with friction, mainly when the interest is on pattern formation. This simplification has also been used in convection of viscoelastic fluids. To describe the nonlinear envelope of the convection cells spatial modulation, it is possible to obtain a non linear partial differential equation by means of the multiple scales approximation [3], as done by Newell and Whitehead [39] and Segel [56]. This equation is called the Newell-Whitehead-Segel (NWS) equation. For a scaled A(X,Y,T), it is:

$$\frac{\partial A}{\partial T} = rA - |A|^2 A + \left(\frac{\partial}{\partial X} - \frac{1}{2}i\frac{\partial^2}{\partial Y^2}\right)^2 A. \tag{2}$$

Here, X, Y and T are the scaled horizontal coordinates and time, respectively. In the absence of space modulation it reduces to the Landau Equation 1. It is used to understand the non linear instability of convection flow. However, it has been found that this equation also appears in the description of many different physical phenomena. The non linear stability of convection rolls depends on the magnitude of the coefficients of the equation. If the possibility of the appearance of square or hexagonal patterns is of interest, then the stability of two coupled or three coupled NWS equations have to be investigated. They are obtained from the coupling of modes having different directions (see [22] and [23]).

The shear stress tensor of Newtonian fluids have a linear constitutive relation with respect to the shear rate tensor. The constitutive equation of that relation has as constant of proportionality the dynamic viscosity of the fluid, that is

$$\tau_{ij} = 2\eta_0 e_{ij}. \tag{3}$$

Here, the shear rate tensor is

$$e_{ij} = \frac{1}{2}\left(\frac{\partial v_i}{\partial x_j} + \frac{\partial v_j}{\partial x_i}\right), \tag{4}$$

Any fluid whose stress tensor has a different constitutive relation, or equation, with respect to the shear rate tensor is called non Newtonian. That relation might have an algebraic or differential form. Here, only natural convection of viscoelastic fluids will be discussed [4, 9] as non Newtonian flows. These fluids are defined by constitutive equations which include complex differential operators. They also include relaxation and retardation times. The physical reason can be explained by the internal structure of the fluids. They can be made of polymer melts or polymeric solutions in some liquids. In a hydrostatic state, the large polymeric chains take the shape of minimum energy. When shear is applied to the melt or solution, the polymeric chains deform with the flow and then they are extended or deformed according to the energy transferred by the shear stress. This also has influence on the applied shear itself and on the shear stress. When the shear stress disappears, the deformed polymeric chains return to take the form of minimum energy, carrying liquid with them. This will take a time to come to an end, which is represented by the so called retardation time. On the other hand, there are cases when shear stresses also take some time to vanish, which is represented by the so called relaxation time. It is possible to find fluids described by constitutive equations with both relaxation and retardation times. The observation of these viscoelastic effects depend on different factors like the percentage of the polymeric solution and the rigidity of the macromolecules.

A simple viscoelastic model is the incompressible second order fluid [10, 16, 34]. Assuming τ_{ij} as the shear-stress tensor, the constitutive equation is:

$$\tau_{ij} = 2\eta_0 e_{ij} + 4\beta e_{ik} e_{kj} + 2\gamma \frac{\mathcal{D} e_{ij}}{\mathcal{D} t}. \tag{5}$$

and

$$\frac{\mathcal{D} P_{ij}}{\mathcal{D} t} = \frac{D P_{ij}}{D t} + P_{ik}\frac{\partial v_k}{\partial x_j} + \frac{\partial v_k}{\partial x_i} P_{kj}, \tag{6}$$

for a tensor P_{ij} and where

$$\frac{D}{Dt} = \frac{\partial}{\partial t} + v_k \frac{\partial}{\partial x_k}, \tag{7}$$

is the Lagrange or material time derivative. The time derivative in Equation 6 is called the lower-convected time derivative, in contrast to the following upper convected time derivative

$$\frac{\mathcal{D} P_{ij}}{\mathcal{D} t} = \frac{D P_{ij}}{D t} - P_{ik}\frac{\partial v_k}{\partial x_j} - P_{jk}\frac{\partial v_k}{\partial x_i}, \tag{8}$$

and to the corrotational time derivative

$$\frac{\mathcal{D}P_{ij}}{\mathcal{D}t} = \frac{DP_{ij}}{Dt} + \omega_{ik}P_{kj} - P_{ik}\omega_{kj}, \tag{9}$$

where the rotation rate tensor is

$$\omega_{ij} = \frac{1}{2}\left(\frac{\partial v_i}{\partial x_j} - \frac{\partial v_j}{\partial x_i}\right). \tag{10}$$

These time derivatives can be written in one formula as

$$\frac{\mathcal{D}P_{ij}}{\mathcal{D}t} = \frac{DP_{ij}}{Dt} + \omega_{ik}P_{kj} - P_{ik}\omega_{kj} - a\left(e_{ik}P_{kj} + P_{ik}e_{kj}\right), \tag{11}$$

where the time derivatives correspond to the upper convected for $a = 1$, the corrotational for $a = 0$ and the lower convected for $a = -1$, respectively [47]. These time derivatives are invariant under a change of reference frame. In Equations 3 and 5 η_0 is the viscosity and in Equation 5 β and γ are material constants. The second order model Equation 5 has limitations in representing fluid motion. It is an approximation for slow motion with small shear rate [4]. Linear and nonlinear convection of second order fluids has been investigated by Dávalos and Manero [12] for solid walls under the fixed heat flux boundary condition. The same fluid has been investigated looking for the possibility of chaotic motion (aperiodic and sensitive to initial conditions [28]) by [58] for the case of free boundaries and fixed temperature boundary condition.

The Maxwell model [4] is used to describe motion where it is possible to have shear stress relaxation. The constitutive equation of this model is:

$$\tau_{ij} + \lambda\frac{\mathcal{D}\tau_{ij}}{\mathcal{D}t} = 2\eta_0 e_{ij}. \tag{12}$$

where λ is the relaxation time. A characteristic of this equation is that for λ small the fluid nearly behaves as Newtonian. For large λ it tends to behave as an elastic solid as can be seen if e_{ij} is considered as the time derivative of the strain. In the limit of very large λ, the approximate equation is integrated in time to get Hook's law, that is, the stress is proportional to the strain. This constitutive equation has three versions, the upper convected, the lower convected and the corrotational Maxwell models, depending on the time derivative selected to describe the fluid behavior. The natural convection of the Maxwell fluid has been investigated by Vest and Arpaci [66] for free-free and solid-solid walls with fixed temperature. Sokolov and Tanner [59] investigated the linear problem of the Maxwell fluid, among other viscoelastic fluids, using an integral form of the stress tensor. The non linear problem has been investigated for free-free boundaries by Van Der Borght et al. [64], using the upper convected time derivative. Brand and Zielinska [5] show that nonlinear traveling waves appear for different Prandtl numbers in a convecting Maxwell fluid with free-free walls. The Prandtl number Pr is the ratio of the kinematic viscosity over the thermal diffusivity. The chaotic behavior of convection of a Maxwell fluid has been investigated by Khayat [29]. The effect of the thickness and thermal conductivity of the walls has been taken into account in the linear convection of a Maxwell fluid by Pérez-Reyes and Dávalos-Orozco [46].

The Oldroyd's fluid model [4, 41] includes, apart from a relaxation time, a retardation time. The linear version of this model is called the Jeffreys model (but the non linear model is sometimes called by this name). The constitutive equation is

$$\tau_{ij} + \lambda \frac{\mathcal{D}\tau_{ij}}{\mathcal{D}t} = 2\eta_0 \left(e_{ij} + \lambda_1 \frac{\mathcal{D}e_{ij}}{\mathcal{D}t} \right). \tag{13}$$

where λ_1 is the retardation time. Notice that when $\lambda_1 = 0$ this contitutive equation reduces to that of a Maxwell fluid. Therefore, a number of papers which investigate the convection in Oldroyd fluids also include results of the Maxwell fluid. When $\lambda = 0$, the equation reduces to that of the second-order fluid with a zero coefficient γ. Linear convection of Oldroyd fluids has been investigated by Green [21], Takashima [62], Kolkka and Ierley [30], Martínez-Mardones and Pérez-García [35] and Dávalos-Orozco and Vázquez-Luis [14] for free upper surface deformation. Nonlinear calculations of the Oldroyd fluid where done first by Rosenblat [55] for free-free boundaries. The non linear problem of solid-solid and solid-free boundaries was investigated by Park and Lee [43, 44]. Nonlinear problems were investigated by Martínez-Mardones et al. for oscillatory and stationary convection [36], to study the stability of patterns in convection [37] and to investigate the convective and absolute instabilities by means of amplitude equations [38].

The following section presents the balance equations suitable for natural convection. Section 3 is an introduction to Newtonian fluids convection. The Sections 4, 5, and 6 correspond to reviews of convection of second-order, Maxwell and Oldroyd fluids, respectively. Finally, some conclusions are given in the last Section 7.

2. Equations of balance of momentum, mass and energy

Here, the basic equations of balance of momentum, mass and energy for an incompressible fluid are presented. In vector form, they are

$$\rho \left[\frac{\partial \mathbf{u}^*}{\partial t^*} + (\mathbf{u}^* \cdot \nabla^*) \, \mathbf{u}^* \right] = -\nabla^* p^* + \nabla^* \cdot \boldsymbol{\tau}^* + \rho \mathbf{g} \tag{14}$$

$$\nabla^* \cdot \mathbf{u}^* = 0 \tag{15}$$

$$\rho C_V \left[\frac{\partial T^*}{\partial t^*} + (\mathbf{u}^* \cdot \nabla^*) \, T^* \right] = X_F \nabla^{*2} T^* \tag{16}$$

The dimensional variables are defined as follows. ρ is the density, $\mathbf{u}^* = (u^*, v^*, w^*)$ is the velocity vector, p^* is the pressure, $\boldsymbol{\tau}^*$ is the stress tensor which satisfies one of the constitutive equations presented above. T^* is the temperature, C_V is the specific heat at constant volume and X_F is the heat conductivity of the fluid. Use is made of $\nabla^* = (\partial/\partial x^*, \partial/\partial y^*, \partial/\partial z^*)$. $\mathbf{g} = (0, 0, -g) = -g\hat{k}$ is the vector of the acceleration of gravity with g its magnitude and $\hat{k}$ a unit vector in the direction opposite to gravity. Equation 15 means that the fluid is incompressible and that any geometric change of a fluid element volume in one direction is reflected in the other the directions in such a way that the volume is preserved according to this equation.

If the thickness and conductivity of the walls are taken into account, the temperature in each wall satisfies the equation

$$\rho_W C_{VW} \frac{\partial T_W^*}{\partial t^*} = X_W \nabla^{*2} T_W^* \tag{17}$$

where T_W is the temperature of one of the walls (T_L for the lower wall and T_U for the upper wall). ρ_W, C_{VW} and X_W are the density, specific heat at constant volume and heat conductivity of one of the walls (ρ_L, C_{VL}, X_{LW} for the lower wall and ρ_U, C_{VU}, X_{UW} for the upper wall). The variables are subjected to boundary conditions. The velocity has two types of conditions: for friction free walls and for solid walls with friction. They are, respectively:

$$\mathbf{n} \cdot \mathbf{u}^* = 0 \quad and \quad \mathbf{n} \cdot \nabla^* u^* = \mathbf{n} \cdot \nabla^* v^* = 0 \quad at \quad z^* = z_1 \; and \; z^* = z_1 + d \; free \; boundary \tag{18}$$

$$\mathbf{u}^* = 0 \quad at \quad z^* = z_1^* \quad and \quad z^* = z_1^* + d \quad solid \quad boundary$$

where $\mathbf{n}$ is the unit normal vector to one of the walls, z_1 is a particular position of the lower wall in the z-axis and d is the thickness of the fluid layer. The conditions in the first line of Equation 18 mean that the fluid can not penetrate the wall and that the wall does not present any shear due to the absence of friction. The condition of the second line means that the fluid sticks to the wall due to friction.

The temperature satisfies the boundary conditions of fixed temperature and fixed heat at the walls, respectively,

$$T^* = T_0 \quad at \quad z^* = z_1^* \quad and \quad z^* = z_1^* + d \quad fixed \quad temperature \tag{19}$$

$$\mathbf{n} \cdot \nabla^* T^* = \frac{q_0}{X_F} \quad at \quad z^* = z_1^* \quad and \quad z^* = z_1^* + d \quad fixed \quad heat \quad flux$$

where q_0 is a constant heat flux normal and through one of the walls.

If the thickness and heat conductivity of the walls are taken into account, the temperature has to satisfy the conditions

$$\begin{aligned} T_L^* &= T_{BL} \quad \text{at} \quad z^* = z_1^* - d_L \\ T_U^* &= T_{AU} \quad \text{at} \quad z^* = z_1^* + d + d_U \\ T_L^* = T^*, \quad X_L \mathbf{n_L} \cdot \nabla^* T^* &= \mathbf{n_L} \cdot \nabla^* T_L^* \quad \text{at} \quad z^* = z_1^* \\ T_U^* = T^*, \quad X_U \mathbf{n_U} \cdot \nabla^* T^* &= \mathbf{n_U} \cdot \nabla^* T_L^* \quad \text{at} \quad z^* = z_1^* + d \end{aligned} \tag{20}$$

where $X_L = X_{LW}/X_F$ and $X_U = X_{UW}/X_F$. T_{BL} and T_{AU} are the temperatures below the lower wall and above the upper wall. d_L and d_U are the thicknesses of the lower and the upper walls, respectively. The normal unit vectors to the upper and lower walls are $\mathbf{n_U}$ and $\mathbf{n_L}$. The two conditions in the third and forth lines of Equation 20 mean the continuity of temperature and the continuity of the heat flux between the fluid and each wall, respectively.

The equations and boundary conditions can be made non dimensional by means of representative magnitudes for each of the dependent and independent variables. For example, the distance is scaled by the thickness of the fluid layer d or a fraction of it, the time is scaled with d^2/κ, where the thermal diffusivity is $\kappa = X_F/\rho_0 C_V$, the velocity with κ/d, the pressure and the stress tensor with $\rho_0 \kappa^2/d^2$. ρ_0 is a representative density of the fluid. The temperature is made non dimensional with a characteristic temperature difference or with a quantity proportional to a temperature difference. The time can also be scaled with d^2/ν, where the kinematic viscosity is $\nu = \eta_0/\rho_0$ and the velocity with ν/d. Then, the pressure and the stress tensor can be scaled in two ways, by means of $\rho_0\kappa\nu/d^2$ or by $\rho_0\nu^2/d^2$. The difference stems on the importance given to the mass diffusion time d^2/ν or to the heat diffusion time d^2/κ.

It is assumed that before a perturbation is applied to the Equations 14 to 16 the system is in a hydrostatic state and that the variables satisfy

$$0 = -\nabla^* P_0^* + \rho_0 \left[1 - \alpha_T \left(T_0^* - T_R^*\right)\right] \mathbf{g} \tag{21}$$

$$\nabla^{*2} T_0^* = 0 \tag{22}$$

The solution of these two equations will be the main pressure P_0 and the main temperature profile T_0^* of the system before perturbation. Here, ρ_0 is a reference density at the reference temperature T_R^* which depends on the boundary conditions. α_T is the coefficient of volumetric thermal expansion of the fluid. These two solutions of Equations 21 and 22 are subtracted from Equations 14 to 16 after introducing a perturbation on the system. In non dimensional form, the equations of the perturbation are

$$\frac{1}{Pr}\left(\frac{\partial \mathbf{u}}{\partial t} + (\mathbf{u}\cdot\nabla)\,\mathbf{u}\right) = -\nabla p + \nabla\cdot\boldsymbol{\tau} + \mathbf{R}\theta\hat{k} \tag{23}$$

$$\nabla\cdot\mathbf{u} = 0 \tag{24}$$

$$\frac{\partial\theta}{\partial t} + (\mathbf{u}\cdot\nabla)\,\theta - \mathbf{u}\cdot\hat{k} = \nabla^2\theta \tag{25}$$

$$\frac{\kappa}{\kappa_W}\frac{\partial\theta_W}{\partial t} = \nabla^2\theta_W \tag{26}$$

The non dimensionalization was based on the heat diffusion time and the scaling of the pressure and shear stress with $\rho_0\kappa\nu/d^2$. $\mathbf{u}$, p, τ, θ and θ_W are the perturbations of velocity, pressure, shear stress, fluid temperature and walls temperature (θ_L and θ_U for the lower and upper walls), respectively. $\mathbf{R} = g\alpha_T\Delta T d^3/\nu\kappa$ is the Rayleigh number and $Pr = \nu/\kappa$ is the Prandtl number. ΔT is a representative temperature difference. κ_W is the thermal diffusivity of one of the walls (κ_L for the lower wall and κ_U for the upper wall).

The last term in the left hand side of Equation 25 appears due to the use of the linear temperature solution of Equation 22. If the temperature only depends on z^* in the form $T_0^* = a_1 z^* + b_1$, this solution is introduced in a term like $\mathbf{u}^*\cdot\nabla^* T_0^*$. Here, a_1 is a constant which is proportional to a temperature difference or an equivalent if the heat flux is used. In these equations, the Boussinesq approximation has been taken into account, that is, in Equations 14 to 16 the density was assumed constant and equal to ρ_0 everywhere except in the term $\rho\mathbf{g}$ where it changes with temperature. The other parameters of the fluid and wall are also assumed as constant. These conditions are satisfied when the temperature gradients are small enough.

The constitutive Equations 3, 5, 12, 13 are perturbed and also have to be made non dimensional. For the perturbation shear stress tensor τ_{ij} and shear rate tensor e_{ij}, they are

$$\tau_{ij} = 2e_{ij}. \tag{27}$$

$$\tau_{ij} = 2e_{ij} + 4Be_{ik}e_{kj} + 2\Gamma \frac{\mathcal{D}e_{ij}}{\mathcal{D}t}. \tag{28}$$

$$\tau_{ij} + L\frac{\mathcal{D}\tau_{ij}}{\mathcal{D}t} = 2e_{ij}. \tag{29}$$

$$\tau_{ij} + L\frac{\mathcal{D}\tau_{ij}}{\mathcal{D}t} = 2\left(e_{ij} + LE\frac{\mathcal{D}e_{ij}}{\mathcal{D}t}\right). \tag{30}$$

where $B = \beta\kappa/\rho\nu d$, $\Gamma = \gamma\kappa/\rho\nu d^2$, $L = \lambda\kappa/d^2$ and $E = \lambda_1/\lambda < 1$.

The boundary conditions of the perturbations in non dimensional form are

$$\mathbf{n}\cdot\mathbf{u} = 0 \quad \textit{and} \quad \mathbf{n}\cdot\nabla u = \mathbf{n}\cdot\nabla v = 0 \quad \textit{at} \quad z = z_1 \quad \textit{and} \quad z = z_1 + 1 \quad \textit{free} \quad \textit{boundary} \tag{31}$$

$$\begin{aligned} \mathbf{u} = 0 \quad &\textit{at} \quad z = z_1 \quad \textit{and} \quad z = z_1 + 1 \quad \textit{solid} \quad \textit{boundary} \\ \theta = 0 \quad &\textit{at} \quad z = z_1 \quad \textit{and} \quad z = z_1 + 1 \quad \textit{fixed} \quad \textit{temperature} \\ \mathbf{n}\cdot\nabla\theta = 0 \quad &\textit{at} \quad z = z_1 \quad \textit{and} \quad z = z_1 + 1 \quad \textit{fixed} \quad \textit{heat} \quad \textit{flux} \end{aligned} \tag{32}$$

$$\begin{aligned} \theta_L = 0 \quad &\text{at} \quad z = z_1 - D_L \\ \theta_U = 0 \quad &\text{at} \quad z = z_1 + 1 + D_U \\ \theta_L = \theta, \quad X_L\mathbf{n_L}\cdot\nabla\theta = \mathbf{n_L}\cdot\nabla\theta_L \quad &\text{at} \quad z = z_1 \\ \theta_U = \theta, \quad X_U\mathbf{n_U}\cdot\nabla\theta = \mathbf{n_U}\cdot\nabla\theta_U \quad &\text{at} \quad z = z_1 + 1 \end{aligned} \tag{33}$$

Here, D_L and D_U are the ratios of the thickness of the lower and upper walls over the fluid layer thickness, respectively. The meaning of the conditions Equation 32 is that the original temperature and heat flux at the boundary remain the same when $\theta = 0$ and $\mathbf{n}\cdot\nabla\theta = 0$. The same can be said from the first two Equations 33, that is, the temperature below the lower wall and the temperature above the upper wall stay the same after applying the perturbation.

3. Natural convection in newtonian fluids

The basics of natural convection of a Newtonian fluid are presented in this section in order to understand how other problems can be solved when including oscillatory and non linear flow. The section starts with the linear problem and later discuss results related with the non linear equations. The system is a fluid layer located between two horizontal and parallel plane walls heated from below or cooled from above. Gravity is in the z-direction. As seen from Equations 23 to 25, the linear equations are

$$\frac{1}{Pr}\frac{\partial\mathbf{u}}{\partial t} = -\nabla p + \nabla^2\mathbf{u} + \mathbf{R}\theta\hat{k} \tag{34}$$

$$\nabla\cdot\mathbf{u} = 0 \tag{35}$$

$$\frac{\partial\theta}{\partial t} - w = \nabla^2\theta \tag{36}$$

In Equation 34 use has been made of Equations 3, 4 and 35. The first boundary conditions used will be those of free-free and fixed temperature at the wall [7]. These are the simplest conditions which show the qualitative behavior of convection in more complex situations.

To eliminate the pressure from the equation it is necessary to apply once the curl operator to Equation 34. This is the equation of the vorticity and its vertical component is independent from the other components of the vorticity vector. Applying the curl one more time, it is possible to obtain an equation for the vertical component of the velocity independent of the other components. The last and the first equations are

$$\frac{1}{Pr}\frac{\partial \nabla^2 w}{\partial t} = \nabla^4 w + R\nabla_{\perp}^2 \theta \tag{37}$$

$$\frac{\partial \zeta_Z}{\partial t} = \nabla^2 \zeta_Z \tag{38}$$

Here, $\nabla_{\perp}^2 = (\partial^2/\partial^2 x, \partial^2/\partial^2 y)$ is the horizontal Laplacian which appears due to the unit vector $\hat{k}$ in Equation 34. The third component of vorticity is defined by $\zeta_Z = [\nabla \times \mathbf{u}]_Z$. The three components of ζ are related with the three different elements of the rotation tensor Equation 10 multiplied by 2.

These Equations 37, 38 and 36 are partial differential and their variables can be separated in the form of the so called normal modes

$$f(x,y,z,t) = F(z)exp\left(ik_x x + ik_y y + \sigma t\right) \tag{39}$$

$F(z)$ is the amplitude of the dependent variable. The wavevector is defined by $\vec{k} = (k_x, k_y)$, k_x and k_y are its x and y-components and its magnitude is $\left|\vec{k}\right| = k$. When the flow is time dependent, $\sigma = s + i\omega$ where s is the growth rate and ω is the frequency of oscillation. Then, using normal modes and assuming that the system is in a neutral state where the growth rate is zero ($s = 0$), the equations are

$$\frac{i\omega}{Pr}\left(\frac{d^2}{dz^2} - k^2\right)W = \left(\frac{d^2}{dz^2} - k^2\right)^2 W - Rk^2\Theta \tag{40}$$

$$i\omega Z = \left(\frac{d^2}{dz^2} - k^2\right)Z \tag{41}$$

$$i\omega\Theta - W = \left(\frac{d^2}{dz^2} - k^2\right)\Theta \tag{42}$$

$$\frac{dW}{dz} + ik_x U + ik_y V = 0 \tag{43}$$

The last equation is the equation of continuity and U and V are the z- dependent amplitudes of the x and y-components of the velocity. Θ, W and Z are the z-dependent amplitudes of the temperature and the vertical components of velocity and vorticity, respectively. If the heat diffusion in the wall is taken into account with a temperature amplitude Θ_W, Equation 26 becomes

$$i\omega\frac{\kappa}{\kappa_W}\Theta_W = \left(\frac{d^2}{dz^2} - k^2\right)\Theta_W \tag{44}$$

In normal modes the boundary conditions Equations 31 to 33 change into those for the amplitude of the variables. They are

$$W = 0 \quad , \quad \frac{d^2W}{dz^2} = 0 \quad , \quad \frac{dZ}{dz} = 0 \quad at \quad z = 0 \quad and \quad z = 1 \quad free \quad boundary \tag{45}$$

$$\begin{aligned} W = 0 \quad , \quad \frac{dW}{dz} = 0 \quad , \quad Z = 0 \quad at \quad z = 0 \quad and \quad z = 1 \quad solid \quad boundary \\ \Theta = 0 \quad at \quad z = 0 \quad and \quad z = 1 \quad fixed \quad temperature \\ \frac{d\Theta}{dz} = 0 \quad at \quad z = 0 \quad and \quad z = 1 \quad fixed \quad heat \quad flux \end{aligned} \tag{46}$$

$$\begin{aligned} \Theta_L = 0 \quad \text{at} \quad z = -D_L \\ \Theta_U = 0 \quad \text{at} \quad z = 1 + D_U \\ \Theta_L = \Theta, \quad X_L \frac{d\Theta}{dz} = \frac{d\Theta_L}{dz} \quad \text{at} \quad z = 0 \\ \Theta_U = \Theta, \quad X_U \frac{d\Theta}{dz} = \frac{d\Theta_U}{dz} \quad \text{at} \quad z = 1 \end{aligned} \tag{47}$$

where the two free boundary conditions $\mathbf{n} \cdot \nabla u = \mathbf{n} \cdot \nabla v = 0$ where replaced by $d^2W/dz^2 = 0$ using the z-derivative of the continuity Equation 43 and the x and y-components of the solid boundary condition $\mathbf{u} = 0$ are replaced by $dw/dz = 0$ using Equation 43. The conditions of the z-component of vorticity Z are obtained from its definition using the x and y-components of the boundary condition $\mathbf{u} = 0$ for a solid wall and the derivative for the free wall. Notice that in the linear problem, in the absence of a source of vorticity (like rotation, for example) for all the conditions investigated here, the vorticity $Z = 0$. Vorticity can be taken into account in the non linear problem (see for example Pismen [49] and Pérez-Reyes and Dávalos-Orozco [48]). Equations 40 and 42 are independent of the vorticity Z. They can be combined to give

$$\left(\frac{d^2}{dz^2} - k^2\right)\left(\frac{d^2}{dz^2} - k^2 - i\omega\right)\left(\frac{d^2}{dz^2} - k^2 - \frac{i\omega}{Pr}\right) W + Rk^2W = 0 \tag{48}$$

The first boundary conditions used will be those of Equations 45 and 46 of free-free and fixed temperature at the wall [7]. These are the simplest conditions, important because they allow to obtain an exact solution of the problem and may help to understand the qualitative behaviour of convection in more complex situations. Using Equations 40 and 42 evaluated at the boundaries, these conditions can be translated into:

$$W = D^2w = D^4w = D^6w = \ldots \quad and \quad \Theta = D^2\Theta = D^4\Theta = D^6\Theta = \ldots \tag{49}$$

where $D = d/dz$. These are satisfied by a solution

$$W = A \sin(n\pi z) \tag{50}$$

Here, n is an integer number and A is the amplitude which in the linear problem can not be determined. In this way, substitution in Equation 48 leads to an equation which can be written as

$$R = \frac{1}{k^2}\left(n^2\pi^2 + k^2\right)\left[\left(n^2\pi^2 + k^2\right)^2 - \frac{\omega^2}{Pr}\right] + \frac{1}{k^2}\left(n^2\pi^2 + k^2\right)^2\left(1 + \frac{1}{Pr}\right) i\omega \tag{51}$$

The Rayleigh number must be real and therefore the imaginary part should be zero. This condition leads to the solution $\omega = 0$. That is, under the present boundary conditions, the flow can not be oscillatory, it can only be stationary. Thus, the marginal Rayleigh number for stationary convection is

$$R = \frac{1}{k^2}\left(n^2\pi^2 + k^2\right)^3 \tag{52}$$

From this equation it is clear that n represents the modes of instability that the system may show. If the temperature gradient is increased, the first mode to occur will be that with $n = 1$. Here, the interest is to calculate the lowest magnitude of R with respect to k because it is expected that the mode $n = 1$ will appear first for the wavenumber that minimizes R. Thus, taking the derivative of R with respect to k and solving for k, the minimum is obtained for

$$k = \pi/\sqrt{2} \tag{53}$$

This critical wavenumber is substituted in Equation 52 to obtain the critical Rayleigh number for free-free walls and fixed temperature

$$R = \frac{27}{4}\pi^4 = 657.51 \tag{54}$$

This may be interpreted as the minimum non dimensional temperature gradient needed for the beginning of linear convection. The space variables were made non dimensional using the thickness of the layer. Therefore, the result of Equation 53 shows that, under the present conditions, the size (wavelength) of the critical convection cells will be $\Lambda = 2\sqrt{2}d$.

Now, the solid-solid conditions and the fixed temperature conditions are used. In this case too, it has been shown that convection should be stationary [7]. The problem is more complicated and numerical methods are needed to calculate approximately the proper value problem for R. From Equation 48 and $\omega = 0$ the equation for W is :

$$\left(\frac{d^2}{dz^2} - k^2\right)^3 W + Rk^2W = 0 \tag{55}$$

Due to the symmetry of the boundary conditions it is possible to have two different solutions. One is even and the other is odd with respect to boundary conditions located symmetrically with respect to the origin of the z-coordinate. That is, when $z_1 = -1/2$. Assume $W \propto exp(mz)$ in Equation 55 to obtain

$$\left(m^2 - k^2\right)^3 + Rk^2 = 0 \tag{56}$$

This is a sixth degree equation for m (or a third degree equation for m^2) which has to be solved numerically. The solutions for Equation 55 can be written as

$$W = A_1e^{m_1z} + A_2e^{-m_1z} + A_3e^{m_2z} + A_4e^{-m_2z} + A_5e^{m_3z} + A_6e^{-m_3z} \tag{57}$$

one of this coeficients A_i has to be normalized to one. The m_i's contain the proper values R and k. Three conditions for W are needed at each wall. They are $W = DW = D^4W - 2k^2D^2W$ at $z = \pm 1/2$. The last one is a result of the use of the condition for Θ in Equation 40 with $\omega = 0$. As an example of the even and odd proper value problem, the evaluation of the conditions of

W at $z = \pm 1/2$ will be presented. They are

$$0 = A_1 e^{\frac{m_1}{2}} + A_2 e^{-\frac{m_1}{2}} + A_3 e^{\frac{m_2}{2}} + A_4 e^{-\frac{m_2}{2}} + A_5 e^{\frac{m_3}{2}} + A_6 e^{-\frac{m_3}{2}} \tag{58}$$

$$0 = A_1 e^{-\frac{m_1}{2}} + A_2 e^{\frac{m_1}{2}} + A_3 e^{-\frac{m_2}{2}} + A_4 e^{\frac{m_2}{2}} + A_5 e^{-\frac{m_3}{2}} + A_6 e^{\frac{m_3}{2}}$$

Addition and subtraction of both conditions give, respectively

$$0 = 2\,(A_1 + A_2)\,cosh\left(\frac{m_1}{2}\right) + 2\,(A_3 + A_4)\,cosh\left(\frac{m_2}{2}\right) + 2\,(A_5 + A_6)\,cosh\left(\frac{m_3}{2}\right) \tag{59}$$

$$0 = 2\,(A_1 - A_2)\,sinh\left(\frac{m_1}{2}\right) + 2\,(A_3 - A_4)\,sinh\left(\frac{m_2}{2}\right) + 2\,(A_5 - A_6)\,sinh\left(\frac{m_3}{2}\right)$$

The same can be done with the other boundary conditions. The important point is that two sets can be separated, each one made of three conditions: one formed by the addition of the conditions and another one made of the subtraction of the conditions, that is, the even and the odd modes of the proper value problem. It has been shown numerically [7] that the even mode gives the smaller magnitude of the marginal proper value of the Rayleigh number. The odd mode gives a far more larger value and therefore it is very stable in the present conditions of the problem. However, there are situations where the odd mode can be the first unstable one (see Ortiz-Pérez and Dávalos-Orozco [42] and references therein). Recently, Prosperetti [51] has given a very accurate and simple formula for the marginal Rayleigh number by means of an improved numerical Galerkin method. That is

$$R = \frac{1}{k^2}\left(\frac{(\pi^2 + k^2)^5 (sinh(k) + k)}{(\pi^2 + k^2)^2 (sinh(k) + k) - 16\pi^2 k cosh(k/2)^2}\right) \tag{60}$$

The marginal curve plotted from this equation gives a minimum $R = 1715.08$ at $k = 3.114$. These critical values are very near to those calculated by means of very accurate but complex numerical methods. The accepted values are $R = 1707.76$ at $k = 3.117$ [7]. From the critical wavenumber it is possible to calculate the size (wavelength) of the cell at onset of convection. That is, $\Lambda = 2\pi d/3.117$, which is smaller than that of the free-free case. This is due to the friction at the walls. Walls friction also stabilizes the system increasing the critical Rayleigh number over two and a half times the value of the free-free case.

Linear convection inside walls with fixed heat flux has been investigated by Jakeman [26]. Hurle et al. [25] have shown that the principle of exchange of instabilities is valid for a number of thermal boundary conditions, that is, oscillatory convection is not possible and $\omega = 0$, including the case of fixed heat flux. Jakeman [26] used a method proposed by Reid and Harris [7, 54] to obtain an approximate solution of the proper value of R. This is a kind of Fourier-Galerkin method [17, 18]. From the expresion obtained, the critical Rayleigh and wavenumber were calculate analytically by means of a small wavenumber approximation. The reason is that it has been shown numerically in the marginal curves, that the wavenumber of the smallest Rayleigh number tends to zero. The critical Rayleigh number for the free-free case is $R = 120 = 5!$ ($k = 0$) and for the solid-solid case $R = 720 = 6!$ ($k = 0$). It is surprising that the critical Rayleigh numbers are less than half the magnitude of those of the fixed temperature case. This can be explained by the form of the temperature boundary condition, that is $D\theta = 0$. From the condition it is clear that the perturbation heat flux can not be dissipated at the wall and therefore the perturbation remains inside the fluid layer.

This makes the flow more unstable and consequently the critical Rayleigh number is smaller. The linear problem was generalized by Dávalos [11] to include rotation and magnetic field, and obtained explicit formulas for the critical Rayleigh number depending on rotation and magnetic field and a combination of both. Notice that the method by Reid and Harris [54] is very effective and it is still in use in different problems of convection. However, as explained above, it has been improved by Prosperetti [51].

The nonlinear problem for the fixed heat flux approximation has been done by Chapman and Proctor [8]. They improved the methods of calculation with respect to previous papers. The method to obtain a nonlinear evolution equation is to make an scaling of the independent and dependent variables taking into account that, if the Rayleigh number is a little far above criticality (weak nonlinearity), the wavenumber of the convection cell is still small. Consequently, the motion will be very slow because the cell is very large. Then, the scaling used in the nonlinear Equations 23 to 25 is as follows

$$R = R_C + \mu^2\epsilon^2, \quad \mu = O(1), \quad \frac{\partial}{\partial x} = \epsilon\frac{\partial}{\partial X}, \quad \frac{\partial}{\partial y} = \epsilon\frac{\partial}{\partial Y}, \quad \frac{\partial}{\partial t} = \epsilon^4\frac{\partial}{\partial \tau} \tag{61}$$

They solve a two-dimensional problem using the stream function by means of which the velocity vector field satisfies automatically Equation 24. The velocity with the scaling is defined as $\mathbf{u} = (\partial\psi/\partial z, 0, -\epsilon\partial\psi/\partial X)$. The stream function is also scaled as $\psi = \epsilon\phi$. The expansion of the functions in terms of ϵ is

$$\theta = \theta_0(X,z,\tau) + \epsilon^2\theta_2(X,z,\tau) + \cdots \qquad \phi = \phi_0(X,z,\tau) + \epsilon^2\phi_2(X,z,\tau) + \cdots \tag{62}$$

The reason for this expansion is that the substitution of the scaling in Equations 23 to 25 only shows even powers of ϵ. The problem is solved in different stages according to the orders of ϵ subjected to the corresponding scaled boundary conditions. The critical value of R is obtained from a solvability condition at $O(\epsilon^2)$. Notice that they locate the walls at $z = \pm 1$ and obtain $R_c = 15/2$ and $R_c = 45$ for the free-free and the solid-solid cases, respectively. If the definition of the Rayleigh number includes the temperature gradient it depends on the forth power of the thickness of the layer. Therefore, the Rayleigh number defined here is sixteen times that defined by Chapman and Proctor [8]. The evolution equation is obtained as a solvability condition at $O(\epsilon^4)$. That is, with $\theta_0 = f(X,\tau)$

$$\frac{\partial f}{\partial \tau} + \frac{\mu^2}{R_C}\frac{\partial^2 f}{\partial X^2} + B\frac{\mu^4}{R_C}\frac{\partial^2 f}{\partial X^4} + C\frac{\partial}{\partial X}\left[\left(\frac{\partial f}{\partial X}\right)^3\right] = 0 \tag{63}$$

This equation is valid for free-free and solid-solid boundary conditions and the constants B and C have to be calculated according to them. Chapman and Proctor found that the patterns of convection cells are rolls but that they are unstable to larger rolls. Therefore, the convection will be made of only one convection roll. An extension of this problem was done by Proctor [50] including the Biot number Bi in the thermal boundary conditions. The Biot number is a non dimensional quantity that represents the heat flux across the interface between the fluid and the wall. The fixed heat flux boundary condition is obtained when the Bi is zero. When Bi is small but finite, the critical convection cells are finite and therefore more realistic.

The effect of the thickness and heat conductivity of the wall on natural convection has been investigated by Cericier et al. [6]. The goal is to be able to obtain more realistic critical values of the Rayleigh number and wavenumber. Here it is necessary to use the thermal conditions Equations 22 and 33 for the main temperature profile and the perturbation of temperature, respectively. The main temperature profile is linear with respect to z but is more complex due to the presence of terms which depend on all the new extra parameters coming from the geometry and properties of the walls. From Equation 33 it is possible to calculate a new condition which only has the amplitude of the fluid temperature perturbation. In the present notation it has the form

$$D\Theta - \frac{q}{X_L \tanh\left(qD_L\right)}\Theta = 0 \quad at \quad z = 0, \qquad D\Theta + \frac{q}{X_U \tanh\left(qD_U\right)}\Theta = 0 \quad at \quad z = 1, \tag{64}$$

where the coefficients of Θ in both conditions might be considered the Biot numbers of each wall. Here, $q = \sqrt{k^2 + i\omega}$. Now there are four new parameters which influence the convective instability, the heat conductivities ratio X_L and X_U and the thicknesses ratio D_L and D_U. Assuming that $X = X_L = X_U$ and $d = D_L = D_U$ the problem has some simplification.

Figure 1 shows results for the case when the properties of both walls are the same. Notice that when X increases the critical values in both figures change from the fixed temperature case to the fixed heat flux case. The results are similar to those of Cericier et al. [6] and were plotted using a formula calculated from a low order Galerkin approximation. It is important to observe that in the middle range of X the thickness of the walls play a relevant roll producing large differences between the critical values, for fixed X.

The problem of surface deformation in convection requires lower conditions for free or solid walls and an upper condition of a free deformable surface. The stationary linear problem was first investigated by V. Kh. Izakson and V. I. Yudovich in 1968 and their work is reviewed in [19]. The stationary problem with rotation and a variety of thermal boundary conditions was investigated by Dávalos-Orozco and López-Mariscal [13]. The problem of oscillating convection was first investigated by Benguria and Depassier [2]. They found that when the wall is solid, due to the restriction $R/PrG < 1$ (discussed presently) it is not possible for oscillatory convection to have a smaller critical Rayleigh number than that of stationary convection with surface deformation. Therefore, only the free wall case presents oscillatory convection as the first unstable one. $G = gd^3/\nu^2$ is the Galileo number, representative of the surface restoration due to the gravitational force. The deformation of the surface is due to a pressure difference which is balanced by the shear stresses at the fluid surface, that is, the dimensional zero stress jump at the surface

$$\left(p^* - p_\infty\right) n_i^* = \tau_{ik}^* n_i^* \quad at \quad z^* = z_1^* + d + \eta^*\left(x, t\right). \tag{65}$$

When the surface is flat the pressure condition is $p^* - p_\infty = 0$ (no pressure jump), where p_∞ is the pressure of the ambient gas whose viscosity is neglected. This problem requires the kinematic boundary condition of the surface deformation which in two-dimensions and in non dimensional form is

$$w = \frac{\partial \eta}{\partial t} + u\frac{\partial \eta}{\partial x} \qquad at \quad z = z_1 + 1 + \eta\left(x, t\right) \tag{66}$$

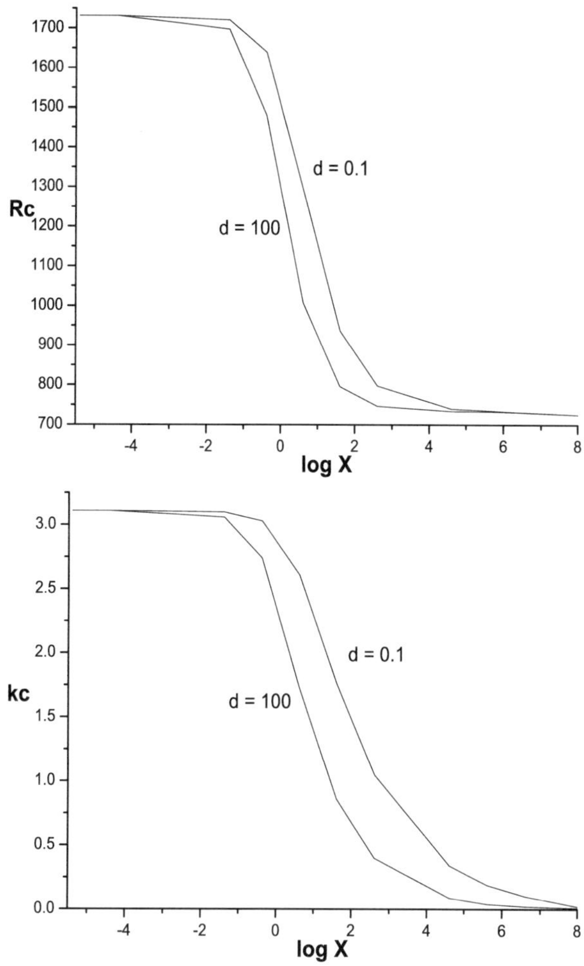

Figure 1. Critical Rayleigh number and wavenumber vs conductivities ratio for two thicknesses ratio

which means that a fluid particle remains on the fluid interface and that the time variation of the convected surface deflection moves with the same vertical velocity as that of the fluid particle. η is an extra dependent variable representing the amplitude of the free surface deformation. The non dimensional normal and tangent vectors are defined as

$$\mathbf{n} = \frac{(-\eta_x, 1)}{\mathcal{N}} \tag{67}$$

$$\mathbf{t} = \frac{(1, \eta_x)}{\mathcal{N}} \tag{68}$$

where $\mathcal{N} = (\eta_x^2 + 1)^{1/2}$ and the subindexes mean partial derivative. Other conditions, like Equations 18 to 20, defined using the normal and tangential vectors to the free deformable surface have to be modified with the definitions given in Equations 67 and 68. Equation 66 has to be multiplied by **n** to obtain the normal stress boundary condition and by **t** to calculate the tangential stress boundary condition. Here, the problem is assumed two-dimensional, but when it is three-dimensional it is necessary to define another tangential vector to the surface, perpendicular to both **n** and **t**. The problem simplifies if $z_1 = -1$ and the boundary conditions at the free surface are set at $z = \eta(x, t)$. For the linear problem $\eta(x, t)$ is assumed small. This gives the possibility to make a Taylor expansion of the variables at the free surface, that is, around $z = 0$, and simplify the boundary conditions. From this expansion, it is found that the fixed heat flux and the fixed temperature conditions remain the same in the linear problem. In two-dimensional flow it is possible to use the stream function. In this case, as a result of the expansion and use of normal modes, the kinematic, tangential stress and normal stress boundary conditions, respectively, are

$$i\omega\eta + ik\Psi = 0 \quad at \quad z = 0 \tag{69}$$

$$\left(D^2 + k^2\right)\Psi = 0 \quad at \quad z = 0 \tag{70}$$

$$\frac{i\omega D^3\Psi}{G} - \frac{i\omega}{G}\left(3k^2 + \frac{i\omega}{\mathrm{Pr}}\right)D\Psi - k^2\,\mathrm{Pr}\,\Psi = 0 \quad at \quad z = 0 \tag{71}$$

Ψ is the amplitude of the stream function. The approximations done here are only valid when the Galileo number satisfies $R/PrG < 1$. The reason is that the density and the temperature perturbations, related by $\rho' = (R/PrG)\theta$, should be $\rho' < \theta$ in order to satisfy the Boussinesq approximation, which, among other things, neglects the variation of density with temperature in front of the inertial term of the balance of momentum equation. Consequently, the approximation is valid when the critical Rayleigh numbers satisfy $R_C < PrG$. In the stationary problem the new parameter is in fact PrG due to the condition Equation 71 (see [19],[13]). However, in the oscillatory problem, Pr is an independent parameter, as seen in Equation 48, but it appears again in the condition Equation 71. Notice that in the limit of $PrG \to \infty$ condition Equation 71 reduces to that of a flat wall. Then, the product PrG has two effects when it is large: 1) it works to guarantee the validity of the Boussinesq approximation under free surface deformation and 2) it works to flatten the free surface deformation.

The problem of oscillatory convection was solved analytically by Benguria and Depassier [2] when it occurs before stationary convection, that is, when the flat wall is free with fixed temperature and the upper deformable surface has fixed heat flux. They found that the cells are very large and took the small wavenumber limit. The critical Rayleigh number is $R_c = 30$ and $k_c = 0$.

Nonlinear waves for the same case of the linear problem, have been investigated by Aspe and Depassier [1] and by Depassier [15]. In the first paper, surface solitary waves of the Korteweg-de Vries (KdV) type were found. In the other one, Depassier found a perturbed Boussinesq evolution equation to describe bidirectional surface waves.

4. Natural convection in second-order fluids

The methods used in natural convection of Newtonian fluids can also be used in non Newtonian flows. Linear and non linear natural convection of second order fluids was investigated by Siddheshwar and Sri Krishna [58]. They assumed the flow is two-dimensional and used the free-free and fixed temperature boundary conditions. Here, the constitutive Equation 5 is used in the balance of momentum equation.

For the linear problem use is made of normal modes. The amplitudes of the stream function and the temperature are assumed of the form $\sin n\pi z$ which satisfies the boundary conditions. As before, the substitution in the governing equations leads to the formula for the marginal Rayleigh number of a second order fluid

$$R = \frac{1}{k^2}\left(n^2\pi^2 + k^2\right)\left[\left(n^2\pi^2 + k^2\right)^2 - \frac{\omega^2}{Pr}\left(1 + Q\left(n^2\pi^2 + k^2\right)\right)\right] \tag{72}$$

$$+\frac{1}{k^2}\left(n^2\pi^2 + k^2\right)^2\left[1 + \frac{1}{Pr} + \frac{Q}{Pr}\left(n^2\pi^2 + k^2\right)\right] i\omega$$

Here, $Q = \gamma/\rho_0 d^2$ is the elastic parameter. The Rayleigh number is real and the imaginary part should be zero. The only way this is possible is that $\omega = 0$. Therefore, the flow can not be oscillatory and this reduces to Equation 52 for the marginal Rayleigh number and to the critical one of a Newtonian fluid for free-free convection, that is, $R_C = 27\pi^4/4$ at $k_C = \pi/\sqrt{2}$. By means of the energy method for the linear problem, Stastna [60] has shown that, in the solid-solid case and fixed temperature at the walls, the convection can not be oscillatory and $\omega = 0$. Therefore, again the linear critical Rayleigh number and wavenumber are the same as those of the Newtonian fluid.

In their paper, Siddheshwar and Sri Krishna [58] also investigated the possibility of chaotic behaviour to understand the role played by the elastic parameter Q. They use the doble Fourier series method of Veronis [65] to calculate, at third order, a nonlinear system of Lorenz equations [32] used to investigate possible chaotic behavior in convection. In particular, the form selected by Lorenz for the time dependent amplitudes of the stream function and the temperature are

$$\Psi(x,z,t) = X(t)\sin kx \sin \pi z \tag{73}$$

$$\Theta(x,z,t) = Y(t)\cos kx \sin \pi z + Z(t)\sin 2\pi z$$

which satisfy the boundary conditions. These are used in the equations to obtain the nonlinear coupled Lorenz system of equations for the amplitudes $X(t)$, $Y(t)$ and $Z(t)$

$$\frac{dX(t)}{dt} = q_1 X(t) + q_2 Y(t), \tag{74}$$

$$\frac{dY(t)}{dt} = q_3 X(t) + q_4 Y(t) + q_5 X(t) Z(t),$$

$$\frac{dZ(t)}{dt} = q_6 Z(t) + q_7 X(t) Y(t),$$

where the q_i $(1 = 1, \cdots, 7)$ are constants including parameters of the problem. According to [58] the elastic parameter Q appears in the denominator of the constants q_1 and q_2. In the system of Equations 74 the variables can be scaled to reduce the number of parameters to three. The results of [58] show that random oscillations occur when the parameters Q and Pr are reduced in magnitude. Besides, they found the possibility that the convection becomes chaotic for the magnitudes of the parameters investigated.

When the walls are solid and the heat flux is fixed, results of the nonlinear convective behaviour of a second-order fluid have been obtained by Dávalos and Manero [12]. They used the method of Chapman and Proctor [8] to calculate the evolution equation that describes the instability. A small wavenumber expansion is done like that of Equation 61 and62. However, in contrast to [8], their interest was in a three-dimensional problem and instead of using the stream function, use was made of a function defined by $\mathbf{u} = \nabla \times \nabla \times \phi \hat{k}$. The boundary condition of this function ϕ at the walls is $\phi = 0$. It is found, by means of the solvability condition at first order, that the critical Rayleigh number is the same as that of the Newtonian case, that is, $R_C = 720$ at $k_C = 0$ and that convection can not be oscillatory. At the next approximation level, the solvability condition gives the evolution equation for the nonlinear convection. The result was surprising, because it was found that the evolution equation is exactly the same as that of the Newtonian fluid, that is, Equation 63 but in three-dimensions. The only difference with Newtonian fluids will be the friction on the walls due to the second order fluid. As explained above, the flow under the fixed heat flux boundary condition is very slow and the convection cell is very large. Therefore, this result may be related with the theorems of Giesekus, Tanner and Huilgol [20, 24, 63] which say that the creeping flow of a second-order incompressible fluid, under well defined boundary conditions, is kinematically identical to the creeping flow of a Newtonian fluid. The results presented here are a generalization of those theorems for three-dimensional natural convection evolving in time.

5. Natural convection in Maxwell fluids

In order to investigate the convection of a Maxwell fluid Equations 12 and 29 have to be used in the balance of momentum equation. The linear problem was investigated by Vest and Arpaci [66] for both free and solid walls and by Sokolov and Tanner [59] who present an integral model for the shear stress tensor which represent a number of non Newtonian fluids, including that of Maxwell. The linear equations in two-dimension use the stream function. In normal modes they are expressed as:

$$\left(D^2 - k^2 - \frac{Ni\omega}{\mathrm{Pr}}\right)\left(D^2 - k^2\right)\Psi = ikNR\Theta \tag{75}$$

$$\left(D^2 - k^2 - i\omega\right)\Theta = ik\Psi \tag{76}$$

where $N = (1 + i\omega L)$ is a complex constant which depends on the viscoelastic relaxation time and the frequency of oscillation.

The combination of these two equations gives:

$$\left(D^2 - k^2 - \frac{Ni\omega}{\mathrm{Pr}}\right)\left(D^2 - k^2 - i\omega\right)\left(D^2 - k^2\right)\Psi + k^2 NR\Psi = 0 \tag{77}$$

The free-free linear problem of [66] is illustrative. Assuming that the amplitudes are proportional to $\sin n\pi z$, the Equation 77 is transformed into a complex algebraic equation for ω

$$-i\omega^3 - \omega^2 A_1 + i\omega A_2 + A_3 = 0 \tag{78}$$

where

$$A_1 = \frac{1}{L}\left[L\left(n^2\pi^2 + k^2\right) + 1\right], \quad A_2 = \left[\frac{1 + Pr}{L}\left(n^2\pi^2 + k^2\right) - \frac{PrRk^2}{n^2\pi^2 + k^2}\right], \tag{79}$$

$$A_3 = \frac{Pr}{L}\left[\left(n^2\pi^2 + k^2\right)^2 - \frac{Rk^2}{n^2\pi^2 + k^2}\right].$$

In Equation 75 the real and imaginary parts have to be zero. Then

$$i\omega\left(-\omega^2 + A_2\right) = 0, \quad -\omega^2 A_1 + A_3 = 0 \tag{80}$$

There are two possibilities. 1) From the first $\omega = 0$ and from the second $A_3 = 0$. 2) From the first $\omega \neq 0$ with $\omega^2 = A_2$ and from the second, after substitution of ω, $A_1 A_2 - A_3 = 0$. From 1), A_3 gives the marginal stationary Rayleigh number for different modes n. The critical value for mode $n = 1$ has already been given above. From 2), the marginal Rayleigh number for oscillatory convection can be calculated for different modes n. Vest and Arpaci show that, when the relaxation time parameter has a magnitude large enough, it is possible to have oscillatory convection as the first unstable one, with R_C smaller than that of stationary convection. Also, they show that an increase of Pr decreases considerably R_C making convection very unstable. A similar behavior at criticality can be found for the solid-solid case. However, the solution is far more complex because it has to be solved numerically ensuring that the proper value of the Rayleigh number is real. The frequency is obtained form the numerical solution of the imaginary part and it is substituted into the real part, which still contains the frequency. The marginal Rayleigh number is obtained from the real part. Variation of the wavenumber leads to the minimum of the Rayleigh number, the critical one, with its corresponding wavenumber and frequency. The conclusions obtained by Vest and Arpaci [66] are that the solid-solid case is more stable than the free-free case but qualitatively the response to the parameters L and Pr is similar.

The problem of a viscoelastic fluid layer with free and deformable surface will be discussed in detail in the section for Oldroyd fluid convection. The Maxwell fluid case is included in that problem.

The effect of the thickness and thermal conductivity of the walls on linear convection of a Maxwell fluid layer was investigated by Pérez-Reyes and Dávalos-Orozco [46]. They found that, by making some algebra, those effects can be included in a kind of Biot number which appears in the thermal boundary conditions of the upper and lower walls. The difference with respect to the Newtonian problem is that here it is necessary to calculated numerically the frequency of oscillation in the same way as explained in the last paragraph for the solid-solid case. The number of parameters in the equations increased and therefore the ratio κ/κ_W which appears in the heat diffusion Equation 26 of the walls is assumed equal to one. Besides, it is supposed that the ratios of conductivities and thicknesses of the upper and lower walls are the same. With this in mind, some results are presented in Fig. 2. Notice that in the figures F is used instead of L, for the relaxation time, and that D is used instead of d. Note that here the curves of R_C increase with X instead of decreasing as in the Newtonian case for both magnitudes of D. Then, in this case it is easier to reach a codimension-two point where stationary convection competes with oscillatory convection to be the first unstable one. In the figure, the dashed lines correspond to stationary convection. This codimension-two point depends on the Prandtl number. When Pr increases there is a magnitude after which oscillatory convection is always the first unstable one (see [46]). In contrast, for other magnitudes of the relaxation time the behavior is similar to that of the Newtonian fluid (see Fig. 1) but with far more smaller magnitudes of R_C, as seen in Fig. 3.

Thus, the oscillatory flow is very unstable for the new magnitude of $F = 100$. It is of interest to observe the different reaction the flow instability has with respect to X and D for both magnitudes of F.

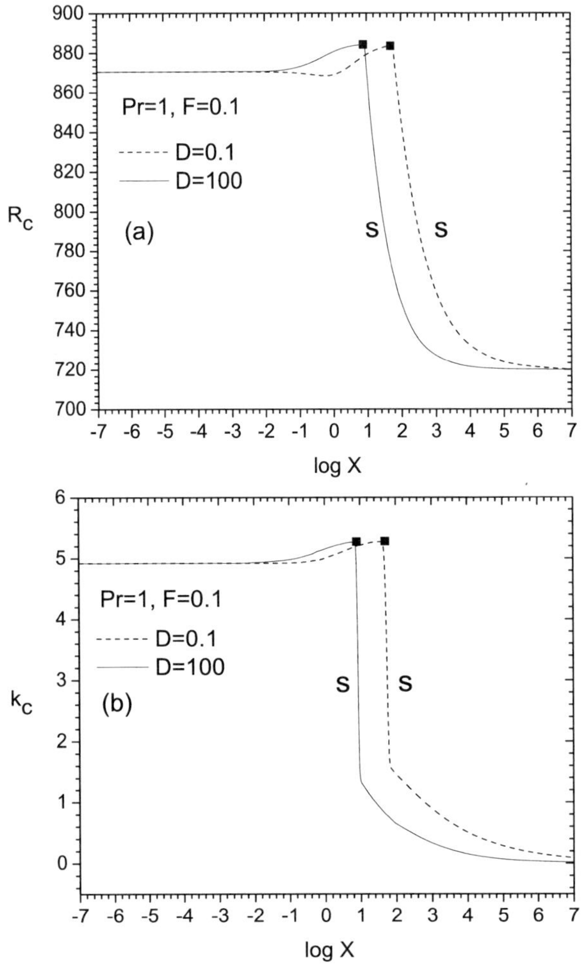

Figure 2. Pr=1 and F=0.1. A) Critical Rayleigh number vs X. B) Critical wavenumber vs X

Nonlinear convection of a Maxwell fluid was investigated by Van Der Borght [64] using the ideal free-free boundary conditions. They calculated the heat dissipation of nonlinear stationary hexagonal convection cells by means of the Nusselt number. It was found that, for a given Rayleigh number, viscoelasticity effects only produce a slightly higher Nusselt number than Newtonian convection. Nonlinear traveling waves in a Maxwell fluid were investigated by Brand and Zielinska [5] using free-free boundary conditions. They obtain one Landau equation for stationary convection and other one for oscillatory convection (see Equation1). They found that standing waves are preferred over traveling waves for $Pr < 2.82$ at a codimension-two point. They also investigated the wave modulation in space by means of an equation similar to Equation 2 but of higher nonlinear order. Irregular and sensitive to initial conditions behavior of a convecting Maxwell fluid was investigated by Khayat [29]

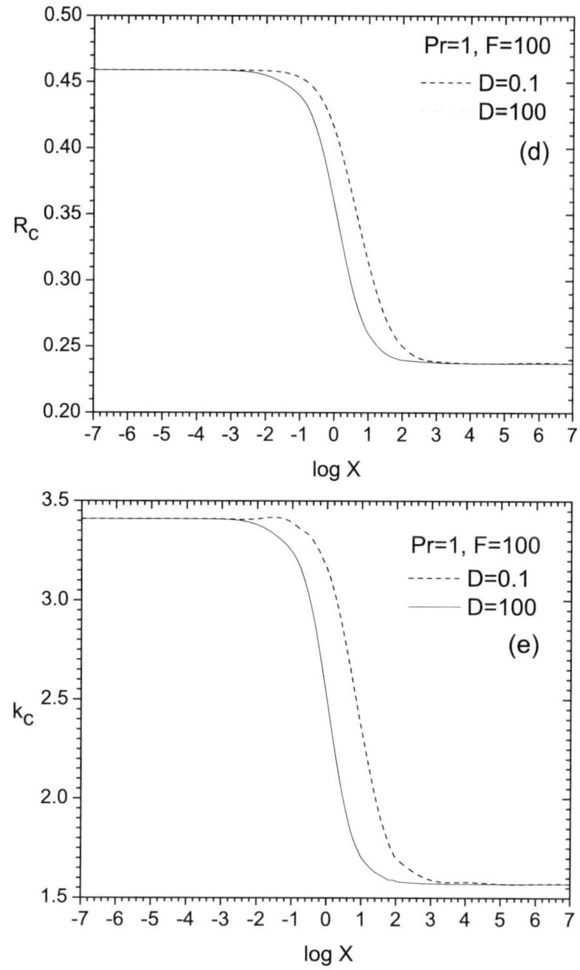

Figure 3. Pr=1 and F=100. A) Critical Rayleigh number vs X. B) Critical wavenumber vs X

using the free-free boundary conditions. The variables are written in the form of Equations 73, in addition to those of the shear and the two normal stresses. In this way, instead of three coupled Lorenz Equations 74, he obtains a system of four coupled equations which include as new time dependent variable, a linear combination of the amplitude of the normal stresses difference and the shear stress. In the limit $L \to 0$, the new system reduces to that of Lorenz. He found that above a critical magnitude of the relaxation time L_C the flow is oscillatory. For an L below the critical one, the route to chaotic motion is different from that of a Newtonian fluid, even in the case where L is very small. Viscoelasticity produces chaos when the Newtonian fluid still is non chaotic.

6. Natural convection in Oldroyd fluids

The Maxwell model of viscoelastic fluids presented above, shows an extreme (violent) mechanical behavior in comparison to other models. Mainly, this occurs when the relaxation time is large and the flow shows a more elastic behavior. The most elementary correction to this model is the Oldroyd fluid model which includes the property of shear rate retardation. That means that the fluid motion relaxes for a time interval even after the shear stresses have been removed. The representative magnitude of that time interval is called retardation time. This characteristic moderates the mechanical behavior of the Oldroyd fluid. The equations of motion require the constitutive Equation 13 or 30 (in non dimensional form). The ratio of the retardation time over the relaxation time E appears as an extra parameter which satisfies $E < 1$ (see [4], page 352). Notice that when $E = 0$ the Maxwell constitutive equation is recovered. Therefore, for very small E the behavior of the Oldroyd fluid will be similar to that of the Maxwell one.

The linear problem requires the same Equations 75 to 77, but here $N = (1 + i\omega L)/(1 + i\omega LE)$. The convection with free-free boundary conditions was investigated by Green [21] who obtained an equation similar to Equation 78. In the same way, from the solution of the real and imaginary parts, it is possible to calculate the marginal Rayleigh number and frequency of oscillation. In this case it is also possible to find a magnitude of L and E where the stationary and oscillatory convection have the same Rayleigh numbers, the codimension-two point. The Prandtl number plays an important role in this competition to be the first unstable one. The solid-solid problem was solved numerically by Takashima [62]. He shows that the critical Rayleigh number is decreased by an increase of L and increased by an increase of E. The numerical results show that an increase of Pr decreases drastically the magnitude of R_C for the Maxwell fluid, but it is not very important when $E > 0$. Oscillatory convection is the first unstable one after a critical magnitude L_C is reached, which depends on the values of Pr and E. For small Pr, L_C is almost the same for any E. However, for large Pr the L_C for the Maxwell fluid is notably smaller than that of the Oldroyd fluid. This fluid was also investigated by Sokolov and Tanner [59] using an integral model. In contrast to the papers presented above, Kolkka and Ierley [30] present results including the fixed heat flux boundary condition. They also give some corrections to the results of Vest and Arpaci [66] and Sokolov and Tanner [59]. The qualitative behavior of convection with fixed heat flux is the same, for both free-free and solid-solid boundaries, but with important differences in the magnitude of R_C. The codimension-two point also occurs for different parameters. Interesting results have been obtained by Martínez-Mardones and Pérez-García [35] who calculated the codimension-two points with respect to L and E for both the free-free and the solid-solid boundary conditions. Besides, they calculated the dependance these points have on the Prandtl number. They show that for fixed E, the L_C of codimension-two point decreases with Pr.

When natural convection occurs with an upper free surface it is every day experience to see that the free surface is deforming due to the impulse of the motion of the liquid coming in the upward direction. The assumption that the free surface is deformable in the linear convection of an Oldroyd viscoelastic fluid was first investigated by Dávalos-Orozco and Vázquez-Luis [14]. Under this new condition, the description of linear convection needs the same Equations 75 to 77 and $N = (1 + i\omega L)/(1 + i\omega LE)$. However, the free boundary conditions have to change because the surface deformation produces new viscous stresses due to viscoelasticity. The problem is assumed two-dimensional and the stream function appears in the boundary conditions of the upper free deformable surface. The mechanical boundary conditions Equations 69 and 70 are the same. However, the normal stress boundary condition

Equation 71 changes into

$$\frac{i\omega D^3\Psi}{GN} - \frac{i\omega}{G}\left(\frac{3k^2}{N} + \frac{i\omega}{\mathrm{Pr}}\right) D\Psi - k^2\,\mathrm{Pr}\,\Psi = 0 \quad at \quad z = 0 \tag{81}$$

which includes the viscoelastic factor N. Note that here the reference frame locates the free surface at $z = 0$, that is $z_1 = -1$. The advantages of doing this were explained above. The thermal boundary conditions remain the same. Numerical calculations were done for free and solid lower walls. In both cases, the fixed temperature boundary condition was used in the lower wall and the fixed heat flux boundary condition was used in the upper surface. In all the calculations the Prandtl number was set equal to $Pr = 1$. The goal was to compare with the paper by Benguria and Depassier [2]. Under these conditions, the results were first compared with those of the Newtonian flat surface solid-free convection ($R_{CS} = 669$, $k_{CS} = 2.09$ see [2]), the Newtonian deformable surface oscillatory solid-free convection ($R_{ON} = 390.8$, $k_{CS} = 1.76$ see [2]) and with the viscoelastic (Oldroyd and Maxwell) flat surface solid-free convection (presented in the figures with dashed lines). The results were calculated for different Galileo numbers G. However, here only some sample results are presented (see [14] for more details).

Figure 4 shows results for the solid-free case with $G = 100$. The dashed lines are extended until the stationary curve $R_{CS} = 669$ to show points of codimension-two. The curves of viscoelastic convection with deformable surface (solid lines) always have smaller R_C than $R_{ON} = 390.8$ and than those of the flat surface (dashed lines). When L increases both solid and dashed curves tend to the same value. Then, surface deformation is irrelevant for very large L. The Maxwell fluid is always the most unstable one. It was found that when L decreases below a critical value, R_C increases in such a way that it crosses above the line R_{ON}, reaches a maximum (around $L = 0.03$, $R_C = 393.19$ for $E = 0.1$ and $R_C = 393.27$ for $E = 0.01, 0.001, 0.0$) and then decreases until it reaches the line R_{ON} again for very small L. This means that there is a range of values of L where $R_C \geq R_{ON} = 390.8$. Then, inside this rage, viscoelastic convection with deformable surface is more stable than that of the Newtonian convection with deformable surface. The result was verified with different numerical methods (see [14]). This phenomenon was explained by means of the double role played by the Galileo number as an external body force on convection (like rotation, see [13]) and as restoring force of the surface deformation.

In Figure 5 shown are the results of the free-free case for $G = 1000$, which represents a larger restoration force of the surface deformation. This results were compared with those of the Newtonian flat surface free-free convection ($R_{CS} = 384.7$, $k_{CS} = 1.76$ see [2]), the Newtonian deformable surface oscillatory free-free convection ($R_{ON} = 30.0$, $k_{CS} = 0.0$ see [2]) and with the viscoelastic (Oldroyd and Maxwell) flat surface free-free convection (presented in the figures with dashed lines). The behavior of the curves is similar but, except for very small E (nearly Maxwell fluid) and large L, it is found that G has no influence on the instability of the free-free convection under the present conditions. The curve of the Maxwell fluid is the more unstable. The jumps found in the curves of k_C are also explained due to the dual role played by G on the instability. The results presented for the solid-free and free-free boundary conditions show the importance that free surface deformation has on the natural convection instability of viscoelastic fluids.

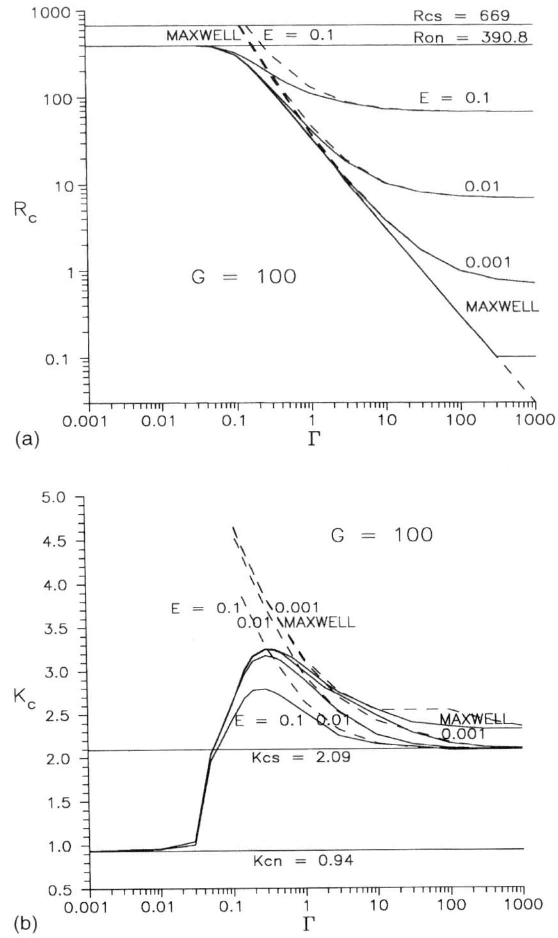

Figure 4. Solid-free case. Curves of criticality for $G = 100$ and different values of E. The upper deformable free surface case has solid curves and the undeformable one has dashed curves. A) R_c against Γ(here L). B) k_c against Γ (here L).

The nonlinear problem for an Oldroyd fluid was investigated by Rosenblat [55] for free-free boundary conditions. He found results for the three time derivatives Equations 6, 8 and 9. The weakly nonlinear approximation is used where the Rayleigh number is very near to the critical one. He found conditions for subcriticality and supercriticality calculating an algebraic quantity K which includes non dimensional parameters of all the fluids investigated. The conclusion for stationary convection is that the corotational Oldroyd model has subcritical bifurcation (and therefore is unstable) and that the upper and lower convected Oldroyd models can not have this bifurcation and are stable, as the Newtonian model. For oscillatory convection the problem is more complex and it is resolved numerically with plots of L vs E. However, the results are reviewed as follows. The corotational Oldroyd model has supercritical bifurcation and is stable. The upper and lower convected Oldroyd models have subcritical bifurcation and are unstable. A system of four coupled differential equations is

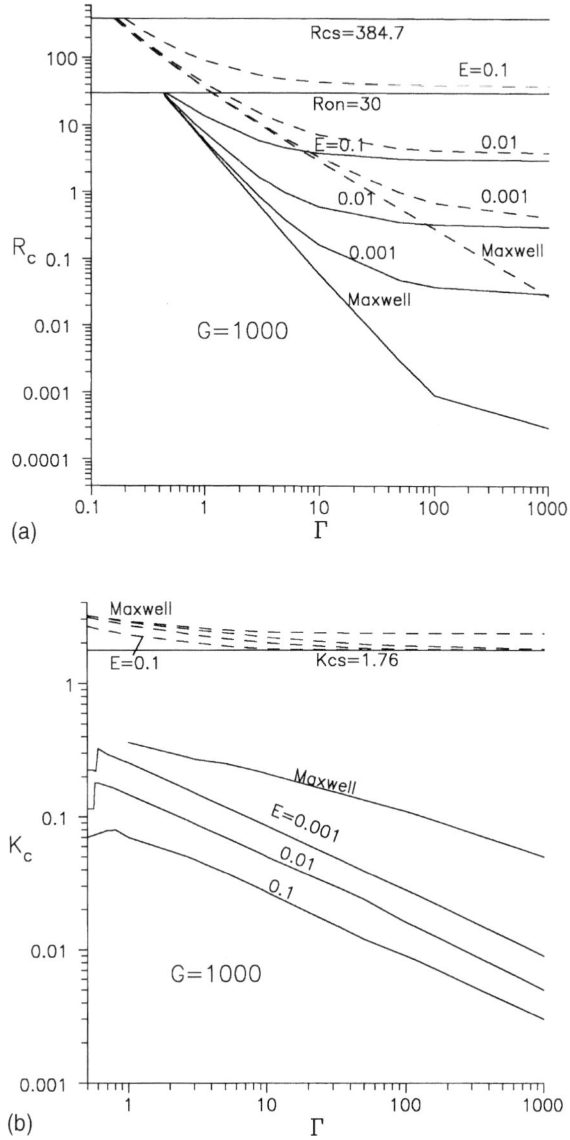

Figure 5. Free-free case. Curves of criticality for $G = 1000$ and different values of E. The upper deformable free surface case has solid curves and the undeformable one has dashed curves. A) R_c against Γ(here L). B) k_c against Γ (here L).

proposed to investigate chaotic behavior which generalize the Lorenz system of Equations 74. He found the possibility of chaotic behavior. The solid-solid and solid-free boundary conditions were used by Park and Lee [43, 44]. Important results are that the amplitude of convection and heat dissipation increase with L and for E small. The rigid walls cause more easily the subcritical bifurcation than free walls for the same viscoelastic parameters.

They conclude that Oldroyd fluid viscoelastic convection is characterized both by a Hopf bifurcation (for very large value of L) and a subcritical bifurcation.

Analytical and numerical methods were used by Martínez-Mardones et al. [36] to calculate the nonlinear critical parameters which lead to stationary convection as well as traveling and standing waves. By means of coupled Landau amplitude equations Martínez-Mardones et al. [37] investigated the pattern selection in terms of the viscoelastic parameters. They fix Pr and E and show that increasing L stationary convection changes into standing waves by means of a subcritical bifurcation. The convective and absolute instabilities for the three model time derivatives of the Oldroyd fluid were investigated by Martínez- Mardones et al. [38]. If the group velocity is zero at say $k = k_0$ and the real part of σ, s, in Equation 39 is positive, it is said that the instability is absolute. In this case, the perturbations grow with time at a fixed point in space. If the perturbations are carried away from the initial point and at that point the perturbation decays with time, the instability is called convective. By means of coupled complex Ginzburg-Landau equations (Landau equations with second derivatives in space and complex coefficients) they investigated problems for which oscillatory convection appears first. Besides, they investigated the effect the group velocity has on oscillatory convection. It is found that the conductive state of the fluid layer is absolutely unstable if $L > 0$ or $E > E_C$ and therefore, when $0 < E < E_C$, the state is convectively unstable. They also show that there is no traveling wave phenomena when passing from stationary convection to standing waves.

7. Conclusions

In this chapter many phenomena have been discussed in order to show the variety of problems which can be found in natural convection of Newtonian and viscoelastic fluids. One of the goals was to show that the different boundary conditions may give results which differ considerably from each other. Sometimes, the results are qualitatively the same and this is taken as an advantage to solve "simpler" problems as those corresponding to the linear and nonlinear equations with free-free boundary conditions. A change in the setting of the problem may produce large complications, as in the case of the free-free boundary conditions, but with one of them being deformable. In this case a new parameter appears, the Galileo number G, which complicates not only the number of numerical calculations, but also the physical interpretation of the results, as explained above. As have been shown, the introduction of viscoelasticity complicates even more the physics of convection. Depending on the boundary conditions, there can be stationary and oscillatory cells in linear convection. Nonlinear convection can be stationary but for other magnitudes of the parameters, traveling and standing waves may appear as the stable fluid motion. The problem is to find the conditions and magnitudes of the viscoelastic parameters when a particular convection phenomenon occurs. This is the thrilling part of viscoelastic convection. It is the hope of the present author that this review may motivate a number of readers to work in this rich area of research.

Acknowledgements

The author would like to thank Joaquín Morales, Cain González, Raúl Reyes, Ma. Teresa Vázquez and Oralia Jiménez for technical support.

Author details

L. A. Dávalos-Orozco
Instituto de Investigaciones en Materiales, Departamento de Polímeros,
Universidad Nacional Autónoma de México,
Ciudad Universitaria, Circuito Exterior S/N,
Delegación Coyoacán, 04510 México D. F., México

8. References

[1] Aspe, H. & Depassier, M. C. (1990). Evolution equation of surface waves in a convecting fluid. *Physical Review A*, Vol. 41, No. 6, 3125 - 3128.
[2] Benguria, R. D. & Depassier, M. C. (1987). Oscillatory instabilities in the Rayleigh-Bénard problem with a free surface. *Physics of Fluids A*, Vol. 1, No. 7, 1123 - 1127.
[3] Bhattacharjee, J. K. (1987). *Convection and Chaos in Fluids*, World Scientific, ISBN 9971-50-224-0 , Singapore.
[4] Bird, R. B., Armstrong, R. C. & Hassager, O. (1987). *Dynamics of Polymeric Liquids, Vol. 1*, Wiley, ISBN 0-471-80245-X, New York.
[5] Brand, H. R. & Zielinska B. J. A. (1986). Tricritical codimension-2 point near the onset of convection in viscoelastic liquids. *Physical Review Letters*, Vol. 57, No. 25, 3167 - 3170.
[6] Cericier, P., Rahal, S., Cordonnier J. & Lebon, G. (1998). Thermal influence of boundaries on the onset of Rayleigh-Bénard convection. *International Journal of Heat and Mass Transfer*, Vol. 41, No. 21, 3309 - 3320.
[7] Chandrasekhar, S. (1981). *Hydrodynamic and Hydromagnetic Stability*, Dover, ISBN 0-486-64071-X, New York.
[8] Chapman, C. J. & Proctor, M. R. E., (1980). Rayleigh-Bénard convection between poorly conducting boundaries. *Journal of Fluid Mechanics*, Vol. 101, No. 4, 759 - 782.
[9] Coleman, B. D. & Noll, W., (1961). Foundations of linear viscoelasticity. *Review of Modern Physics*, Vol. 33, No. 2, 239 - 249.
[10] Coleman, B. D. & Markovitz, H., (1964). Normal stress effects in second-order fluids. *Journal of Applied Physics*, Vol. 35, No. 1, 1 - 9.
[11] Dávalos, L. A. O. (1984). Magnetoconvection in a rotating fluid between walls of low thermal conductivity. *Journal of the Physical Society of Japan*, Vol. 53, No. 7, 2173 - 2176.
[12] Dávalos, L. A. O. & Manero O. (1986). Thermoconvective instability of a second order fluid. *Journal of the Physical Society of Japan*, Vol. 55, No. 2, 442 - 445.
[13] Dávalos-Orozco, L. A. & López Mariscal P. G. (1995). Natural convection in a rotating fluid layer with deformable free surface. *Geophysical and Astrophysical Fluid Dynamics*, Vol. 80, No. 1, 75 - 102.
[14] Dávalos-Orozco, L. A. & Vázquez-Luis, E. (1999). Natural convection of a viscoelastic fluid with deformable free surface. *Journal of Non-Newtonian Fluid Mechanics*, Vol. 85, No. 2, 257 - 271.
[15] Depassier, M. C. (2006). Evolution equation for bidirectional surface waves in a convecting fluid. *Physics of Fluids*, Vol. 18, No. 107102, 1 - 6.
[16] Dunn, J. E. & Rajagopal, K. R. (1995). Fluids of differential type: Critical review and thermodynamic analysis. *International Journal of Engineering Science*, Vol. 33, No. 5, 689 - 729.

[17] Finalyson, B. A. (1968). The Galerkin Method applied to convective instability problems. *Journal of Fluid Mechanics*, Vol. 33, No. 1, 201 - 208.

[18] Finlayson, B. A. (1972). *The Method of Weighted Residuals and Variational Principles*, Academic Press, ISBN 978-0-122-57050-6, New York.

[19] Gershuni, G. Z. & Zhukhovitskii, E. M. (1976). *Convective Stability of Incompressible Fluids*, Keter Publishing House Jerusalem Ltd., ISBN 0-7065-1562-5, Jerusalem.

[20] Giesekus, H. (1963). Die simultane Translations- und Rotationsbewegun einer Kugel in einer elastoviscosen flussigkeit. *Rheologica Acta*, Vol. 03, No. 1, 59 - 71.

[21] Green III, T. (1968). Oscillating convection in an elasticoviscous liquid. *Physics of Fluids*, Vol. 11, No. 7, 1410 - 1412.

[22] Hoyle, R. B. (1998). Universal instabilities of rolls, squares and hexagones, In: *Time-Dependent Nonlinear Convection*, Tyvand, P. A., (Ed.), 51 - 82, Computer Mechanics Publications, ISBN 1-85312-521-0, Southampton.

[23] Hoyle, R. B. (2006). *Pattern Formation, An Introduction to Methods*, Cambridge University Press, ISBN 978-0-521-81750-9, Cambridge.

[24] Huilgol, R. R. (1973). On the solution of the Bénard problem with boundaries of finite conductivity. *SIAM Journal of Applied Mathematics*, Vol. 24, No. 2, 226 - 233.

[25] Hurle, D. T. J., Jakeman, E. & Pike E. R. (1967). On the solution of the Bénard problem with boundaries of finite conductivity. *Proceeding of the Royal Society of London A*, Vol. 296, No. 1447, 469 - 475.

[26] Jakeman, E (1968). Convective instability in fluids of high thermal diffusivity. *Physics of Fluids*, Vol. 11, No. 1, 10 - 14.

[27] Jeffreys, H. (1926). The stability of a layer of fluid heated from below. *Philosophical Magazine Series 7*, Vol. 2, No. 10, 833 - 844.

[28] Kapitaniak, T. (2000). *Chaos for Engineers*, Springer-Verlag, ISBN 3-540-66574-9, Berlin.

[29] Khayat, R. E. (1995). Fluid elasticity and the transition to chaos in thermal convection. *Physical Review E*, Vol. 51, No. 1, 380 - 399.

[30] Kolkka, R. W. & Ierley, G. R. (1987). On the convected linear instability of a viscoelastic Oldroyd B fluid heated from below. *Journal of Non-Newtonian Fluid Mechanics*, Vol. 25, No. 2, 209 - 237.

[31] Landau, L. D. & Lifshitz, E. M. (1987). *Fluid Mechanics*, Pergamon Press, ISBN 0-08-033933-6 , New York.

[32] Lorenz, E. N. (1963). Deterministic non- periodic flows. *Journal Atmospheric Science*, Vol. 20, No. 2, 130 - 141.

[33] Malkus, W. V. R. & Veronis, G. (1958). Finite amplitude cellular convection. *Journal of Fluid Mechanics*, Vol. 4, No. 3, 225 - 260.

[34] Markovitz, H. & Coleman, B. D. (1964). Incompressible Second-Order Fluids, In: *Advances in Applied Mechanics Volume 8*, Dryden, H. L. & von Kármán, T., (Ed.), 69 - 102, Academic Press, ISBN 978-0120020089, London.

[35] Martínez-Mardones, J. & Pérez-García, C. (1990). Linear instability in viscoelastic fluid convection. *Journal of Physics: Condensed Matter*, Vol. 2, No. 5, 1281 - 1290.

[36] Martínez-Mardones, J., Tiemann, R., Zeller, W. & Pérez-García, C. (1994). Amplitude equation in polymeric fluid convection. *International Journal of Bifurcation and Chaos*, Vol. 4, No. 5, 1347 - 1351.

[37] Martínez-Mardones, J., Tiemann, R., Walgraef, D. & Zeller, W.(1996). Amplitude equations and pattern selection in viscoelastic convection. *Physical Review E*, Vol. 54, No. 2, 1478 - 1488.
[38] Martínez-Mardones, J., Tiemann, R., Walgraef, D.(1999). Convective and absolute instabilities in viscoelastic fluid convection. *Physica A*, Vol. 268, No. 1, 14 - 23.
[39] Newell, A. C. & Whitehead, J. A. (1969). Finite bandwidth, finite amplitude convection. *Journal of Fluid Mechanics*, Vol. 38, No. 2, 279 - 303.
[40] Nield, D. A. (1964). Surface tension and buoyancy effects in cellular convection. *Journal of Fluid Mechanics*, Vol. 19, No. 3, 341 - 352.
[41] Oldroyd, J. G. (1950). On the formulation of rheological equations of state. *Proceedings of the Royal Society of London A*, Vol. 200, No. 1063, 523 - 541.
[42] Ortiz-Pérez A. S. & Dávalos-Orozco, L. A. (2011). Convection in a horizontal fluid layer under an inclined temperature gradient. *Physics of Fluids*, Vol. 23, No. 084107, 1 - 11.
[43] Park, H. M. & Lee, H. S. (1995). Nonlinear hydrodynamic stability of viscoelastic fluids heated from below. *Journal of Non-Newtonian Fluid Mechanics*, Vol. 60, No. 1, 1 - 26.
[44] Park, H. M. & Lee, H. S. (1996). Hopf bifurcation of viscoelastic fluids heated from below. *Journal of Non-Newtonian Fluid Mechanics*, Vol. 66, No. 1, 1 - 34.
[45] Pearson, J. R. A. (1958). On convection cells induced by surface tension. *Journal of Fluid Mechanics*, Vol. 4, No. 5, 489 - 500.
[46] Pérez- Reyes, I. & Dávalos-Orozco, L. A. (2011). Effect of thermal conductivity and thickness of the walls in the convection of a viscoelastic Maxwell fluid layer. *International Journal of Heat and Mass Transfer*, Vol. 54, No. 23, 5020 - 5029.
[47] Petrie, C. J. S. (1979). *Elongational Flows*, Pitman Publishing Limited, ISBN 0-273-08406-2 , London.
[48] Pérez- Reyes, I. & Dávalos-Orozco, L. A. (2012). Vertical vorticity in the non- linear long wavelength instability of a viscoelastic fluid layer. To be submitted.
[49] Pismen, L. M. (1986). Inertial effects in long-scale thermal convection. *Physics Letters A*, Vol. 116, No. 5, 241 - 244.
[50] Proctor, M. R. E. (1981). Planform selection by finite-amplitude thermal convection between poorly conducting slabs. *Journal of Fluid Mechanics*, Vol. 113, 469 - 485.
[51] Prosperetti, A. (2011). A simple analytic approximation to the Rayleigh-Bénard stability threshold. *Physics of Fluids*, Vol. 23, No. 124101, 1 - 8.
[52] Rajagopal, K. R., Ruzicka, M. and Srinivasa, A. R. (1996) On the Oberbeck-Boussinesq approximation. *Mathematical Models and Methods in Applied Sciences*, Vol. 6, No. 8, 1157 - 1167.
[53] Lord Rayleigh (1916). On convective currents in a horizontal layer of fluid when the higher temperature is on the under side. *Philosophical Magazine Series 6*, Vol. 32, No. 192, 529 - 546.
[54] Reid, W. H. & Harris, D. L. (1958). Some further results on the Bénard problem. *Physics of Fluids*, Vol. 1, No. 2, 102 - 110.
[55] Rosenblat, S. (1986). Thermal convection in a viscoelastic liquid. *Journal of Non-Newtonian Fluid Mechanics*, Vol. 21, No. 2, 201 - 223.
[56] Segel, L. A. (1969). Distant side-walls cause slow amplitude modulation of cellular convection. *Journal of Fluid Mechanics*, Vol. 38, No. 1, 203 - 224.
[57] Segel, L. A. & Stuart, J. T. (1962). On the question of the preferred mode in cellular thermal convection. *Journal of Fluid Mechanics*, Vol. 13, No. 2, 289 - 306.

[58] Siddheshwar, P. G. & Sri Krishna C. V. (2002). Unsteady non-linear convection in a second- order fluid. *International Journal of Non-Linear Mechanics*, Vol. 37, No. 2, 321 - 330.

[59] Sokolov, M. & Tanner, R. I., (1972). Convective stability of a general viscoelastic fluid heated from below. *Physics of Fluids*, Vol. 15, No. 4, 534 - 539.

[60] Stastna, J. (1985). Convection and overstability in a viscoelastic fluid. *Journal of Non-Newtonian Fluid Mechanics*, Vol. 18, No. 1, 61 - 69.

[61] Stuart, J. T. (1964). On cellular patterns in thermal convection. *Journal of Fluid Mechanics*, Vol. 18, No. 4, 481 - 498.

[62] Takashima, M (1972). Thermal instability of a viscoelastic fluid layer. I *Journal of the Physical Society of Japan*, Vol. 33, No. 2, 511 - 518.

[63] Tanner, R. I., (1966). Plane Creeping Flows of Incompressible Second-Order Fluids. *Physics of Fluids*, Vol. 9, No. 8, 1246 - 1247.

[64] Van Der Borght, R., Murphy, J. O. & Steiner J. M. (1974). A theoretical investigation of finite amplitude thermal convection in non-Newtonian fluids. *Zeitschrift für Angewandte Mathematik und Mechanik*, Vol. 36, No. 3, 613 - 623.

[65] Veronis, G. (1966). Large amplitude Bénard convection. *Journal of Fluid Mechanics*, Vol. 26, No. 1, 49 - 68.

[66] Vest, C. M. & Arpaci, V. S. (1969). Overstability of a viscoelastic fluid layer heated from below. *Journal of Fluid Mechanics*, Vol. 36, No. 3, 613 - 623.

Chapter 2

Turbulent Flow of Viscoelastic Fluid Through Complicated Geometry

Takahiro Tsukahara and Yasuo Kawaguchi

Additional information is available at the end of the chapter

http://dx.doi.org/10.5772/52049

1. Introduction

Viscoelastic liquids with very small amounts of polymer/surfactant additives can, as well known since B.A. Toms' observation in 1948, provide substantial reductions in frictional drag of wall-bounded turbulence relative to the corresponding Newtonian fluid flow. Friction reductions of up to 80% compared to the pure water flow can be occasionally achieved with smooth channel/pipe flow of viscoelastic surfactant solution [11, 54]. This friction-reducing effect, referred to as turbulent drag reduction (DR) or Toms effect, has been identified as an efficient technology for a large variety of applications, e.g. oil pipelines [25] and heating/cooling systems for buildings [43], because of major benefits in reducing energy consumption.

It has been known that long, high-molecular-weight, flexible polymers or rod-like micelle networks of surfactant are particularly efficient turbulence suppressor, so that those solutions lead to different turbulent states both qualitatively and quantitatively, resulting in dramatic DRs. One of promising additives, which may allow their solutions to induce DR, is a cationic surfactant such as "cetyltrimethyl ammonium chloride (CTAC)" under appropriate conditions of surfactant chemical structure, concentration, counter-ion, and temperature to form micellar networks in the surfactant solution. Those resulting micro-structures give rise to viscoelasticity in the liquid solution. The properties and characteristics of the viscoelastic fluids measured even in simple shear or extensional flows are known to exhibit appreciably different from those of the pure solvent. From a phenomenological perspective, their turbulent flow is also peculiar as is characterized by extremely elongated streaky structures with less bursting events. Therefore, the viscoelastic turbulence has attracted much attention of researchers during past 60 years. Intensive analytical, experimental, and numerical works have been well documented and many comprehensive reviews are available dealing with this topic: [cf., 18, 19, 26, 35, 51, , and others].

Although the mechanism of DR is still imperfectly understood, but some physical insights have emerged. In particular, with the aid of recent advanced supercomputers, direct

INTECH
open science | open minds

© 2012 The Author(s). Licensee InTech. This chapter is distributed under the terms of the Creative Commons Attribution License http://creativecommons.org/licenses/by/3.0), which permits unrestricted use, distribution, and reproduction in any medium, provided the original work is properly cited.

numerical simulations (DNSs) of viscoelastic fluid as well as the Newtonian fluid have been increasingly performed [e.g., 1, 7, 17, 41, 44]. Some progresses in the model of DR and in the understanding of modulated turbulent structures have been made by L'vov et al. [20] and Roy et al. [39]. Later, Kim et al. [16] carried out DNS to examine interactions between the coherent structures and the fluid viscoelasticity. They reported a dependency of the vortex-strength threshold for the auto-generation of new hairpin vortices in the buffer layer on the viscoelasticity. Most of DNS studies in the literature are performed on flows over smooth wall surface and other simple flow configurations, such as channel flow, boundary layer, isotropic turbulence, and shear-driven turbulence.

As well as smooth turbulent flows in plane channel and pipe, the turbulent flow through complex geometries has both fundamental scientific interest and numerous practical applications: such flows are associated with the chemical, pharmaceutical, food processing, and biomedical engineering, where the analysis and designing for their pipe-flow systems are more difficult than for its Newtonian counterpart. This is mainly because severe limitations in the application of ideal and Newtonian flow theories to these relevant flow problems. Most of the previous work presented in the literature concerning this subject has been done with flows either through sudden expansion or over backward-facing step. The flow even in such relatively simple cases of complex geometries exhibits important features that partain to complex flows containing flow separation, reattachment, and often an extremely high level of turbulence. A better understanding of viscoelastic-fluid behavior and turbulent flow properties of those flows should lead to both the design and the development of hydrodynamically more efficient processes in various pipe-flow systems and to an improved quality control of the final products. Consequently, in situations of both practical and fundamental importance, we have investigated the detailed mechanism and efficiency of DR for viscoelastic turbulent flow through roughened channel, or an orifice flow, that is one of canonical flows involving separation and reattachment. The goal of a series of our works is to better understand the physics of viscoelastic turbulent flow in complicated flow geometry.

The following subsections give a brief introduction to the preceding studies that motivated us to further investigate the viscoelastic turbulent orifice flow and describe the more specific purpose of the study reported in this chapter.

1.1. Related studies

As far as we know there exist no other DNS studies on the viscoelastic turbulent orifice flow than those carried out by authors' group recently. However, there are a few experimental and numerical works on sudden expansion and backward-facing step owing to their geometrical simplicity. Table 1 summarizes several earlier works.

As for the Newtonian fluid, Makino et al. [22, 23] carried out DNSs of the turbulent orifice flow, and investigated also the performance of heat transfer behind the orifice. They reported several differences in turbulent statistics between the orifice flow and other flows of the sudden expansion and the backward-facing step. Recently, the authors' group investigated the viscoelastic fluid in the channel with the same rectangular orifice using DNS [46, 49]. We found phenomenologically that the fluid viscoelasticity affected on various turbulent motions in just downstream of the orifice and attenuated spanwise vortices.

By means of experiments, we confirmed the turbulence suppression in the region behind the orifice and analyzed the flow modulation with respect to the turbulent structures by

Configuration	Author(s)	Method	Expansion ratio
Orifice	Present	Sim.	1:2
	Tsurumi et al. [50]	Exp.	1:2
Sudden expansion	Pak et al. [29]	Exp.	1:2, 3:8
	Castro & Pinho [3]	Exp.	1:1.54
	Escudier & Smith [8]	Exp.	1:1.54
	Poole & Escudier [32, 33]	Exp.	1:2, 1:4
	Oliveira [28]	Sim.	1:2
	Manica & De Bortoli [21]	Sim.	1:3
	Dales et al. [5]	Exp.	1:1.5
	Poole et al. [34]	Sim.	1:3
Backward-facing step	Poole & Escudier [31]	Exp.	1:1.43

Table 1. Relevant previous studies on viscoelastic turbulent flow: Exp., experiment; Sim., numerical simulation.

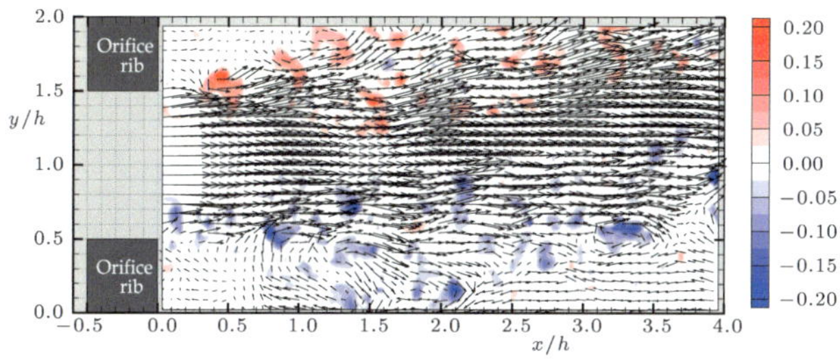

(a) Water

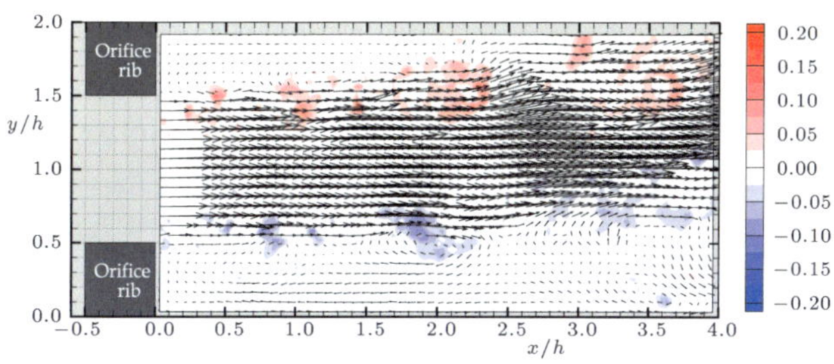

(b) CTAC, 150 ppm

Figure 1. Snapshots of flow fields behind the orifice, taken by PIV measurement: vector, (u, v); contour, the swirling strength $\lambda_{ci}\omega_z/|\omega_z|$ (positive, anti-clockwise rotation; negative, clockwise). The main flow direction is from left to right. Cited from [50].

using PIV (particle image velocimetry) [50]. Figure 1 shows the instantaneous velocity vectors in a plane of interest. Also shown is the contour of swirling strength, by which the vortex core can be extracted by plotting iso-surface of $\lambda_{ci} > 0$, the imaginary part of complex conjugate eigenvalue of velocity-gradient tensor in the two-dimensional plane, and the rotational direction be evaluated by the sign of spanwise rotation ω_z. As can be seen in the figure, the sudden expansion of the orifice leads to generation of strong separated shear layers just behind the orifice-rib edges. This shear layer enhances turbulence dominantly

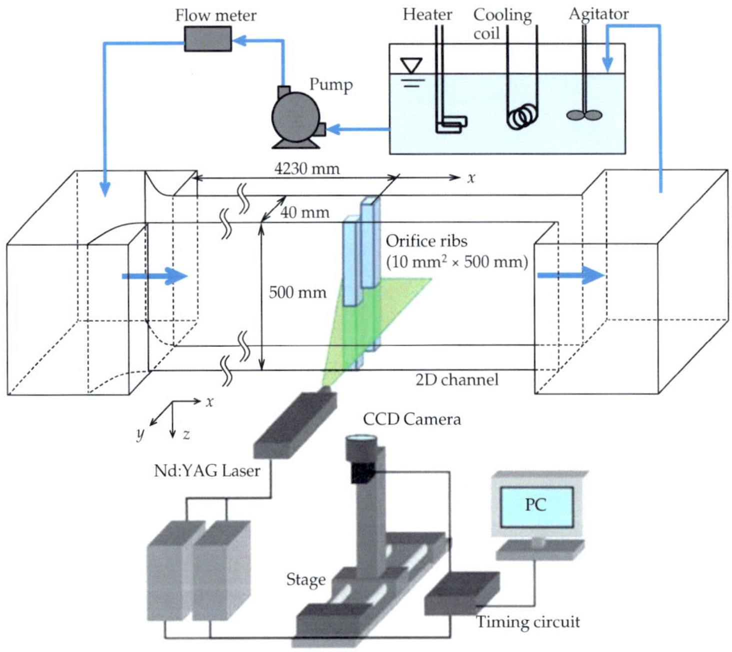

Figure 2. Outline of experimental apparatus with PIV system.

in the water flow, while the viscoelastic flow seems rather calm. Here, the viscoelastic fluid they employed was the CTAC solution with 150 ppm of weight concentration. A schematic of the experimental set up is depicted in Fig. 2. The Reynolds number based on the actual bulk mean velocity passing the orifice were $Re_m = 8150$ for water and 7840 for the viscoelastic fluid (CTAC solution), which were obtained under the same pumping power. It is interesting to note that the hydrodynamic drag throughout the channel including the orifice is rather increased in the viscoelastic flow despite the presence of turbulence-suppression phenomenon. We conjectured that, in the experiment, any DR did not apparently occur because an increment of the skin friction by an extra shear stress due to viscoelasticity exceeded a decrement of the Reynolds shear stress. It might be difficult to determine the individual contribution of either turbulence, viscosity, or viscoelasticity in such an experimental study. To achieve clearer pictures of the role of viscoelasticity and turbulence modulations affecting on DR, we should re-examine the viscoelastic turbulent orifice flow with emphasis on the viscoelastic force (stress) exerted on the fluctuating flow motion.

1.2. Purpose

In the present study, we will focus on an instantaneous field of the viscoelastic turbulent flow past the rectangular orifice and discuss mainly the interaction between the turbulent fluid motion and the (polymer/surfactant) additive conformation field, i.e. the balance of the inertia, viscous, and viscoelastic forcing terms in the governing momentum equation. We have made some preliminary studies which have shown that this flow exhibits a change in the augmentation of the local heat transfer dependently on the streamwise distance from the orifice [49]. Therefore, we propose in this chapter that this streamwise variation of

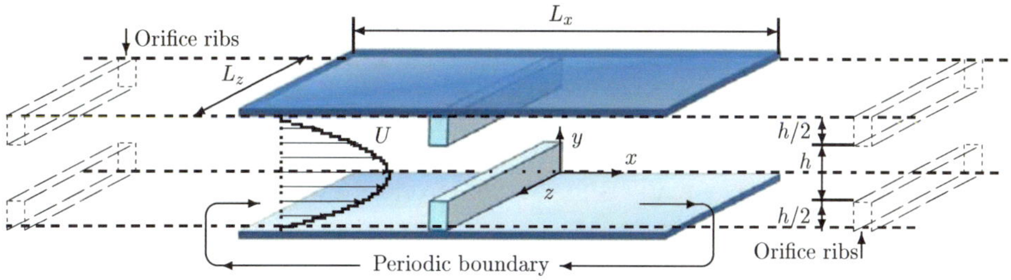

Figure 3. Configuration of the roughened-channel flow for the simulation, where a sequence of regularly-spaced, rectangular, orifices is considered.

the heat-transfer augmentation would be deeply related to the turbulence-viscoelasticity interaction, and suggest its scenario.

We performed DNSs without any turbulence model but with the Giesekus' viscoelastic-fluid model, valid for a polymer/surfactant solution, which is generally capable of reducing the turbulent frictional drag in a smooth channel. The geometry considered here is periodic orifices with the 1:2 expansion ratio.

2. Problem formulation

In this section, the equations governing incompressible viscoelastic-fluid flows are presented in their dimensional and non-dimensional forms. Rheological properties relating to a model we employed here to calculate the polymer/surfactant, or the fluid-viscoelasticity, contribution to the extra-stress tensor are also described.

2.1. Flow configuration

Prior to introducing the equations, let us depict the configuration of the computational domain in Fig. 3. In the three dimensional Cartesian coordinate system, x, y, and z indicate the streamwise, wall-normal, and spanwise directions, respectively. The main flow is driven by the streamwise mean pressure gradient. The flow that we analyzed by DNS was assumed to be fully-developed turbulent flow through an obstructed channel, of height $L_y = 2h$, with periodically repeating two-dimensional orifices (i.e., transverse rectangular orifices): namely, in the simulations, the periodic boundary condition was adopted in the x direction as well as the z direction to allow us to demonstrate an infinite channel and regularly-spaced obstructions Note that, by contrast with the above-mentioned experiment, where transient flows past only one orifice were studied, we numerically investigated the fully-developed flows through a sequence of orifices. As illustrated in Fig. 3, the transverse orifices are placed in every L_x in the x direction.

The height of each rib is chosen as $0.5h$—the channel half height is h—and thus the expansion ratio of the orifice is 1:2, that is equivalent to the experimental condition but the thickness in the x direction is small as $0.1h$. These present conditions relating to the orifice installation are the same as those studied by Makino et al. [22]. The no-slip boundary condition is used on all the wall surfaces including the faces of the orifice.

The domain size in the streamwise direction ($L_x = 12.8h$) was not sufficiently long that the effects of the orifice on the flow approaching next one could be neglected. The domain size along the spanwise direction was chosen as $L_z = 6.4h$, which was confirmed to be adequate based on the convergence of the spanwise correlation to almost zero at this domain size.

2.2. Governing equations

In this study, the governing equations for the three velocity components $\mathbf{u} = \{u, v, w\}$ and pressure p take the form:

$$\nabla \cdot \mathbf{u} = 0, \tag{1}$$

$$\rho \frac{\mathrm{D}\mathbf{u}}{\mathrm{D}t} = -\nabla p + \eta_s \nabla^2 \mathbf{u} + \nabla \cdot \boldsymbol{\tau}_{\mathrm{p}}, \tag{2}$$

with t the time, ρ the fluid density, and η_s the Newtonian-solvent viscosity. These fluid properties are assumed to be constant, irrespective of the flow fields. In the last term, there exists an additional stress-tensor component $\boldsymbol{\tau}_{\mathrm{p}}$, which is related with kinematic quantities by an appropriate constitutive equation. A model which has proved effective in reproducing a power-law region for viscosity and normal-stress coefficients as well as a reasonable description of the elongational viscosity for viscoelastic surfactant solutions was proposed by Giesekus [10]. This model assumes that the extra stress $\boldsymbol{\tau}_{\mathrm{p}}$ due to additives in the solution satisfies

$$\boldsymbol{\tau}_{\mathrm{p}} + \lambda \overset{\nabla}{\boldsymbol{\tau}}_{\mathrm{p}} + \alpha \frac{\lambda}{\eta_{\mathrm{a}}} \left(\boldsymbol{\tau}_{\mathrm{p}} \cdot \boldsymbol{\tau}_{\mathrm{p}} \right) = 2\eta_{\mathrm{a}} \mathbf{S}, \tag{3}$$

where λ is the relaxation times, $\overset{\nabla}{\boldsymbol{\tau}}_{\mathrm{p}}$ is the upper convected derivative of the stress tensor, and $\mathbf{S}$ is the deformation tensor. The parameter η_{a} has dimensions of viscosity (but note that η_{a} represents the additive contribution to the zero-shear-rate solution viscosity: $\eta_0 = \eta_{\mathrm{s}} + \eta_{\mathrm{a}}$). For the mobility parameter representing magnitude of the non-linearity of the fluid elasticity, $\alpha = 0.001$ was given as our previous studies [45, 46, 52, 53].

From several kinds of the non-Newtonian fluid model, such as FENE-P model and Oldroyd-B model, we chose the Giesekus model to properly resolve the evolution of extra stress due to the deformation of macromolecules in the surfactant solution. One can find in the literature [14, 42] that measured rheological properties of the surfactant solution agree well with those of a Giesekus fluid.

To re-write Equations (2) and (3) in their non-dimensional forms, we should introduce a dimensionless conformation tensor $\mathbf{c} = c_{ij}$ given by an explicit function of

$$\boldsymbol{\tau}_{\mathrm{p}} = \frac{\eta_{\mathrm{a}}}{\lambda} \left(\mathbf{c} - \mathbf{I} \right). \tag{4}$$

and derive the dimensionless forms as follows:

$$\frac{\partial u_i^+}{\partial t^*} + u_j^+ \frac{\partial u_i^+}{\partial x_j^*} = -\frac{\partial p^+}{\partial x_i^*} + \frac{\beta}{Re_{\tau 0}} \frac{\partial^2 u_i^+}{\partial x_j^{*2}} + \frac{1-\beta}{We_{\tau 0}} \frac{\partial c_{ij}}{\partial x_j^*} + F_i, \tag{5}$$

and

$$\frac{\partial c_{ij}}{\partial t^*} + u_k^+ \frac{\partial c_{ij}}{\partial x_k^*} - \frac{\partial u_i^+}{\partial x_k^*} c_{kj} - c_{ik} \frac{\partial u_j^+}{\partial x_k^*} + \frac{Re_{\tau 0}}{We_{\tau 0}} \left[c_{ij} - \delta_{ij} + \alpha \left(c_{ik} - \delta_{ik} \right) \left(c_{kj} - \delta_{kj} \right) \right] = 0. \tag{6}$$

Quantities with a superscript, $\square^+$, indicate that they are normalized by the friction velocity $u_{\tau 0}$, that is given by the force balance between the wall shear stress and the mean pressure gradient through the computational volume in the case of the plane channel without any orifice nor obstruction, i.e.,

$$u_{\tau 0} = \sqrt{\frac{\tau_{w0}}{\rho}} = \sqrt{-\frac{\delta}{\rho} \cdot \frac{\Delta p}{L_x}}. \tag{7}$$

Here, τ_{w0} is the averaged wall shear stress for a smooth plane channel flow and Δp is the time-averaged pressure drop between $x = -L_x/2$ and $L_x/2$. The other superscript, or $\square^*$, represents non-dimensionalization by the channel half width: e.g., $x^* = x/h$. As for the three non-dimensional parameters of $Re_{\tau 0}$, $We_{\tau 0}$, and β, their definitions and specific values will be described in Section 2.3

In order to mimic the solid body of the orifice in the fluid flow, the direct-forcing immersed boundary method [9, 24] was applied on the surface of the orifice ribs and their inside. The additional term of F_i in Equation (5) represents the body force vector per unit volume for this method.

2.3. Rheological and flow parameters

We executed two simulations of the orifice flows for either viscoelastic fluid or Newtonian fluid, for comparison. All present DNSs were run under a constant pressure drop: $\Delta p/L_x$ was constant and it enabled us to define a constant friction Reynolds number $Re_{\tau 0} = \rho u_{\tau 0} h/\eta_0$. In this work, $Re_{\tau 0}$ was fixed at 100 for each simulation. The friction Weissenberg number representing the ratio between the relaxation time and the viscous time scale was chosen to be $We_{\tau 0} = \rho \lambda u_{\tau 0}^2/\eta_0 = 20$ in the viscoelastic flow. The viscosity ratio of the solvent viscosity to the solution viscosity at a state of zero shear stress was taken as $\beta = \eta_s/\eta_0 = 0.8$. These rheological conditions would provide a noticeable drag-reducing effect to turbulent flows in a smooth channel. Actually, our previous DNS result that pertained to the similar condition ($Re_{\tau 0} = 150$, $We_{\tau 0} = 10$, and $\beta = 0.8$) provided a moderate drag reduction more than 10% [45]. As for the Newtonian fluid, these parameters corresponds to $We_{\tau 0} = 0$ and $\beta = 1$, which leads to Equation (5) in the common form for the Newtonian fluid.

In our previous works [46, 49], while the friction Reynolds number and the Weissenberg number were ranged, respectively, from 100 to 200 and from 0 to 30, the drag reduction rates of 15–20% were achieved in the viscoelastic flows. Unfortunately, to the authors' knowledge, there has never been any other DNS study of viscoelastic turbulent orifice flow, partly due to numerical difficulty, namely, the Hadamard instability [12] in viscoelastic-flow calculations.

3. Numerical procedure

3.1. spatial discretization

The finite difference method was adopted for the spatial discretization. Two different grid numbers of $256 \times 128 \times 128$ and $128 \times 128 \times 128$ in (x, y, z) were used for the Newtonian and the viscoelastic flows, respectively, since the Newtonian turbulent flow is basically accompanied by finer eddies that need to be resolved. For the coarser mesh, the spatial resolutions were $\Delta x^+ = 10$ and $\Delta z^+ = 5$ with the computational domain size of $L_x \times L_y \times L_z = 12.8h \times 2h \times 6.4h$ and were reasonable to capture flow variations and small-scale eddies behind the orifice. According to the results shown later, the orifice flow of the viscoelastic fluid indeed has exhibited relatively-large turbulent structures and offered reasonable statistics. In

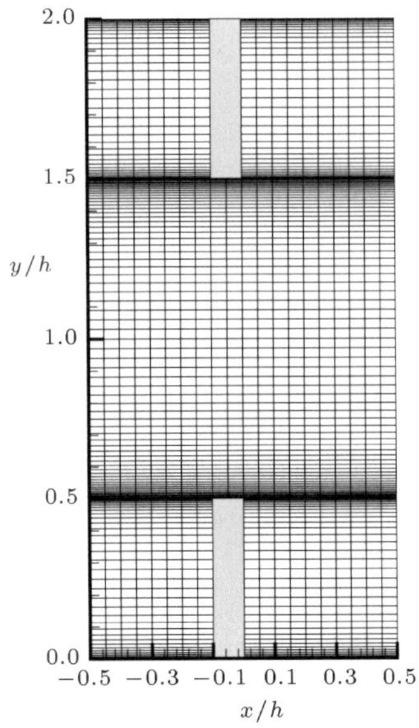

Figure 4. Computational grids with emphasis on the orifice ribs, viewed from the spanwise direction.

the wall-normal direction, the density of the computational mesh is not uniform so that dense grids appear at the height of the orifice rib and near the walls, as shown in Fig. 4. Using a hyperbolic tangent stretching factor, we employed the resolutions of $\Delta y^+ = 0.31$–3.01.

In the x and z directions, the central difference scheme with the 4th-order accuracy was employed, while the 2nd-order accuracy was in the y direction. It should be noted, however, that the 'minmod' flux-limiter scheme as a TVD (total-variation diminishing) method was adopted to the non-linear term in Equation (6): details of this numerical method can be found in the authors' papers [48, 52].

3.2. Time integration

The SMAC (simplified marker and cell) method was applied for coupling between Equations (1) and (5); and the time advancement was carried out by the 3rd-order Runge-Kutta (RK) scheme, but the 2nd-order Crank-Nicolson scheme was used for the viscous term in the y direction, although of course several alternatives to these methods for the coupling and time integration may be available. Regarding the issue to ensure proper methods, we preliminarily tested a variety of combinations with the same flow geometry and conditions and evaluated their availability with emphasis on the orifice rib. Before showing this verification result, let us note again that we employed the immersed-boundary technique for the orifice ribs, or the rigid solid phase in fluid circumstance. The idea of this technique, firstly proposed by Peskin [30], is to employ a regular Cartesian grid, but to apply additional suitable momentum source within the domain in order to satisfy the requisite conditions at the interface and inside of the solid phase [36]. In the present simulation with the Cartesian grid, the velocities defined either at the surface or the inside of the orifice ribs were required to be zero. Hence, we should appropriately calculate the additional term F_i in Equation (5) to drive those velocities to the desired value, when compute the set of the governing equations. We examined two coupling methods—the SMAC method

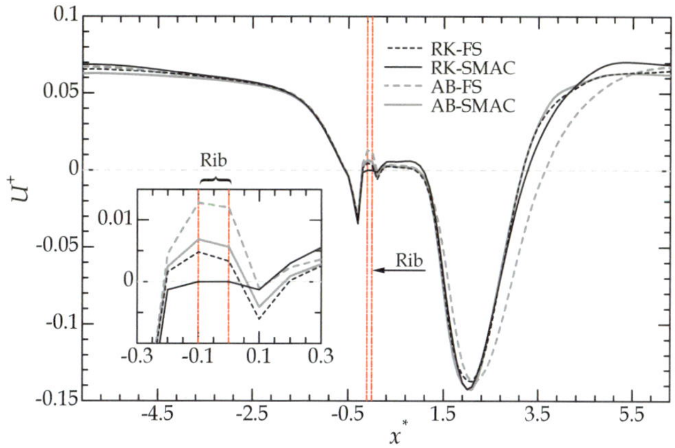

Figure 5. Dependence on numerical methods: streamwise distribution of a velocity in the vicinity of the upper wall for the Newtonian-fluid flow at $Re_{\tau 0} = 100$.

and the fractional-step (FS) method—and two time-integration scheme—the RK scheme and the 2nd-order Adams-Bashforth (AB) scheme. Figure 5 compares the results obtained by the different combinations of them. We choose to show only the curves for the near-wall distribution of the mean streamwise velocity, U^+ at $y^* = 0.0053$ ($y^+ = 0.53$) from the upper wall. Focusing on the location of the rib, you can find considerable variability of the value dependently on the scheme combination: see the inset of Fig. 5. As might be expected, the RK scheme with higher accuracy gave near-zero U at the rib, indicating a better performance than the AB scheme. Moreover, the marked reduction in the U pertaining to the SMAC method can be also clearly seen. It reflects the fact that, when combined with appropriate choice of coupling method and time-integration scheme, this immersed-boundary method leads to a reasonable simulation.

One may observe other locations where scheme-dependent variability seems to be significant. For instance, the reattachment point at which the near-wall U changes its sign was apparently found to vary according to numerical schemes, as seen in Figure 5. This might be true, but a large deviation of the reattachment point by the combination of AB and SMAC is attributed to a straightening phenomena in the mean flow, which is essentially bended to one wall by the Coanda effect. Although not shown in figure, the same verification of U in the core region was also done, but revealed no remarkable variation between different methods at every streamwise position. It implies that the scheme-dependency can be ignorable except for the near-wall region at the orifice. The undesirable non-zero U through the rib in the case of AB-FS is thought of as a main reason for the weaken Coanda effect and the straightened flow.

In the following, instantaneous flow fields and several turbulence statistics obtained by the DNS with RK-SMAC are shown.

4. Results

4.1. Instantaneous flow field: Kelvin-Helmholtz and turbulent eddies

A major difference between the present study and published works on the smooth channel is related to the streamwise variation of the flow state and the main turbulence-producing area. Although the flow past an orifice is one of the simplest reattaching flows, the flow field is still very complex compared to the smooth channel flow. The contracted flow passing the orifice detaches at its leading edge, forming a separated shear layer. Even if the the contracted flow is laminar-like, transition begins soon after separation unless the Reynolds number is very low as $Re_m < 400$ for the backward-facing step flow [2, 27] and the same orifice flow [22]. The present bulk Reynolds numbers as low as $Re_m = 579$ and 646 for the Newtonian and viscoelastic flows, respectively, exceed this critical value and are in the transitional regime. As a consequence of the increase in Re_m, the viscoelastic flow is determined as it offers a lower drag by about 20%, which corresponds to the 'drag reduction rate.' Most of drag-reduced turbulent flows over smooth wall differ from the Newtonian flows in the same general way [51]: for instance, the fluid viscoelasticity inhibits transfer of kinetic energy from the streamwise to the wall-normal velocity fluctuations, and vorticity fluctuations inducing production of turbulence in near-wall region disappear in the highly drag-reduced flow, even as the bulk Reynolds number is raised from that for the Newtonian flow with the same pressure drop. In these contexts, the instantaneous vortex structures both within the strong shear layer just downstream of the orifice and in the downstream after the reattachment point should be of interest for investigation of viscoelastic-fluid behaviors. Actually, the DNS study on the turbulent heat transfer [49] demonstrated a heat-transfer augmentation in the region of $x > 6h$ (i.e., area after the reattachment) even with drag reduction, when compared with the Newtonian case: we will discuss again regarding this phenomenon in Section 5.3.

Figure 6 presents an instantaneous snapshot of eddies in each of the Newtonian flow and viscoelastic flow, viewed three-dimensionally with emphasis on the orifice downstream. Here, eddies are visualized by the second invariant of the fluctuating velocity-gradient tensor:

$$II' = \frac{\partial u_i'}{\partial x_j} \frac{\partial u_j'}{\partial x_i}. \tag{8}$$

Additionally, the contours in the figures show the instantaneous streamwise velocity ($u = U + u'$) distribution in an arbitrary x-y plane and its distribution near the bottom wall, revealing the adverse flow region just behind the orifice and high and low momentum patches on the wall surface below eddies. The orifice flows presented in Fig. 6 are highly unsteady and turbulent in region behind the orifice for both fluids. However, the viscoelastic flow seems to involve turbulent structures very similar to those in the Newtonian flow, but the number of eddies is drastically reduced. The spanwise vortices, especially Kelvin-Helmholtz (K-H) vortices, in the strong shear layer released from the orifice edge are less remarkable, which is qualitatively consistent with the experimental observation mentioned in Section 1. This vortex suppression phenomenon is expected to be induced by the viscoelasticity.

It is interesting to note that the near-wall streaks becomes highly intermittent but still occurs in the region far downstream of the reattachment point. The mean reattachment point locates at $x = 4.3h$ on the lower-side wall surface. Because of the bulk Reynolds number as low as 650, it is naturally expected that no apparent turbulent eddies should not be observed in far-downstream region away from the orifice, where the flow would be laminar similar to the smooth channel flow at the same Reynolds number. However, as can be seen in the figure, the velocity distributions both of the Newtonian and the viscoelastic flows are far from those in the laminar state. In particular, the viscoelastic flow exhibits larger vortices far from the orifice: elongated longitudinal vortical structures are observed intermittently at $x = 5$–$7h$, as given in Fig. 6(b). Those large-scale structures may induce velocity fluctuations and also

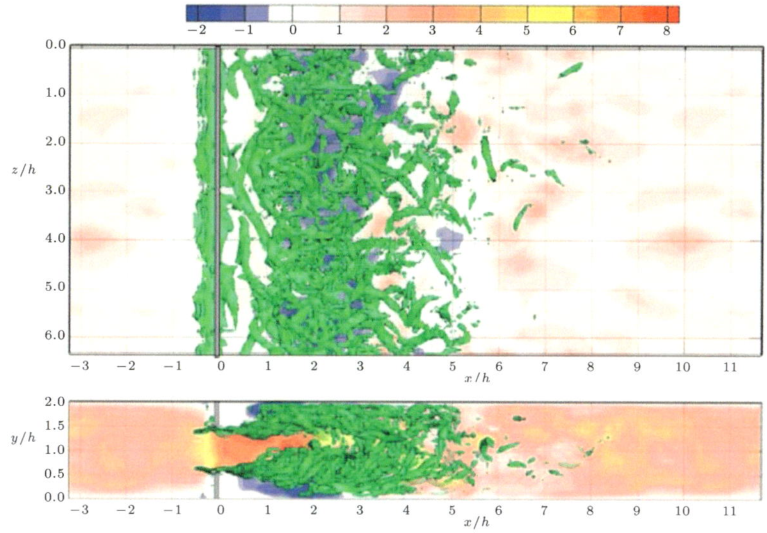

(a) Newtonian fluid

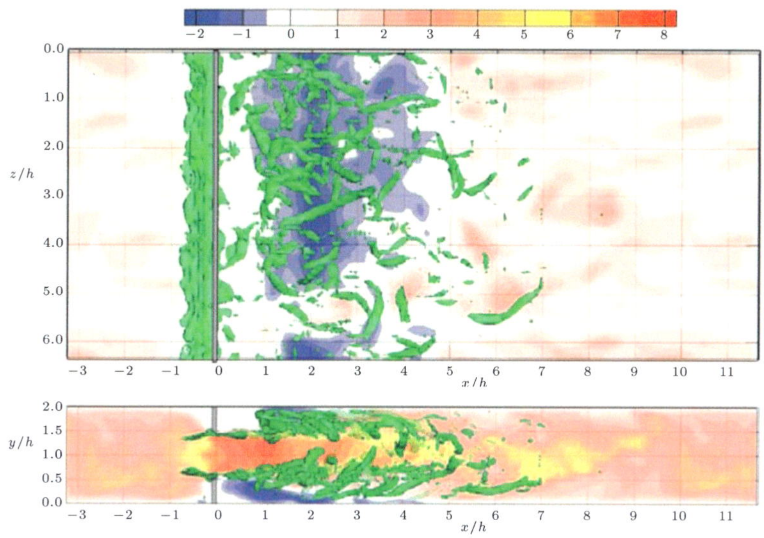

(b) Viscoelastic fluid

Figure 6. Visualized instantaneous flow fields of the obstructed turbulent channel flows. Green iso-surfaces indicate negative regions of the second invariant of the deformation tensor, representing vortical structures. The contours show instantaneous streamwise velocity distribution in an arbitrary x-y plane and in the x-z plane at $y^* = 0.05$ from the lower wall.

significant transports of momentum/heat between the near-wall region and the core region. As demonstrated later (in Section 4.2), the near-wall sweep/ejection motions that pertain to longitudinal vortex are prone to be encouraged by the viscoelastic force. Therefore, it can be conjectured that instability due to the viscoelastic process causes velocity fluctuations as well as vortices in the far downstream region.

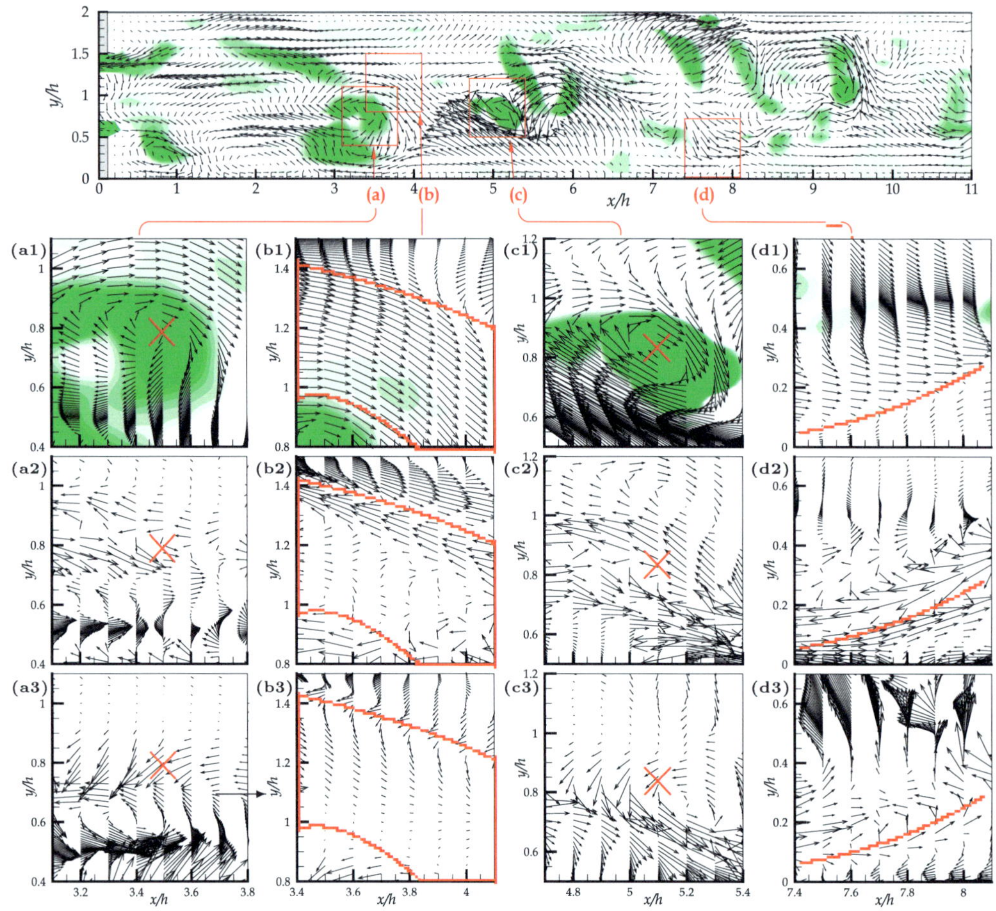

Figure 7. Instantaneous velocity-vector field of the viscoelastic flow, viewed in an arbitrary x-y plane. The main flow moves from left to right. Green contour denotes $II' \leq -0.005$. The small areas surrounded by red borders are shown in enlarged views of the following figures. (a–d) Enlarged views of the instantaneous field of the viscoelastic flow: (a1–d1), same as the top figure, the vector of u' and w' and the contour of $II' \leq -0.005$; (a2–d2) the vector denotes the force contributed by the viscous term (F_x, F_y); (a3–d3) the force by the viscoelastic term (E_x, E_y). The position of (a) focuses on a spanwise vortical motion near the orifice, and the symbol of ($\times$) indicates the vortex center: (b), another location in the core region without determinate vortical motion; (c), another vortical motion far from the orifice: (d), a near-wall turbulent motion (sweep and ejection) downstream of the reattachment point.

Regarding the facts that the K-H vortices as well as subsequent eddies were quickly damped but the quasi-streamwise vortices away from the orifice were sustained in the viscoelastic flow, we will consider these vortical motions in the frame of x-y plane and investigate their relationships to the conformation-stress (polymer or surfactant-micellar network stress) field.

4.2. Viscoelastic force exerted on fluid motions

As indicated in Fig. 7, four different small areas are chosen and compared with the vector patterns of the viscous and the viscoelastic body forcing terms in the governing equation for the fluctuating velocity, namely, the second and third terms in the right-hand side of Equation (5) for u' and v':

$$\text{viscous force} \quad F_x = \frac{\beta}{Re_{\tau 0}} \frac{\partial^2 u'^+}{\partial x_j^{*2}} \ (\text{in } x), \quad F_y = \frac{\beta}{Re_{\tau 0}} \frac{\partial^2 v'^+}{\partial x_j^{*2}} \ (\text{in } y); \tag{9}$$

$$\text{viscoelastic force} \quad E_x = \frac{1-\beta}{We_{\tau 0}} \frac{\partial c_{xj}}{\partial x_j^*} \ (\text{in } x), \quad E_y = \frac{1-\beta}{We_{\tau 0}} \frac{\partial c_{yj}}{\partial x_j^*} \ (\text{in } y). \tag{10}$$

In Fig. 7(a), we extract a spanwise swirling motion that persist to a K-H vortex in a separated shear layer. As clearly described in Fig. 7(a1), the velocity vectors present a clockwise vortical motion and the contour of II' also implies the existence of an eddy. The distributions of viscous and viscoelastic force vectors are displayed in the successive figures, (a2) and (a3), respectively. It is clear that, the both force vectors show an anti-clockwise pattern that is opposite in direction to the fluid swirling, although the center of the viscoelastic force pattern deviate slightly from the center of the flow votex. The viscous force (F_x, F_y) is intensified (either positively or negatively) where the velocity gradient of du/dy drastically changes. The viscous force inherently inhibits the flow vortical motion.

As for a non-rotating fluid motion, shown in Figs. 7(b1) and (b2), the behavior of the viscous force field is similar to the rotating case. However, when compared to the distribution of (E_x, E_y), the flow in the core region without the shear of du/dy is found to be somewhat stimulated by the viscoelastic force: see the consistency in the direction of the force and velocity in the region enclosed by the red line in (b3). It may be relevant to the earlier findings that, away from the orifice ($x > 4.5h$), the mean velocity in the core region of the viscoelastic flow became significantly larger than that of the Newtonian flow: cf. Fig. 2 in the paper of [46] and Fig. 6 of [33]. The cause of the accelerated core flow is probably related to the extensional viscosity, the magnitude of which is accentuated by high levels of viscoelasticity but not varied for the Newtonian fluid. With the high extensional viscosity, the flow motion is hard to alter in the longitudinal direction. This effect should be responsible also for the attenuation of the Coanda effect that would cause the asymmetry in the mean flow past the orifice: for details to Section 6.

Next, let us consider the region away from the orifice. It is clearly seen in Fig. 6(c) that both (F_x, F_y) and (E_x, E_y) vectors oppose the velocity vector of (u', v') with respect to a spanwise vortex, as similar to the trend observed in (a). Considering the locations of upwelling and downwelling flows associated with the vortex, the distribution of (E_x, E_y) shows the viscoelastic force directly counteracting the fluid motions. The weakening of the spanwise vortices may also be attributed to this effect. It may be interesting to note that the anti-correlated swirling vector pattern of (E_x, E_y) is shifted slightly downstream with respect to the center of the swirling fluid motion. Such slight discordances between the velocity and viscoelastic force in terms of the rotational center are frequently observed not only in Fig. 6(a), but also other viscoelastic flows [15]. Further investigations are needed to clarify its cause and importance for the vortex retardation by viscoelasticity.

In the downstream of the reattachment point, quasi-streamwise vortices become more common than spanwise vortices, as seen in Fig. 6(b). Figure 7(d1) shows an impingement of the ejection (Q2) and the sweep (Q4) motions, the so-called 'bursting,' which generally occurs in the buffer layer of the wall turbulence. It is demonstrated that a quasi-streamwise vorticity

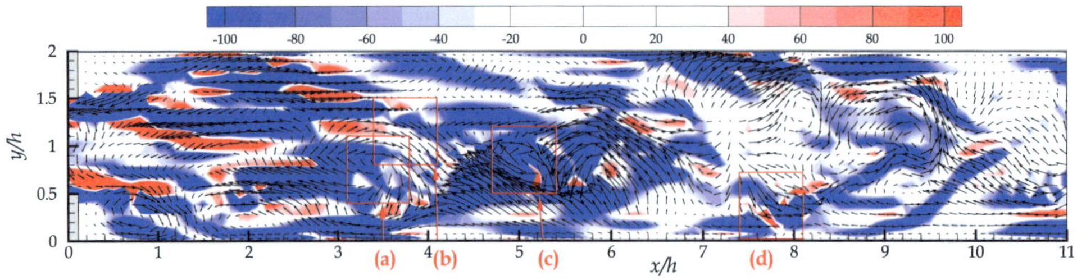

(a) Velocity (u, v) *versus* viscous force (F_x, F_y)

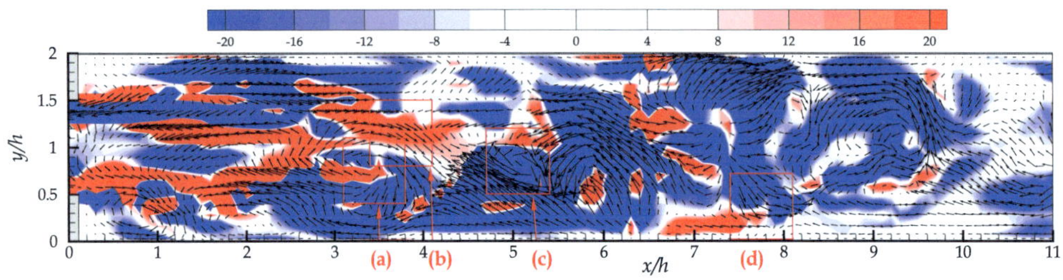

(b) Velocity (u, v) *versus* viscoelastic force (E_x, E_y)

Figure 8. Instantaneous distributions of the inner product of the velocity vector and the vector of either viscous force or the viscoelastic force, viewed in the same x-y plane and at the same instance with Fig. 7: (a) $\mathbf{u} \cdot \mathbf{F}$, (b) $\mathbf{u} \cdot \mathbf{E}$. The arrows represent the velocity vectors $\mathbf{u}$.

induces negative streamwise velocity fluctuations (below the red line in the figure), which results in a low-speed streak, and blowing down of high momentum fluid to the wall (above the red line). The vector fields of the viscous force and the viscoelastic force around them are shown in (d2) and (d3), respectively. Negative F_x is detected above the red line, where positive u' is induced, while positive F_x is observed very close to the wall. It is worth to note that this trend is not, however, consistent with the viscoelastic force (E_x, E_y) which has the almost same sign with (u', v') except for far from the wall, indicating that the viscoelastic force assists flow in some extent. This is presumably consistent with positive correlation between E_x and u' in the vicinity of the wall, as reported for the turbulence on smooth wall [1, 16].

4.3. Alignment between flow and force vectors

In order to see the variation in the relationship between the fluid motions and the viscoelastic force behaviors, we examine here the alignment between the flow vector and individual force. Figures 8(a) and (b) show the contour of the following inner product of two vectors—the velocity and either force of viscosity and viscoelasticity:

$$\mathbf{u} \cdot \mathbf{F} = |\mathbf{u}||\mathbf{F}| \cos \theta_F, \quad (11)$$

$$\mathbf{u} \cdot \mathbf{E} = |\mathbf{u}||\mathbf{E}| \cos \theta_E. \quad (12)$$

If the vectors of the flow and the viscous/viscoelastic force are parallel and have the same sign (θ_F, $\theta_E \approx 0$), the contour is given with red in the contour. If they have the opposite sign (θ_F, $\theta_E \approx \pi$), the contour becomes blue. Note that, when either velocity or force vector is negligible or their vectors align in perpendicular, the inner product should approach zero.

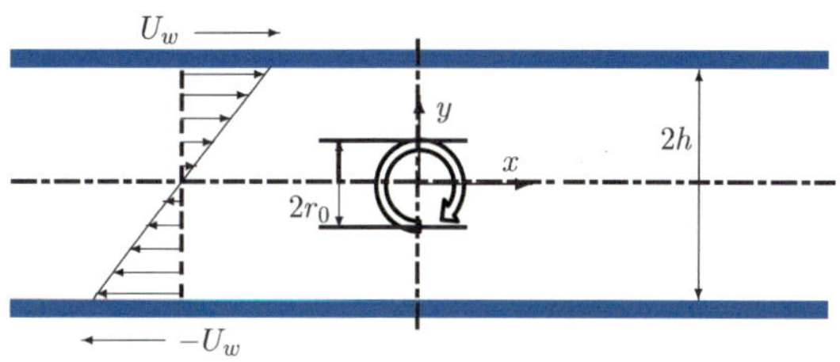

Figure 9. Initial field of laminar plane Couette flow with an immersed Rankine vortex.

As discussed in the preceding section, the plot of $\mathbf{u} \cdot \mathbf{E}$ in Fig. 8(b) shows that the viscoelastic force is anti-correlated with the spanwise vorticity just behind the orifice and in recirculation zone except for the core region. On the other hands, some consistency is observed in the near-wall region, but away from the reattachment point ($x > 6h$), where the sweep flows associated with the quasi-streamwise vortex motion are indeed confirmed to be stimulated by the viscoelastic force. As for the $\mathbf{u} \cdot \mathbf{F}$, Fig. 8(a) indicates that the viscous force inherently inhibits characterized fluid motions that we focus on here. The present concept of modification in vortical structures are qualitatively similar to those observed in the smooth wall-bounded turbulence [e.g., 6, 16].

Despite the rather phenomenological insight revealed by the above study, much further qualitative assessment should be required before understanding of the turbulent vortex modulation in the orifice flow for viscoelastic fluid is achieved.

5. Discussions

5.1. Simple test case: response to Rankine vortex in Couette flow

It would be instructive to examine the behavior of the viscoelastic body force in a simple test case, in the absence of any turbulent disturbance and downstream propagation. In this section, we investigate a localized spanwise eddy in a wall-bounded simple shear flow. In particular, the objective field is an incompressible plane Couette flow, which is driven by the relative movement of two parallel walls with the velocity of $\pm U_w$ (in x). The flow state is assumed to be basically laminar with a Reynolds number as low as $Re = \rho U_w h/\eta_0 = 60$, so that two-dimensional simulations have been performed both for Newtonian fluid and viscoelastic fluid. We focus on structures initially consisting of a Rankine-like vortex with its axis parallel to the z axis, no radial velocity ($u_r = 0$), and the tangential velocity of

$$u_\theta = \begin{cases} \Gamma r/(2\pi r_0^2) & (0 \le r \le r_0) \\ \Gamma/(2\pi r) & (r_0 < r) \end{cases} . \tag{13}$$

Here, (r, θ) and (u_r, u_θ) are the radial and circumferential coordinates and velocities that pertain to the vortex, respectively, r_0 the radius of the vortex, and Γ its circulation. While the gap between the walls is $2h$, the Rankine-line vortex with the diameter of $2r_0 = 0.4h$ was superimposed on the laminar Couette flow. The vortex center was set at the channel center, $y = 0$, so that the vortex would stay in position because of the net-zero bulk velocity. Figure 9 shows diagram of the flow configuration and the vortex, which has the rotational direction same with that of the mean flow vorticity. The governing equations and the relevant numerical scheme we used in this section were identical with those already introduced in Section 3. The

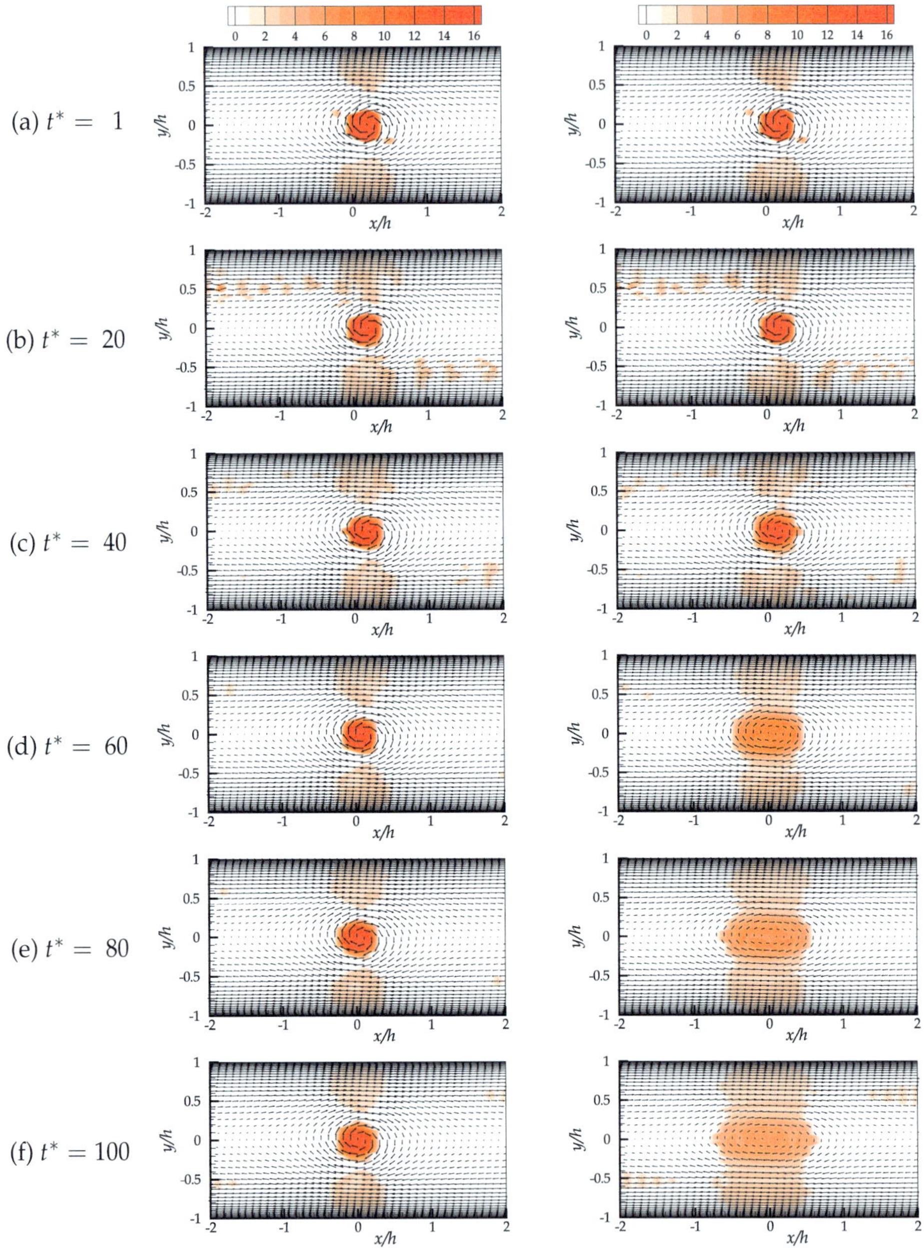

Figure 10. Temporal variation of a Rankin-like vortex in plane laminar Couette flow: (left-side column) Newtonian fluid, (right) viscoelastic fluid. Contour denotes the spanwise vorticity.

rheological parameters related to the Giesekus model were chosen as $We = \rho\lambda u_{\theta\max}^2/\eta_0 = 720$ ($u_{\theta\max}$ is comparable to U_w), $\beta = 0.8$, and $\alpha = 0.001$. The conformation tensor was initially given as zero at every point.

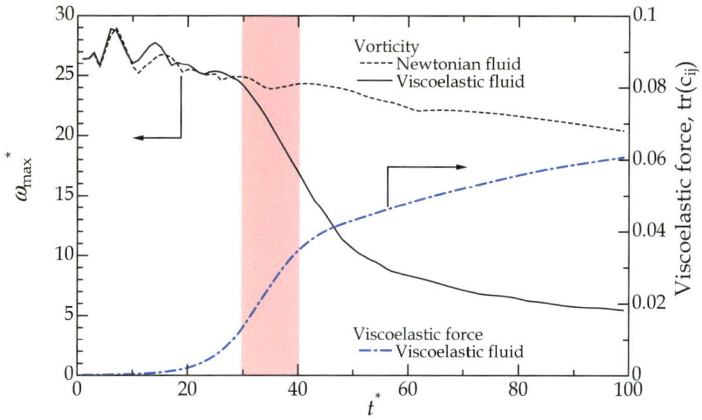

Figure 11. Time developments of the maximum vorticity at the vortex core and of the magnitude of the viscoelastic force.

The calculated vorticity and velocity fields show a rapid destruction, or decay, of the vortex in the viscoelastic fluid, which corresponds to the attenuation of spanwise K-H vortex shedding from the orifice edge. Figure 10 displays its decay process of the (clockwise) vortex for each fluid as a function of time, $t^* = tU_w/(2h)$. In the figure, the general flow pattern is characterized well by the swirling velocity vectors and seems to be not varied significantly in times, but their magnitude and the vorticity are remarkably reduced for $t^* = 40$–60 in the viscoelastic fluid flow. One may also observe that the vortex in this fluid is elongated along the mean flow, while the near-circular vortex stays in shape for the Newtonian fluid. This distortion is probably due to the high extensional viscosity of the viscoelastic fluid.

Figure 11 shows the temporal variation of the maximum vorticity (at the vortex center), ω_{max}, for each fluid and the included within the graph is the trace of viscoelastic stress, $c_{xx} + c_{yy} + c_{zz}$, as an indicator of viscoelastic force magnitude. In the initial stage of development, ω_{max} fluctuates remarkably maybe because of the artificiality of the given initial flow field with immersed vortex, irrespective of the fluid. After that, both fluid flows are settled similarly for a while. From $t^* = 20$, the magnitude of viscoelastic-force becomes increased at an accelerated rate that exhibits some sort of peak during $t^* = 30$–40. A consequence of increased viscoelastic force is that the vortex has been attenuated significantly, as seen in the visualization of Fig. 10 and in Fig. 11. From $t^* = 50$, both of ω_{max} and $\mathrm{tr}(c_{ij})$ take on somewhat moderate attitude: ω_{max} gradually decreases as slowly as that for the Newtonian fluid; and $\mathrm{tr}(c_{ij})$ increases linearly, at least until $t^* = 100$. It is conjectured that the viscoelasticity acts to resist flow and obtain elastic energy from the kinetic energy of the vortex: this phenomenon is thought to occur as a delayed response with a lag that should be relevant to the fluid relaxation time λ.

To investigate further details of the relationship between the flow structure and the fluid viscoelasticity, we study the viscoelastic-force distribution as the way in which the turbulent orifice flow was analyzed in Section 4.3. Figure 12 shows streamlines for the viscoelastic flow at the same instances in time as given in Fig. 10. The background color map shows the inner product of the velocity vector and the viscoelastic body force, $\mathbf{u} \cdot \mathbf{E}$, as similar to the manner in Fig. 8. As already mentioned, the streamlines are practically unchanged or slightly distorted into the shape of an ellipse. It is interesting to note that the inner-product

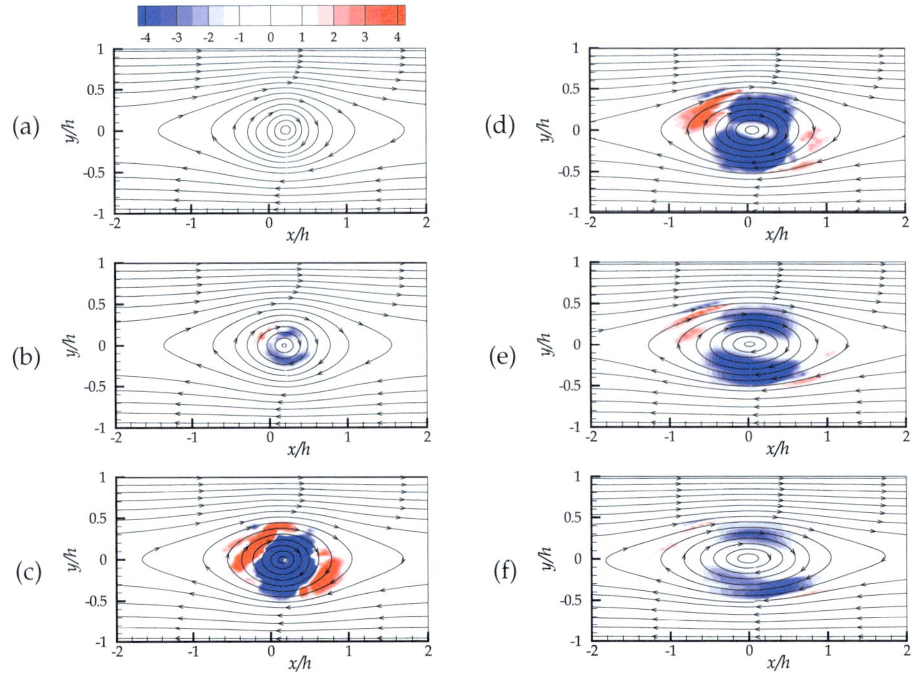

Figure 12. Streamlines and contour of the inner product of the velocity vector **u** and the viscoelastic force vector **E** at the same instance with Fig. 10 for the viscoelastic fluid.

distribution drastically alters in time and implies a mutual relation between the flow and viscoelasticity. If emphases are placed on the top-left and bottom-right parts with respect to the vortex center and on the top-right and bottom-left parts, a flow contraction and expansion, respectively, occur in gaps between each wall and the vortex. We find that the viscoelastic force in Fig. 12 assists flow in regions of strong extension (contraction) area around the vortex, where $\mathbf{u} \cdot \mathbf{E} > 0$, corresponding to red contour (see online version). On the other hand, most other parts of the vortex are found to be exerted resisting force mainly in regions of extension as well as the vortex core. Both these observations might be consistent with the trends observed in the orifice flow discussed earlier: that is, the wake past the orifice contraction would be sustained, whereas the expanding motion, or entrainment to the wall, be rather inhibited in viscoelastic fluid. These viscoelastic-fluid reaction can be confirmed to intensify during a finite time, in particular, $t^* = 30$–50 in the case of the present condition.

Although the vortex ranges in terms of size and magnitude and the relaxation time are not equivalent to those for the orifice flow discussed in Section 4, the concept of spanwise-vortex suppression should be, at least qualitatively, valid for those turbulent flows.

5.2. Vortex structures behind rib

Based on the discussions presented above, we propose a scenario of development of vortex structures and the difference between the two fluids. Figure 13 illustrates diagrams with emphasis on the Kelvin-Helmholtz vortices and successive longitudinal vortices. In the Newtonian fluid flow, the K-H vortices, which emanate from the edge of a rib and align

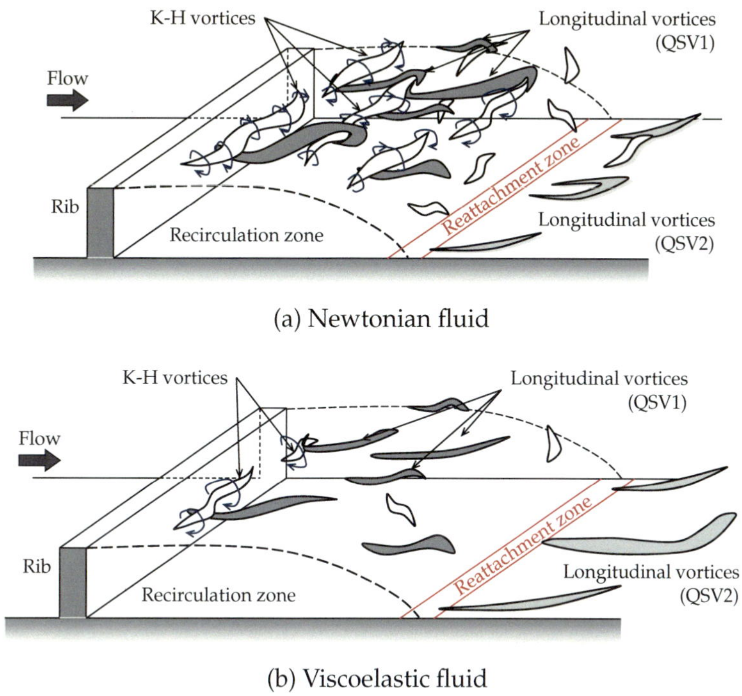

Figure 13. Conceptual scenario of the development of vortices in turbulent orifice flow for Newtonian fluid (a) and viscoelastic fluid (b).

parallel to the rib, propagate downstream with developing and inducing small eddies. Then, an intensive turbulent production arises above the reattachment zone. Moreover, once the three-dimensional disturbance reaches some finite amplitude, it produces a bending of spanwise K-H vortices and gives rise to additional eddies, the so-called *rib* vortices (labelled as QSV1 in the figure), extending in the streamwise direction, which bridge a sequence of the K-H vortices, as in the mixing layer [40]. Comte et al. [4] named this vortex pattern as a vortex-lattice structure, which was actually confirmed also in the present Newtonian orifice flow. In the downstream of the reattachment, quasi-streamwise vortices (QSV2) are expected to be dominant, as in the smooth channel flow. Basically, QSV1 and QSV2 may not the same structure in terms of generation process: the QSV1 should be generated in the separated shear layer and dissipated around the reattachment zone, while the QSV2 may be somewhat intensified structures of those observed in the smooth turbulent channel flow.

As demonstrated in Sections 4 and 5.1, spanwise vortices tend to be preferentially suppressed by the viscoelasticity, so that the K-H vortices rapidly decay, as schematically shown in Fig. 13(b). Accordingly, the longitudinal vortices (QSV1) become dominant structure but sparse even in a region above the recirculation and reattachment zone. The shift downstream of the reattachment zone occurs by the effect of viscoelasticity, as in agreement with the earlier experiments [29, 33]. This increase in the reattachment length inherently results in an expanse of the separated shear layer, i.e., the area of intensive turbulent production, in the downstream region. However, this expanded production area would not significantly contribute in regard to the net turbulent kinetic energy. For reference, the two-dimensional budget for the transport equation of the turbulent kinetic energy is presented in Section 6. In

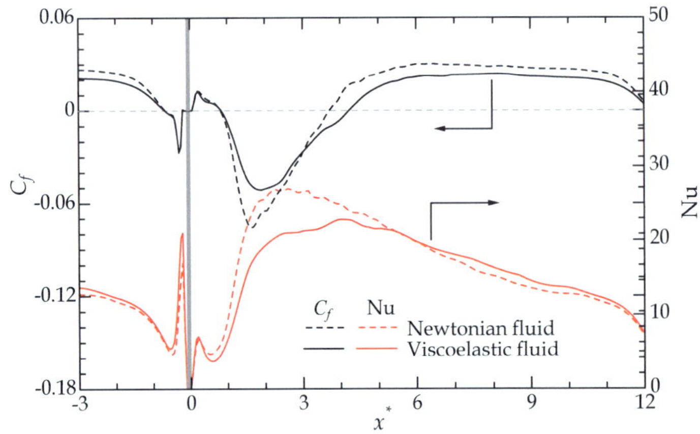

Figure 14. Comparison of Newtonian and viscoelastic flows in terms of streamwise distribution of local skin-frictional coefficient and Nusselt number. The orifice locates at $x^* = 0$–0.1. Cited from [49].

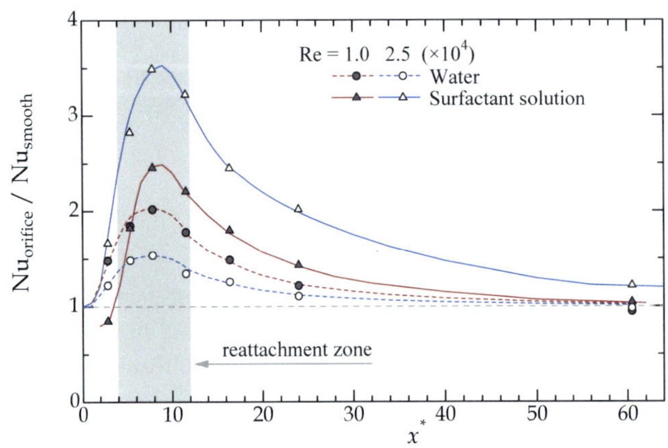

Figure 15. Experimental result for streamwise variations of the ratio between the local Nusselt number for the orifice flow and that for the smooth channel flow. The Reynolds number used here is based on the bulk mean velocity, the channel width $2h$, and the solvent kinematic viscosity. The orifice rear surface locates at $x^* = 0$. Cited from [13].

the downstream of the reattachment zone, much elongated QSV2 occurs and sustains for a longer period compared to that in the Newtonian flow. This is also a phenomenon affected by the fluid viscoelasticity, which is prone to assist some sort of elongational flows. It may be concluded that the viscoelastic flow would avoid rapid transition into turbulence just behind the orifice, whereas that flow be accompanied by long-life longitudinal vortices far downstream of the orifice.

5.3. Heat-transfer augmentation by orifice

In the context of above discussion on flow structures, their variations due to viscoelasticity are generally expected to significantly influence on heat and mass transfers. Better understanding of the thermal fields in the viscoelastic turbulent flow through complicated geometry is practically important for their applications, such as heat exchanger working with coolant of polymer/surfactant solution liquids [37, 38].

For the purpose of briefly describing the phenomena characterized by heat-transfer augmentation/reduction in the viscoelastic orifice flow, the local Nusselt-number (Nu) profile as a function of x^* is plotted in Fig. 14. Here, a constant temperature difference between the top and bottom walls was adopted as for the thermal boundary condition, the other fluid conditions were same as those given in Section 2, and we numerically solved the energy equation for passive scalar with a constant Prandtl number of $Pr = 1.0$ in the absence of any temperature dependency. For details, please see our recent paper [49]. Contained within Fig. 14 is the local skin-frictional coefficient C_f. In a some extent behind the orifice, C_f is broadly negative until the reattachment point locating around $x^* = 4$, at which $C_f = 0$. In this region, both C_f and Nu are decreased in the viscoelastic flow, because turbulent motions as well as the K-H vortices are damped, as concluded in Section 5.2. This phenomenon corresponds to what is termed either DR (drag reduction) or HTR (heat-transfer reduction). It is noteworthy that, for the viscoelastic flow, Nu locally exceeds that for the Newtonian flow, while C_f keeps a lower value: see a range of $x^* = 6$–10 in Fig. 14. This implies a feasibility of highly-efficient heat exchanger with ribs that provides simultaneously heat-transfer enhancement and less momentum loss. We may presume that this paradoxical phenomenon is caused by the mutual interference between QSV2 and the fluid viscoelasticity.

We have also experimentally confirmed the fact that an orifice in the channel flow would significantly promote the heat transfer in its downstream, especially for the case of viscoelastic liquid [13]. The CTAC solution with 150 ppm was used as the test fluid. One of the channel walls was heated to maintain a constant temperature. As seen in Fig. 15, the installation of an orifice induced a drastic increase of Nu in and after the reattachment zone. This effect was found to be more pronounced for higher Reynolds numbers.

6. Conclusions

The effects of viscoelastic force on vortical structures in turbulent flow past the rectangular orifice have been numerically investigated. We confirmed that the viscoelastic force tended to play a role in the attenuation of spanwise vortices behind the orifice. As found in the viscoelastic turbulence through a smooth channel by Kim et al. [16], the counter viscoelastic force reduces the spanwise vortex strength by opposing the vortical motions, which may result in the suppression of the auto-generation of new spanwise vortices and intensive turbulence behind the orifice. On the other hands, in the downstream of the reattachment zone, the flows associated with the quasi-streamwise vortex motion are stimulated by the viscoelastic force. This may lead to longer life-time of longitudinal vortex and the heat-transfer augmentation in far downstream of the orifice, as compared to the Newtonian counterpart. It can be concluded that turbulent kinetic energy is transferred to the elastic energy through the vortex suppression, and the opposite exchange from the elastic energy to the turbulent kinetic energy occurs apart from the orifice.

Appendix: Several turbulence statistics

The streamlines of the mean flow for the viscoelastic fluid we dealt with in the present study and several turbulence statistics are given in Fig. 16. The turbulent intensities of $u'^+_{\rm rms}$, $v'^+_{\rm rms}$,

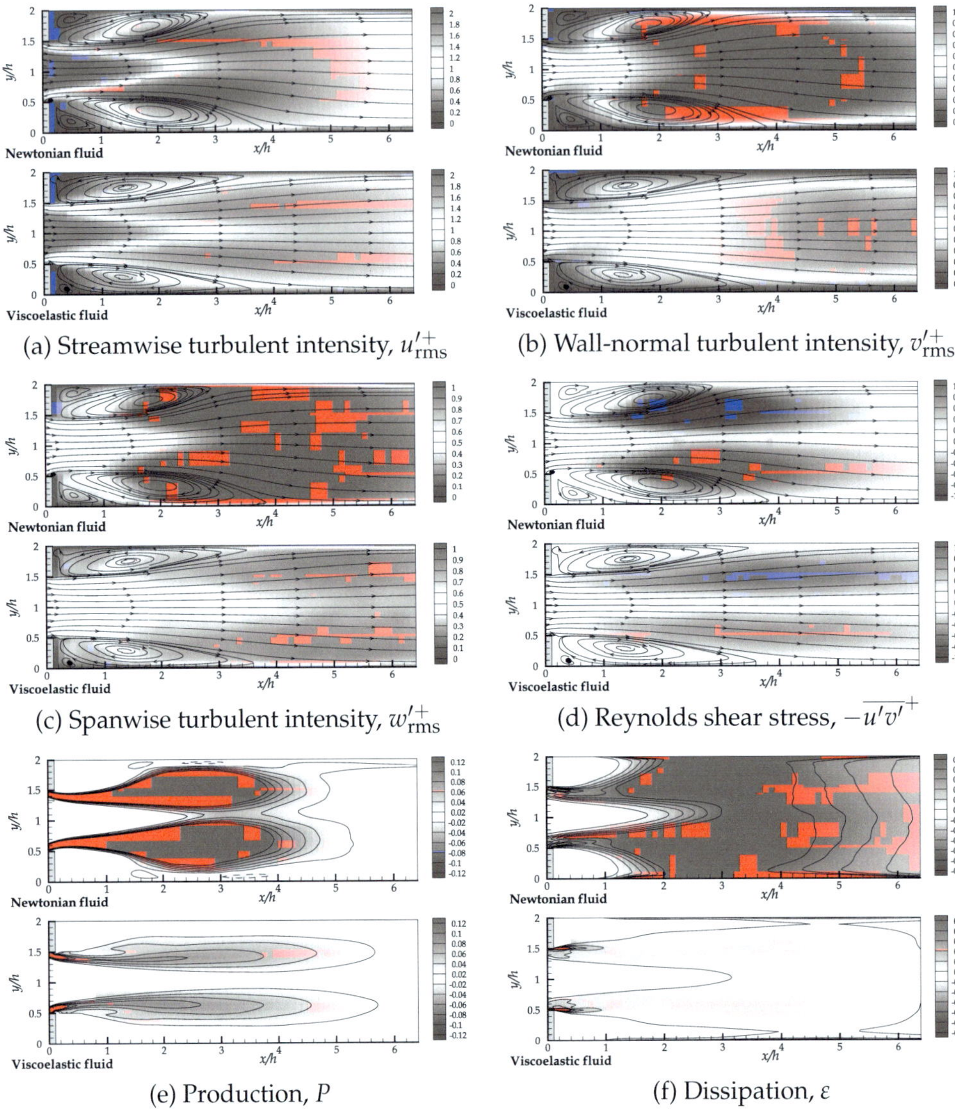

Figure 16. Mean-flow streamlines (a–d) and contours of various turbulence statistics: (a–c) turbulent intensities, (d) Reynolds shear stress, and (e, f) production and dissipation of turbulent kinetic energy. Note that ranges of accompanying color bars are different in each figure. For each statistic, the Newtonian flow and the viscoelastic flow are presented in the upper and lower figures, respectively. The budget terms in (e, f) are non-dimensionalized by $\rho u_{\tau 0}^4/\eta_0$.

and w'^{+}_{rms} represent the root-mean-square of velocity fluctuation in the x, y, and z direction, respectively. Only a major Reynolds shear stress of $-\overline{u'v'}^{+}$ is also shown in Fig. 16(d).

The balance equation for the turbulent kinetic energy $k = \overline{u_i'u_i'}/2$ in a fully-developed flow can be expressed as

$$\frac{\mathrm{d}k}{\mathrm{d}t} = P - \varepsilon + D^p + D^t + D^\nu + A + E = 0, \qquad (14)$$

where

$$\text{production,}\quad P = -\overline{u_i' u_k'} \frac{\partial U_i}{\partial x_k}; \tag{15}$$

$$\text{dissipation,}\quad \varepsilon = \frac{\eta_s}{\rho} \overline{\frac{\partial u_i'}{\partial x_k} \frac{\partial u_i'}{\partial x_k}}; \tag{16}$$

D^p, D^t, and D^v, the diffusion terms by pressure, turbulence, and viscous, respectively; A, the advectional contribution; and E, the viscoelastic contribution. An overbar and capital letter represent average values: U, the streamwise mean velocity. Figures 16(e) and (f) show in-plane distributions of P and ε in a range of $x^* \in [0, 6.4]$ and $y^* \in [0, 2]$.

From the results given in Fig. 16, we have obtained the following insights: (1) the viscoelastic fluid would provide rather symmetric streamlines with respect to the channel center, while the asymmetry due to the Coanda effect occurs clearly despite the symmetric geometry, (2) the intensive turbulence region as well as the turbulence-producing area at the separated shear layer are shifted downstream in the viscoelastic fluid, (3) the suppression of the Kelvin-Helmholtz vortices results in a significant reduction in v'^{+}_{rms} just behind the orifice, and (4) the region where relatively high dissipation occurs shifted far downstream as similar to the turbulence-producing area.

For further information, one may refer to [46], although a detailed discussion including analysis of the energy/stress budget will be presented in the near future.

Acknowledgements

All of the present DNS computations were carried out with the use of supercomputing resources of Cyberscience Center at Tohoku University and those of Earth-Simulator Center of JAMSTEC (Japan Agency for Marine-Earth Science and Technology). We also gratefully acknowledge the assistances of our Master's course student at Tokyo University of Science: Mr. Tomohiro Kawase for his exceptional works in DNS operation; and Mr. Daisei Tsurumi and Ms. Shoko Kawada for their contribution to the experimental part of this study. The authors would like to thank Prof. H. Kawamura, President of Tokyo University of Science, Suwa, for stimulating discussions.

This chapter is a revised and expanded version of a paper entitled "Numerical investigation of viscoelastic effects on turbulent flow past rectangular orifice," presented at the 22nd International Symposium on Transport Phenomena [47].

Author details

Takahiro Tsukahara and Yasuo Kawaguchi
Tokyo University of Science, Japan

7. References

[1] de Angelis, E., Casciola, C.M., & Piva, R. (2002). DNS of wall turbulence: Dilute polymers and self-sustaining mechanisms, *Computers & Fluids*, Vol. 31, 495–507.

[2] Armaly, B.F., Durst, F., Pereira, J.C.F., & Schönung, B. (1983). Experimental and theoretical investigation of backward-facing step flow, *Journal of Fluid Mechcanics*, Vol. 127, 473–496.
[3] Castro, O.S. & Pinho, F.T. (1995). Turbulent expansion flow of low molecular weight shear-thinning solutions, *Experiments in Fluids*, Vol. 20, 42–55.
[4] Comte, P., Lesieur, M., & Lamballais, E. (1992). Large- and small-scale stirring of vorticity and a passive scalar in a 3-D temporal mixing layer, *Physics of Fluid A*, Vol. 4, 2761–2778.
[5] Dales, C., Escudier, M.P., & Poole, R.J. (2005). Asymmetry in the turbulent flow of a viscoelastic liquid through an axisymmetric sudden expansion, *Journal of Non-Newtonian Fluid Mechanics*, Vol. 125, 61–70.
[6] Dubief, Y., White, C.M., Terrapon, V.E., Shaqfeh, E.S.G., Moin, P., & Lele, S.K. (2004). On the coherent drag-reducing and turbulence-enhancing behaviour of polymers in wall flows, *Journal of Fluid Mechanics*, Vol. 514, 271–280 .
[7] Dimitropoulos, C.D., Sureshkumar, R., Beris, A.N., & Handler, R.A. (2001). Budgets of Reynolds stress, kinetic energy and streamwise enstrophy in viscoelastic turbulent channel flow, *Physics of Fluids*, Vol. 13, 1016–1027.
[8] Escudier, M.P. & Smith, S. (1999). Turbulent flow of Newtonian and shear-thinning liquids through a sudden axisymmetric expansion. *Experiments in Fluids*, Vol. 27, 427–434.
[9] Fadlun, E.A., Verzicco, R., Orlandi, P., & Mohd-Yusof, J. (2000). Combined immersed-boundary finite-difference methods for three-dimensional complex flow simulations, *Journal of Computational Physics*, Vol. 161, 35–60.
[10] Giesekus, H. (1982). A simple constitutive equation for polymer fluids based on the concept of deformation-dependent tensorial mobility, *Journal of Non-Newtonian Fluid Mechanics*, Vol. 11, 69–109.
[11] Gyr, A. & Bewersdorff, H.-W. (1995). *Drag reduction of turbulent flows by additives*, Kluwer Academic Publisher, ISBN 978-90-481-4555-3, Dordrecht.
[12] Joseph, D.D. (1990). *Fluid dynamics of viscoelastic liquids*, Springer-Verlag, ISBN 978-0387971551, New York.
[13] Kawada, S., Tsurumi, D., Kawase, T., Tsukahara, T., & Kawaguchi, Y. (2012). Experimental study on heat transfer augmentation in viscoelastic turbulent channel flow by two-dimensional orifice, In: *Turbulent, Heat and Mass Transfer 7*, Begell House, Inc., in press.
[14] Kawaguchi, Y., Wei, J.J., Yu, B., & Feng, Z.P. (2003). Rheological characterization of drag-reducing cationic surfactant solution: shear and elongational viscosities of dilute solutions, *Proceedings of ASME/JSME 2003 4th Joint Fluids Summer Engineering Conference* , pp. 721–728, Honolulu, Hawaii, USA, July 6–10, 2003.
[15] Kim, K., Li, C.-F., Sureshkumar, R, Balachandar, S., & Adrian, R.J. (2007). Effects of polymer stresses on eddy structures in drag-reduced turbulent channel flow, *Journal of Fluid Mechcanics*, Vol. 584, 281–299.
[16] Kim, K., Adrian, R.J., Balachandar, S., & Sureshkumar, R. (2008). Dynamics of hairpin vortices and polymer-induced turbulent drag reduction, *Physical Review Letter*, Vol. 100, 134504, 4 pp.
[17] Li, C.-F., Sureshkumar, R., & Khomami, B. (2006). Influence of rheological parameters on polymer induced turbulent drag reduction, *Journal of Non-Newtonian Fluid Mechcanics*, Vol. 140, 23–40.
[18] Li, F.-C., Yu, B., Wei, J.J. & Kawaguchi, Y. (2012). *Turbulent drag reduction by surfactant additives*, John Wiely & Sons, Inc., ISBN 978-1-118-18107-2, Singapore.
[19] Lumley, J.L. (1969). Drag reduction by additives, *Annual Review of Fluid Mechanics*, Vol. 1, 367–384.

[20] L'vov, V.S., Pomyalov, A., Procaccia, I., & Tiberkevich, V. (2004). Drag reduction by polymers in wall-bounded turbulence, *Physical Review Letter*, Vol. 92, 244503, 4 pp.
[21] Manica, R. & De Bortoli, A.L. (2004). Simulation of sudden expansion flows for power-law fluids. *Journal of Non-Newtonian Fluid Mechanics*, Vol. 121, 35–40.
[22] Makino, S., Iwamoto, K., & Kawamura, H. (2008a). Turbulent structures and statistics in turbulent channel flow with two-dimensional slits, *International Journal Heat and Fluid Flow*, Vol. 29, 602–611.
[23] Makino, S., Iwamoto, K., & Kawamura, H. (2008b). DNS of turbulent heat transfer through two-dimensional slits, *Progress in Computational Fluid Dynamics*, Vol. 8, 397–405.
[24] Mohd-Yusof, J. (1998). Development of immersed boundary methods for complex geometries, *Center for Turbulence Research Annual Research Briefs 1998*, NASA Ames, Stanford University, 325–336.
[25] Motier, J.F. & Carrier, A.M. (1989). Recent studies in polymeric drag reduction in commercial pipelines. *Drag reduction in fluid flows: techniques for friction control*, eds. Sellin, R.H.J. & Moses, R.T., Halsted Press, New York, 197–204.
[26] Nadolink, R.H. & Haigh, W.W. (1995). Bibliography on skin friction reduction with polymers and other boundary-layer additives, *Applied Mechanics Reviews*, Vol. 48, 351–460.
[27] Nie, J.H. & Armaly, B.F. (2004). Reverse flow regions in three-dimensional backward-facing step flow. *International Journal of Heat and Mass Transfer*, Vol. 47, 4713–4720.
[28] Oliveira, P.J. (2003). Asymmetric flows of viscoelastic fluids in symmetric planar expansion geometries. *Journal of Non-Newtonian Fluid Mechanics*, Vol. 114, 33–63.
[29] Pak, B., Cho, Y.I., & Choi, S.U.S. (1990). Separation and reattachment of non-newtonian fluid flows in a sudden expansion pipe. *Journal of Non-Newtonian Fluid Mechanics*, Vol. 37, 175–199.
[30] Peskin, C.S. (1977). Numerical analysis of blood flow in the heart. *Journal of Computational Physics*, Vol. 25, 220–252.
[31] Poole, R.J. & Escudier, M.P. (2003a). Turbulent flow of non-Newtonian liquids over a backward-facing step: Part II. Viscoelastic and shear-thinning liquids, *Journal of Non-Newtonian Fluid Mechanics*, Vol. 109, 193–230.
[32] Poole, R.J. & Escudier, M.P. (2003b). Turbulent flow of a viscoelastic shear-thinning liquid through a plane sudden expansion of modest aspect ratio, *Journal of Non-Newtonian Fluid Mechanics*, Vol. 112, 1–26.
[33] Poole, R.J. & Escudier, M.P. (2004). Turbulent flow of viscoelastic liquids through an asymmetric sudden expansion, *Journal of Non-Newtonian Fluid Mechanics*, Vol. 117, 25–46.
[34] Poole, R.J., Alves, M.A., Oliveira, P.J., & Pinho, F.T. (2007). Plane sudden expansion flows of viscoelastic liquids, *Journal of Non-Newtonian Fluid Mechanics*, Vol. 146, 79–91.
[35] Procaccia, I., L'vov, V.S., & Benzi, R. (2008). Colloquium: Theory of drag reduction by polymers in wall-bounded turbulence, *Reviews of Modern Physics*, Vol. 80, 225–247.
[36] Prosperetti, A. & Tryggvason, G. (2007). *Computational methods for multiphase flow*, Cambridge University Press, ISBN 978-0-521-84764-3, Cambridge.
[37] Qi, Y., Kawaguchi, Y., Lin, Z., Ewing, M., Christensen, R.N., & Zakin, J.L. (2001). Enhanced heat transfer of drag reducing surfactant solutions with fluted tube-in-tube heat exchanger, *International Journal Heat and Mass Transfer*, Vol. 44, 1495–1505.
[38] Qi, Y., Kawaguchi, Y., Christensen, R.N., & Zakin, J.L. (2003). Enhancing heat transfer ability of drag reducing surfactant solutions with static mixers and honeycombs, *International Journal Heat and Mass Transfer*, Vol. 46, 5161–5173.

[39] Roy, A., Morozov, A., van Saarloos, W., & Larson, R.G. (2006). Mechanism of polymer drag reduction using a low-dimensional model, *Physical Review Letter*, Vol. 97, 234501, 4pp.
[40] Schmid, P.J. & Henningson, D.S. (2001). *Stability and Transition in Shear Flows*, Springer-Verlag, ISBN 978-0-387-98985-3, New York.
[41] Sureshkumar, R. & Beris, A.N. (1995). Effect of artificial stress diffusivity on the stability of numerical calculations and the flow dynamics of time-dependent viscoelastic flows, *Journal of Non-Newtonian Fluid Mechanics*, Vol. 60, 53–80.
[42] Suzuki, H., Ishihara, K., & Usui, H. (2001). Numerical study on a drag reducing flow with surfactant additives, *Proceedings of 3rd Pacific Rim Conference on Rheology*, Paper No. 019, Vancouver, Canada, 8–13 July, 2001.
[43] Takeuchi, H. (2012). Demonstration test of energy conservation of central air conditioning system at the Sapporo City Office Building., *Synthesiology, English edition*, Vol. 4, 136–143.
[44] den Toonder, J.M.J., Hulsen, M.A., Kuiken, G.D.C., & Nieuwstadt, F.T.M. (1997). Drag reduction by polymer additives in a turbulent pipe flow: numerical and laboratory experiments, *Journal of Fluid Mechanics*, Vol. 337, 193–231.
[45] Tsukahara, T., Ishigami, T., Yu, B., & Kawaguchi, Y. (2011a). DNS study on viscoelastic effect in drag-reduced turbulent channel flow, *Journal of Turbulence*, Vol. 12, No. 13, 13 pp.
[46] Tsukahara, T., Kawase, T., & Kawaguchi, Y. (2011b). DNS of viscoelastic turbulent channel flow with rectangular orifice at low Reynolds number, *International Journal of Heat and Fluid Flow*, Vol. 32, 529–538.
[47] Tsukahara, T., Kawase, T., & Kawaguchi, Y. (2011c). Numerical investigation of viscoelastic effects on turbulent flow past rectangular orifice, In: *Proceedings of the 22nd International Symposium on Transport Phenomena*, Paper #129 (USB), 7pp., Delft, The Netherlands, 8–11 November 2011.
[48] Tsukahara, T. & Kawaguchi, Y. (2011). Turbulent heat transfer in drag-reducing channel flow of viscoelastic fluid, *Evaporation, Condensation and Heat transfer* (ed., A. Ahsan), InTech, Rijeka, Croatia, pp. 375-400.
[49] Tsukahara, T. & Kawaguchi, Y. (2012). DNS on turbulent heat transfer of viscoelastic fluid flow in a plane channel with transverse rectangular orifices, *Progress in Computational Fluid Dynamics*, in press.
[50] Tsurumi, D., Kawada, S., Kawase, T., Tsukahara, T., & Kawaguchi, Y. (2012). Experimental analysis of turbulent structure of viscoelastic fluid flow in downstream of two-dimensional orifice. In: *Turbulent, Heat and Mass Transfer 7*, Begell House, Inc., in press.
[51] White, C.M. & Mungal, M.G. (2008). Mechanics and prediction of turbulent drag reduction with polymer additives, *Annual Review Fluid Mechacnics*, Vol. 40, 235–256.
[52] Yu, B. & Kawaguchi, Y. (2004). Direct numerical simulation of viscoelastic drag-reducing flow: a faithful finite difference method, *Journal of Non-Newtonian Fluid Mechanics*, Vol. 116, 431–466.
[53] Yu, B., Li, F., & Kawaguchi, Y. (2004). Numerical and experimental investigation of turbulent characteristics in a drag-reducing flow with surfactant additives, *International Journal of Heat and Fluid Flow*, Vol. 25, 961–974.
[54] Zakin, J.L., Myska, J., & Chara, Z. (1996). New limiting drag reduction and velocity profile asymptotes for nonpolymeric additives systems, *AIChE Journal*, Vol. 42, 3544–3546.

Chapter 3

Microscopic Formulation of Fractional Theory of Viscoelasticity

B.N. Narahari Achar and John W. Hanneken

Additional information is available at the end of the chapter

http://dx.doi.org/10.5772/51493

1. Introduction

Viscoelasiticity refers to the phenomenon in which a material body, when deformed exhibits both elastic (akin to solids)and viscous (akin to liquids) behavior. The body stores mechanical energy (elastic behavior) and dissipates it simultaneously(viscous behavior). Linear theory of viscoelasticity treats the body as a linear system which when subjected to an excitation responds with a response function. If the excitation is a stress, the response is a strain and if the excitation is a strain, the response is a stress. Mechanical models involving a spring-mass connected to a dashpot have been used to explain the elastic and viscous behavior.The mathematical structure of the theory and the spring-dash-pot type of mechanical models used and the so called Standard Linear Solid have all been only too well known [1-4]. In recent years methods of fractional calculus have been applied to develop viscoelastic models especially by Caputo and Mainardi [5,6] , Glockle and Nonenmacher [7], and Gorenflo and Mainardi [8]. A recent monograph by Mainardi[9] gives extensive list of references to the literature connecting fractional calculus, linear viscoelasticity and wave motion.All these works treat the phenomenon of viscoelasticity as a macroscopic phenomenon exhibited by matter treated as an elastic continuum albeit including a viscous aspect as well. It should be recognized that matter has an atomic structure and is fundamentally discrete in nature. A microscopic approach would recognize this aspect and a theoretical model would yield the results for the continuum as a limit.

It is well established that lattice dynamics provides a microscopic basis for understanding a host of phenomena in condensed matter physics, including mechanical, thermal, dielectric and optical phenomena, which are macroscopic, generally described from the continuum point of view. Since the pioneering work of Born and Von Karman [10], lattice dynamics has developed into a veritable branch of condensed matter physics. Lucid treatment of various

INTECH
open science | open minds

© 2012 The Author(s). Licensee InTech. This chapter is distributed under the terms of the Creative Commons Attribution License http://creativecommons.org/licenses/by/3.0), which permits unrestricted use, distribution, and reproduction in any medium, provided the original work is properly cited.

aspects of lattice dynamical theory of condensed matter can be found in renowned treatises [11-13]. Of particular interest to the present paper is the work of Askar [14] on the lattice dynamical foundations of continuum theories of elasticity, piezoelectricity, viscoelasticity and plasticity.

The objective of the chapter is to develop a lattice dynamical model based on the methods of fractional calculus so as to provide a microscopic basis for the theory of viscoelasticity. The plan of the chapter is as follows. First a brief review of the mechanical models of viscoelasticity is given. This is followed by an account of conventional lattice dynamical theory of viscoelaticity based on a model of linear chain of coupled oscillators with dissipative elements is given [14]. In the next section, lattice dynamical methods are extended to the model of linear chain of coupled fractional oscillators (requiring no additional dissipative elements) developed by the authors [15]. The response to sinusoidal forcing of the linear chain of fractional oscillators starting from a quiescent state is studied in the continuum limit and expressions for phase velocity, absorption and dispersion and specific dissipation function are derived. The results are discussed and numerical applications are presented in the final section

2. Mechanical models of viscoelasticity

According to the linear theory of viscoelasticity, the body may be considered to be a system responding linearly to an excitation. A fundamental aspect of the theory both from a mathematical and a physical point of view is the response of the system to an excitation usually applied as a Heaviside step function. If a unit step of stress $\sigma(t)$, is applied the material responds by undergoing a strain $\varepsilon(t)$. The test is called creep test and the material function characterizing the response is called the creep compliance and is denoted by J(t). If on the other hand, if a unit strain is applied and the system responds with a stress $\sigma(t)$, the test is called a relaxation test and the material function characterizing the system response, G(t), is called the *relaxation modulus.* The creep compliance and the relaxation modulus are defined through linear hereditary integrals of the Stieltjes type:

$$\begin{aligned} \varepsilon(t) &= \int_{0^-}^{t} J(t-\tau)d\sigma(\tau) \\ \sigma(t) &= \int_{0^-}^{t} G(t-\tau)d\varepsilon(\tau) \end{aligned} \tag{1}$$

Simple mechanical models consisting of springs and dashpots incorporating elastic and dissipative functions have been advanced to explain the behaviour of viscoelastic solids and a good summary of the models of viscoelasticity can be found in the works of Caputo and Mainardi [5,6], Gorenflo and Mainardi[8], and Mainardi[9]. A simple model constituted by a spring in parallel with a dashpot, known as the Kelvin-Voigt model is characterized by the constitutive relation

$$\sigma(t) = m\varepsilon(t) + b\frac{d\varepsilon}{dt} \tag{2}$$

and by the material functions

$$\begin{aligned} J(t) &= \frac{1}{m}\left[1 - e^{-t/\tau_\varepsilon}\right] && \text{(a)} \\ G(t) &= m + b\delta(t) && \text{(b)} \end{aligned} \tag{3}$$

where $\tau_\varepsilon = b/m$ is referred to as the retardation time.

A model consisting of a spring in series with a dashpot known as the Maxwell model is characterized by the constitutive relation

$$\sigma(t) + a\frac{d\sigma}{dt} = b\frac{d\varepsilon}{dt} \tag{4}$$

and by the material functions

$$\begin{aligned} J(t) &= \frac{a}{b} + \frac{t}{b} && \text{(a)} \\ G(t) &= \frac{b}{a}e^{-t/\tau_\sigma} && \text{(b)} \end{aligned} \tag{5}$$

where $\tau_\sigma = a$ is referred to as the relaxation time.

Another model introduced by Zener [4], known as the Standard Linear Solid model is a combination of the above two models and is characterized by the constitutive relation

$$\left[1 + a\frac{d}{dt}\right]\sigma(t) = \left[1 + \frac{d}{dt}\right]\varepsilon(t) \tag{6}$$

and by the material functions

$$\begin{aligned} J(t) &= J_g + \chi_+\left[1 - e^{-t/\tau_\varepsilon}\right] && \text{(a)} \\ G(t) &= G_e + \chi_- e^{-t/\tau_\sigma} && \text{(b)} \end{aligned} \tag{7}$$

Here J_g and G_e are known as the glass compliance and the equilibrium modulus respectively, and χ_+ and χ_- are the limiting values of creep for $t \to \infty$ and of relaxation for $t = 0$ respectively.

More recently methods of fractional calculus have been used [5-8] to generalize these models leading to the so-called Scott-Blair model characterized by the operator equation

$$\left[1 + a\frac{d^\alpha}{dt^\alpha}\right]\sigma(t) = \left[1 + \frac{d^\alpha}{dt^\alpha}\right]\varepsilon(t), 0 < \alpha \leq 1 \tag{8}$$

The details of the characteristic material functions and their behavior with respect to time are discussed in detail in reference [8].

3. Lattice dynamical theory of viscoelasticity

3.1. The elastic continuum

Consider a one dimensional string of mass density ϱ per unit length and E the Young's modulus. If the string is homogeneously stressed and released, elastic forces set up vibrations in the string corresponding to elastic waves propagating in the string. The equation governing this propagation is given by 9a and these waves propagate with a velocity given by 9b

$$\begin{aligned} &\rho\frac{\partial^2 u}{\partial t^2} = E\frac{\partial^2 u}{\partial x^2} \qquad \text{(a)} \\ &c^2 = \frac{E}{\rho} \qquad \text{(b)} \end{aligned} \tag{9}$$

The conventional model of lattice dynamics consists of a chain of N identical masses m connected by springs of spring constant k, shown in Figure 1. The equation of motion for the nth mass is given in the standard notation by

$$\begin{aligned} &m\frac{d^2u_n}{dt^2} = -k(2u_n - u_{n+1} - u_{n-1}) \\ \text{or } &\frac{d^2u_n}{dt^2} = -\omega_0^2(2u_n - u_{n+1} - u_{n-1}) \end{aligned} \tag{10}$$

where u_n is the displacement from the equilibrium position of the nth mass. The right hand side represents the elastic restoring force on the nth atom. The atoms n=1 and n= N define the boundaries.

The frequency of oscillation is given by $\omega_0^2 = k / m$.

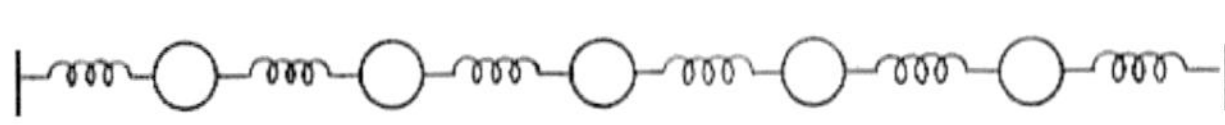

Figure 1. Linear chain of coupled oscillators

The elastic behavior of a continuous chain in equation (9a) can be obtained from the equation of motion of the linear chain of masses connected to each other by springs considered above in equation (9). In the limit when the number of masses N→∞ and the separation between the masses a→0, such that the product Na→ L, a finite length, the linear chain of coupled oscillators reduces to a continuous loaded string and is referred to as the long wavelength limit.

This process of taking the continuum limit is illustrated by considering the term on the right hand side of equation (10):

the factor

$$\omega_0^2\left((2u_n(t)-u_{n+1}(t)-u_n(t)\right)$$

can be written as

$$a^2\omega_0^2\left(\frac{1}{a}\left(\frac{(u_{n+1}(t)-u_n(t)}{a}-\frac{(u_n(t)-u_{n-1}(t)}{a}\right)\right)$$

In the limit of $N\rightarrow\infty$ and $a\rightarrow 0$, the chain of atoms can be treated as a continuum and $u_n(t)\rightarrow u(x,t)$, a continuous function. The term in the brackets can be written as $\frac{\partial^2 u}{\partial x^2}$ and $a^2\omega_0^2$ can be expressed as $=a^2\frac{k}{m}=\frac{ka}{m/a}$. In the limit $ka\rightarrow E$ and $m/a\rightarrow\rho$ The term $a^2\omega_0^2$ reduces to the square of the wave velocity. Thus the equation of motion in the limit can be written as $\frac{\partial^2 u}{\partial t^2}=c^2\frac{\partial^2 u}{\partial x^2}$ yielding the equation of motion of a continuum, equation (9a)

3.2. Viscoelasticity: Lattice dynamical approach

In order to develop a theory of viscoelasticity, the model of linear chain of atoms has been generalized to include dissipative effects [14] by incorporating dashpots either in parallel to the springs as shown in figure 2 to yield the lattice dynamical version of the Kelvin-Voigt model, or the dashpots in series with the springs as shown in figure 3 to yield the lattice dynamical version of the Maxwell model.

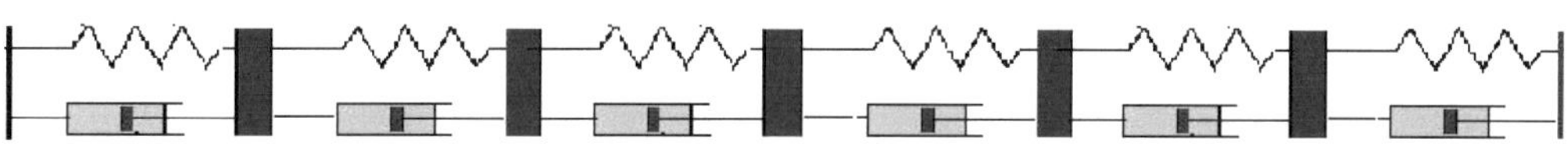

Figure 2. Kelvin-Voigt model for viscoelasticity

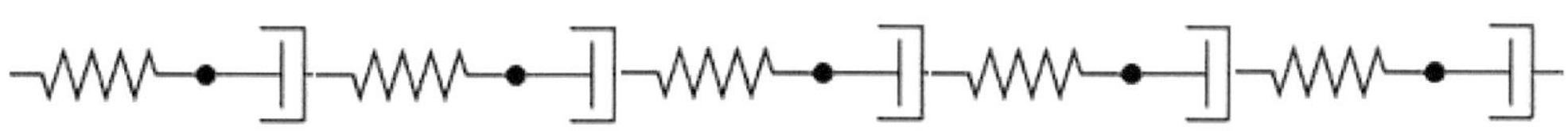

Figure 3. Maxwell model for viscoelasticity

It has been shown [14] that the models in the continuum limit lead to the stress-strain relations

$$\begin{aligned} \sigma &= E\varepsilon + \zeta \frac{d\varepsilon}{dt} && \text{(a)} \\ \frac{d\varepsilon}{dt} &= \frac{\sigma}{E} + \frac{1}{E}\frac{d\sigma}{dt} && \text{(b)} \end{aligned} \tag{11}$$

respectively for the two models. Harmonic waves are found to propagate with decaying amplitude indicating dissipation. The details can be found in reference [14].

4. Fractional calculus approach

In this paper, the lattice dynamical approach is generalized by using the methods of fractional calculus, leading to the so-called linear chain of coupled fractional oscillators. However, this generalization does not require the dissipative elements, namely the dashpots to be explicitly considered, for the dissipation is intrinsic to the fractional oscillators [16-19]. As the method of approach is quite different, further reference will be made not to the work of Askar [14], but to the work of the authors on a linear chain of coupled fractional oscillators [19] and the approach is outlined below.

4.1. Linear chain of fractional oscillators

Instead of starting with equation (10), we consider the integral equation of motion for the nth mass interacting with only its nearest neighbors, which can be written without loss of generality as:

$$u_n(t) = u_n(0) + \dot{u}_n(0)t - \omega_0^2 \int_0^t \left(2u_n(\tau) - u_{n+1}(\tau) - u_{n-1}(\tau)\right)(t-\tau)d\tau \quad 2 \le n \le N-1 \tag{12}$$

Here $u_n(0)$ and $\dot{u}_n(0)$ refer to the displacement from equilibrium and velocity of the nth mass at t=0. For the masses at the ends of the chain, the equations of motion are given by

$$u_1(t) = u_1(0) + \dot{u}_1(0)t - \omega_0^2 \int_0^t \left(2u_1(\tau) - u_2\right)(t-\tau)d\tau + \int_0^t f(\tau)(t-\tau)d\tau \tag{13}$$

and

$$u_N(t) = u_N(0) + \dot{u}_N(0)t - \omega_0^2 \int_0^t \left(2u_N(\tau) - u_{N-1}(\tau)\right)(t-\tau)d\tau, \tag{14}$$

respectively. Of course, it is $\omega_0^2 = k/m$, in the usual notation. The last term in equation (13) indicates a periodic forcing on the end atom.

The integrals on the right hand side of equations (11) -(13) are generalized to fractional integrals of order α (defined by equation (19) below) to yield the equations of motion of a chain of coupled fractional oscillators. However, as has been well known that a fractional oscillator behaves like a damped harmonic oscillator [15-18], the oscillations of the linear chain of fractional oscillators can be sustained only if it is driven by an external force. Hence a periodic force, taken to be sinusoidal without loss of generality, acting only on the first member of the chain is included. Thus the equations of motion are given by

$$u_n = u_n(0) + \dot{u}_n(0)t - \frac{\omega_0^\alpha}{\Gamma(\alpha)}\int_0^t \left(2u_n(\tau) - u_{n+1}(\tau) - u_{n-1}(\tau)\right)(t-\tau)^{\alpha-1}d\tau,\ 2 \le n \le N-1 \quad (15)$$

$$u_1 = u_1(0) + \dot{u}_1(0)t - \frac{\omega_0^\alpha}{\Gamma(\alpha)}\int_0^t \left(2u_1(\tau) - u_2(\tau)\right)(t-\tau)^{\alpha-1}d\tau + \frac{1}{\Gamma(\alpha)}\int_0^t f(\tau)(t-\tau)^{\alpha-1}d\tau \quad (16)$$

$$u_N = u_N(0) + \dot{u}_N(0)t - \frac{\omega_0^\alpha}{\Gamma(\alpha)}\int_0^t \left(2u_n(\tau) - u_{N-1}(\tau)\right)(t-\tau)^{\alpha-1}d\tau \quad (17)$$

The last term on the right hand side of equation (16) represents the effect of the periodic forcing as already noted.

4.2. Formal solution

The set of equations (15) through (17) can be solved formally by using the Laplace transform technique [20-24]. In the standard notation, the Laplace transform is defined by

$$L\{u_n(t)\} = \tilde{u}_n(s) = \int_0^\infty e^{-st}u_n(t)dt. \quad (18)$$

The Laplace transform of a fractional integral of order α, $(\mathrm{Re}\,\alpha > 0)$ defined by [20] and can be evaluated by considering the fractional integral as a convolution of two functions (19b),

$$I_{0+}^\alpha f(t) = \frac{1}{\Gamma(\alpha)}\int_0^t f(\tau)(t-\tau)^{\alpha-1}d\tau \quad \text{(a)}$$

$$\phi(t) = \frac{t^{\alpha-1}}{\Gamma(\alpha)} \text{ and } f(t) \quad \text{(b)} \quad (19)$$

i.e.,

$$I_{0+}^\alpha f(t) = \frac{1}{\Gamma(\alpha)}\int_0^t f(\tau)(t-\tau)^{\alpha-1}d\tau = \phi(t) * f(t),, \quad (20)$$

and using the formula for the Laplace transform of the convolution. Thus

$$L\{I_{0+}^\alpha f(t)\} = L\{\phi(t) * f(t)\} = \tilde{\phi}(s)\tilde{f}(s); \quad (21)$$

But as discussed in [23],

$$\tilde{\phi}(s) = L\{\frac{t^{\alpha-1}}{\Gamma(\alpha)}\} = s^{-\alpha}, \text{Re}\,\alpha > 0 \tag{22}$$

Taking Laplace transforms on both sides of the set of equations (15) through (17) yields

$$\tilde{u}_n(s) = u_n(0)s^{-1} + \dot{u}_n(0)s^{-2} - \omega_0^{\alpha} s^{-\alpha}\left(2\tilde{u}_n(s) - \tilde{u}_{n+1}(s) - \tilde{u}_{n-1}(s)\right),\ 2 \le n \le N-1 \tag{23}$$

$$\tilde{u}_1(s) = u_1(0)s^{-1} + \dot{u}_1(0)s^{-2} - \omega_0^{\alpha} s^{-\alpha}\left(2\tilde{u}_1(s) - \tilde{u}_2(s)\right) + \tilde{f}(s)s^{-\alpha} \tag{24}$$

$$\tilde{u}_N(s) = u_N(0)s^{-1} + \dot{u}_N(0)s^{-2} - \omega_0^{\alpha} s^{-\alpha}\left(2\tilde{u}_N(s) - \tilde{u}_{N-1}(s)\right) \tag{25}$$

The set of equations can be rewritten in the following matrix form:

$$\tilde{A}(s)\tilde{U}(s) = U(0)s^{-1} + \dot{U}(0)s^{-2} + \tilde{F}(s) \tag{26}$$

where the N x N matrix $\tilde{A}(s)$ is given by

$$\tilde{A}(s) = \begin{pmatrix} 1+2\omega_0^{\alpha}s^{-\alpha} & -\omega_0^{\alpha}s^{-\alpha} & 0 & 0 & . & . & 0 \\ -\omega_0^{\alpha}s^{-\alpha} & 1+2\omega_0^{\alpha}s^{-\alpha} & -\omega_0^{\alpha}s^{-\alpha} & 0 & 0 & . & . \\ . & -\omega_0^{\alpha}s^{-\alpha} & 1+2\omega_0^{\alpha}s^{-\alpha} & -\omega_0^{\alpha}s^{-\alpha} & 0 & 0 & . \\ . & . & . & . & . & . & . \\ . & . & . & . & . & . & 0 \\ . & . & . & . & -\omega_0^{\alpha}s^{-\alpha} & 1+2\omega_0^{\alpha}s^{-\alpha} & -\omega_0^{\alpha}s^{-\alpha} \\ 0 & 0 & . & . & 0 & -\omega_0^{\alpha}s^{-\alpha} & 1+2\omega_0^{\alpha}s^{-\alpha} \end{pmatrix} \tag{27}$$

and the N x 1 column vectors are given by

$$\tilde{U}(s) = \begin{pmatrix} \tilde{u}_1(s) \\ \tilde{u}_2(s) \\ \tilde{u}_3(s) \\ . \\ . \\ \tilde{u}_{N-1}(s) \\ \tilde{u}_N(s) \end{pmatrix} \quad U(0) = \begin{pmatrix} u_1(0) \\ u_2(0) \\ u_3(0) \\ . \\ . \\ u_{N-1}(0) \\ u_N(0) \end{pmatrix} \quad \dot{U}(0) = \begin{pmatrix} \dot{u}_1(0) \\ \dot{u}_2(0) \\ \dot{u}_3(0) \\ . \\ . \\ \dot{u}_{N-1}(0) \\ \dot{u}_N(0) \end{pmatrix} \quad \text{and } \tilde{F}(s) = \begin{pmatrix} \tilde{f}(s)s^{-\alpha} \\ 0 \\ 0 \\ . \\ . \\ 0 \\ 0 \end{pmatrix} \tag{28}$$

respectively. It may be noted that among the four column vectors, only the first and the last refer to Laplace transformed quantities, but the middle two refer to constants describing the initial conditions. The solution to Eq. (25) can be formally written as

$$\tilde{U}(s) = \tilde{A}^{-1}(s)\left(U(0)s^{-1} + \dot{U}(0)s^{-2} + \tilde{F}(s)\right) \tag{29}$$

Then the set of linear equations can be explicitly written down and the inverse Laplace transform applied to these equations on both sides term by term to obtain the displacements as functions of time. This formal procedure appears to be straight enough, however, it is not possible to carry out further simplifications in closed form of the expressions in equation. (28). Nevertheless, this set of equations can be numerically solved for specific values of N, the number of oscillators in the chain. Such calculations have been carried out and the details of numerical applications may be found in [19]. In the next section, the continuum limit of these equations is studied.

5. The continuum limit

It would appear to be a straight forward procedure to extend the same considerations to the equations of motion of a chain of coupled fractional oscillators given in equations (15)-(17).

However, it is not so straight forward and due caution has to be exercised in extending such a procedure. The reason is the occurrence of time derivatives of fractional order. Douglas has shown [25] that in the context of fractals, inhomogeneity in space results in the appearance of fractional order time derivatives and inhomogeneity in time implies fractional order derivatives of space. Since fractional order time dependence is being investigated in the present work, the question arises whether inhomogenneity in space can be ignored and whether space derivatives can be taken. The question is one of scale. The long wavelength limit implies that the lengths involved are very much larger than the separation distance between masses. On this scale the inhomogeneity in space can be ignored and space derivative can be interpreted as an average of some sort and a hand waving justification can be made. The limiting form of the equations (12) through (14) can be considered.

Thus the following factor in equation (15)

$$-\omega_0^\alpha\left((2u_n(\tau)-u_{n+1}(\tau)-u_{n-1}(\tau)\right)$$

can be written as

$$a^2\omega_0^\alpha\left(\frac{1}{a}\left(\frac{(u_{n+1}(\tau)-u_n(\tau)}{a}-\frac{(u_n(\tau)-u_{n-1}(\tau)}{a}\right)\right) \tag{30}$$

and

$$a^2\omega_0^\alpha=a^2\frac{k}{m}=\frac{ka}{m/a} \tag{31}$$

In the limit a→ 0, $u_n(\tau)$ becomes a continuous variable $u(x,\tau)$ and the expression in parenthesis in (30) reduces to $\frac{\partial^2 u(x,\tau)}{\partial^2 x}$ and $ka\rightarrow\kappa,\ \ m/a\rightarrow\rho$ represent the tension and the mass density respectively. The expression in (30) can be written in terms of a quantity c_0^α, where c_0 has the dimensions of 'velocity'. Assuming that at t=0 the displacement at the free end is subject to sinusoidal forcing, $u(0,t)=f(t)=A\sin(\omega t)$, the equation of motion reduces in the continuum limit to

$$u(x,t)=\frac{c_0^\alpha}{\Gamma(\alpha)}\int_0^t \frac{\partial^2 u(x,\tau)}{\partial^2 x}(t-\tau)^{\alpha-1}d\tau \tag{32}$$

with the initial conditions

$$u(x,0)=0 \text{ and } \dot{u}(x,0)=0 \text{ for } x>0 \text{ and } u(0,t)=f(t).$$

Taking the Laplace transform on both sides of equation. (32) yields

$$\tilde{u}(x,s)=\frac{c_0^\alpha}{s^\alpha}\frac{d^2\tilde{u}(x,s)}{dx^2} \tag{33}$$

The equation. (33) can be rewritten as

$$\frac{d^2\tilde{u}(x,s)}{dx^2}-\frac{s^\alpha}{c_0^\alpha}\tilde{u}(x,s)=0 \text{ for } x\neq 0 \text{ and } \tilde{u}(0,s)=\tilde{f}(s) \tag{34}$$

This is an ordinary differential equation and can be solved to yield

$$\tilde{u}(x,s)=\tilde{f}(s)e^{-\sqrt{\frac{s^\alpha}{c_0^\alpha}}x} \tag{35}$$

Substituting for $\tilde{f}(s)=\frac{A\omega}{s^2+\omega^2}$ corresponding to a sinusoidal forcing $f(t)=A\sin(\omega t)$,

The eq. (35) can be inverted as a Bromwich integral

$$u(x,t)=\frac{A\omega}{2\pi i}\int_{Br}\frac{e^{st}e^{-\sqrt{\frac{s^\alpha}{c_0^\alpha}}x}ds}{(s^2+\omega^2)} \tag{36}$$

The Bromwich integral in eq.(36) can be evaluated by appealing to the theory of complex variables [22-24]. By considering a Hankel-Bromwich path, it can be evaluated as the sum of two contributions:

$$u(x,t)=u_{tr}(x,t)+u_{st}(s,t) \tag{37}$$

representing a transient part and a steady state part respectively.

The transient part, $u_{tr}(x,t)$, arises from the Hankel loop consisting of the small circle and two lines parallel to the negative x-axis (as shown in figure 1in [17]) and is given by

$$u_{tr}(x,t)=\int_0^{\infty} e^{-rt}K_{\alpha}(r,\omega,c_0^{\alpha},x)dr \tag{38}$$

with

$$K_{\alpha}(r,\omega,c_0^{\alpha},x)=\frac{A\omega}{\pi(r^2+\omega^2)}e^{-(\frac{r}{c_0})^{\alpha/2}x\cos(\pi\alpha/2)}\sin\left((\frac{r}{c_0})^{\alpha/2}x\sin(\pi\alpha/2)\right) \tag{39}$$

The steady part, $u_{st}(x,t)$, arises from the residues of the poles of the integrand in eq. (36) at $s=\pm i\omega$ and is given by

$$u_{st}(x,t)=Ae^{-\left(\frac{\omega}{c_0}\right)^{\alpha/2}x\cos(\pi\alpha/4)}\sin\omega(t-\frac{x}{\omega}\left(\frac{\omega}{c_0}\right)^{\alpha/2}\sin(\pi\alpha/4)) \tag{40}$$

This completes the formal analysis of the response to sinusoidal forcing of a chain of fractional oscillators in the continuum limit. Numerical results will be presented in the next section.

6. Results and discussion

The results of the last section in equations (36)-(40) are new. Applications of fractional calculus to oscillation and dissipation problems and solutions to integral and fractional order differential equations can be found in Gorenflo and Mainardi [8] Impulse response functions are also discussed in their work. The same methodology has been extended to sinusoidal forcing of a linear chain in the present work. The response of a chain of fractional oscillators in the continuum limit, when subject to a sinusoidal forcing starting from a quiescent state consists of a transient part and a steady part. The transient part decays in time and approaches zero as $t\to\infty$. In the limit $\alpha\to 2$ the transient part vanishes entirely. Furthermore, it exhibits attenuation as a function of distance from the end as indicated by the spatial dependence of the kernel in eq. (39). No simple closed form expressions can be obtained, the only recourse is through numerical integration. Closed form solutions are not available even for simple mechanical models such as the standard linear solid, which are not based on fractional calculus, for the reason that the solutions are deemed mathematically cumbersome as has been discussed by Caputo and Mainardi [5]. These authors have applied methods of fractional calculus to the linear theory of viscoelasticity based on mechanical models and have discussed the propagation of viscoelastic waves[6]. Their work is macroscopic in its approach and is not based on the continuum limit of lattice dynamics. However, some concepts developed by these authors are quite insightful and the results of the present study can be related to these concepts as follows. Equation (38) can be rewritten by changing the variable $r \rightarrow 1/\tau$ as

$$u_{tr} = \frac{A}{c_0^{\alpha}} \int_0^{\infty} R(\tau) e^{-t/\tau} d\tau \tag{41}$$

to bring out clearly the decay in time. It is obvious that the decay in time is characterized by a distribution of relaxation times. The relaxation spectrum is given by

$$R(\tau) = \frac{\omega}{\pi(1+\omega^2\tau^2)} e^{-\frac{x}{(c_0\tau)^{\alpha/2}}\cos(\pi\alpha/2)} \times \sin\left(\frac{x}{(c_0\tau)^{\alpha/2}}\sin(\pi\alpha/2)\right) \tag{42}$$

It may be noted that the relaxation spectrum depends also on the distance x, which appears in a similarity combination $x/(c_0\tau)^{\alpha/2}$ and that the factor in the exponent, $\cos(\pi\alpha/2)$, may become negative. Hence caution should be exercised in a strict interpretation in terms of relaxation times. The relaxation spectrum is displayed in figure 4 for several values of the parameter α. The values of the other parameters have been chosen to be $x = 1.0$, $c_0 = 1.0$ and $\omega = 1.88$ for this display.

The steady state solution given by equation (40) can be written as

$$u_{st}(x,t) = \frac{A}{c_0^{\alpha}} e^{-\delta x} \sin\omega(t - x/v) \tag{43}$$

explicitly showing that it as an attenuated wave, propagating with a velocity (phase velocity)

$$v = \frac{\omega c_0^{\alpha/2}}{\omega^{\alpha/2}\sin(\pi\alpha/4)} \tag{44}$$

and characterized by an absorption parameter

$$\delta = \left(\frac{\omega}{c_0}\right)^{\alpha/2} \cos(\pi\alpha/4) \tag{45}$$

In the limit $\alpha \rightarrow 2$, $v \rightarrow c_0$ and $\delta \rightarrow 0$ as can be seen from equations (44) and (45).

Phase velocity as a function of frequency for several values of α is shown in Fig. 5. The variation of the absorption parameter as a function of the driving frequency is shown in Fig.6.

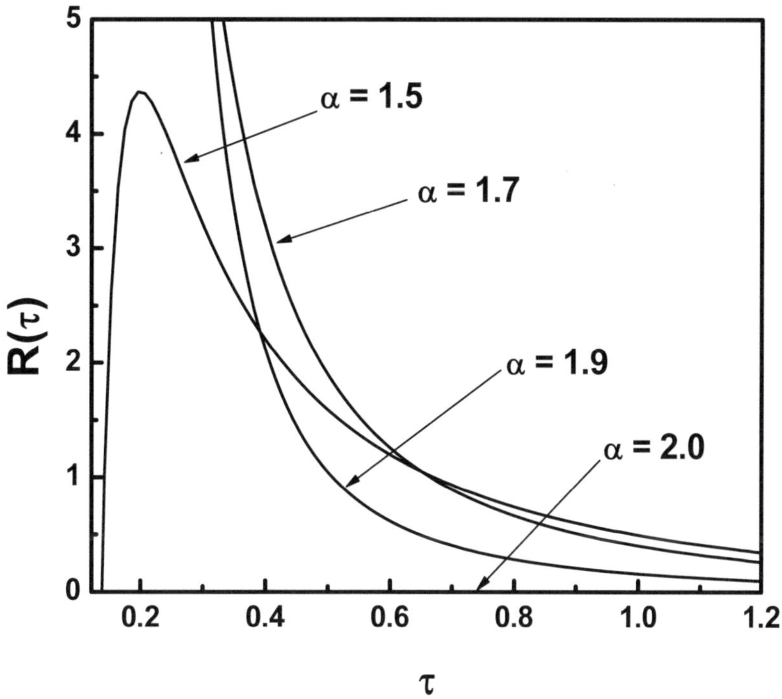

Figure 4. Relaxation Spectrum for different values of α

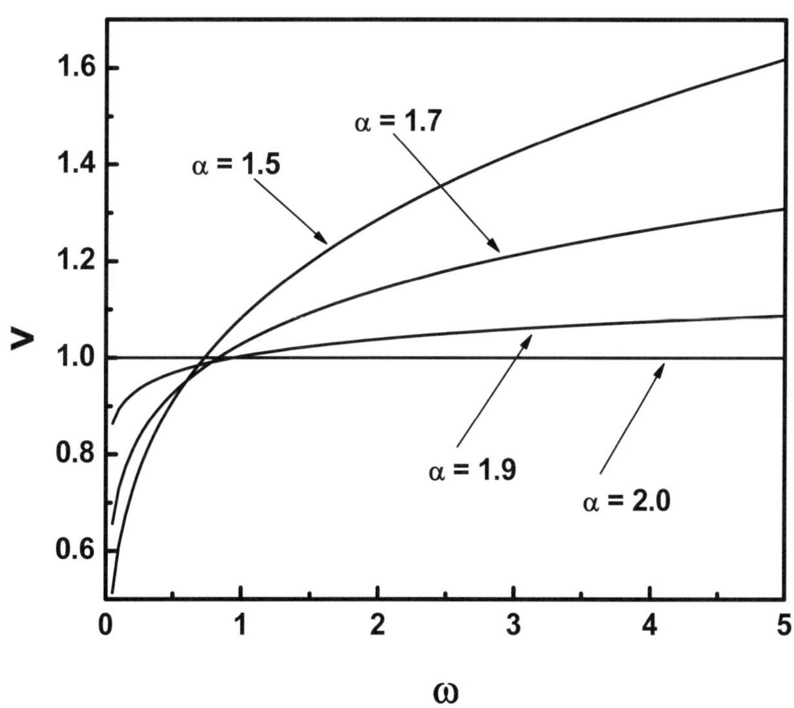

Figure 5. Phase velocity as a function of driving frequency (in arbitrary units).

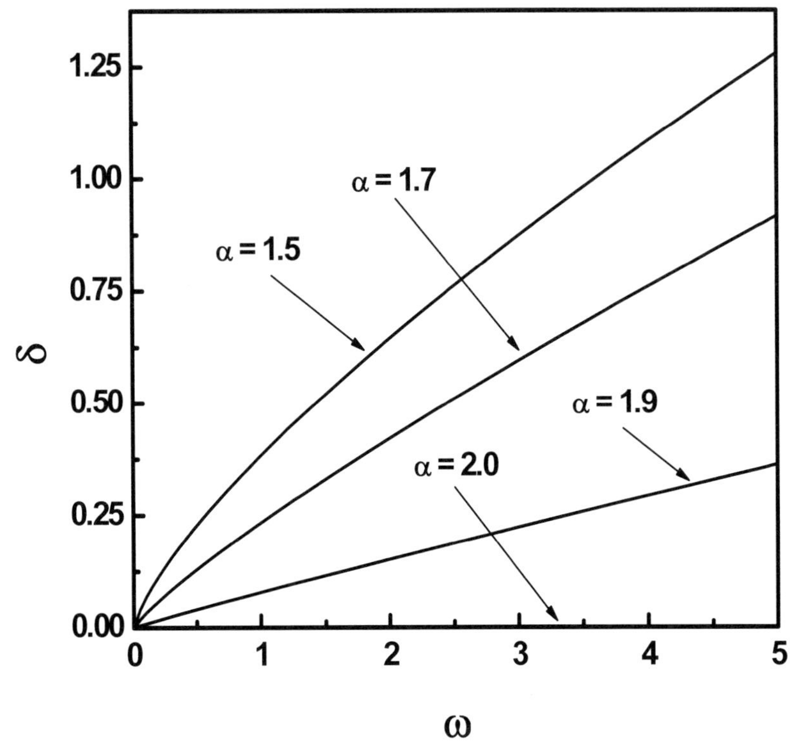

Figure 6. Absorption parameter δ as a function of driving Frequency (in arbitrary units).

According to equation. (40), the amplitude of the response wave decays with distance as $\sim Ae^{-\left(\frac{\omega}{c_0}\right)^{\alpha/2} x\cos(\pi\alpha/4)}$ and hence the energy as $A^2 e^{-2\left(\frac{\omega}{c_0}\right)^{\alpha/2} x\cos(\pi\alpha/4)}$. The specific dissipation function can be defined in terms of the distance over which the energy decays by a factor e^{-1}, which is given by

$$x = \frac{c_0^{\alpha/2}}{2\omega^{\alpha/2}\cos(\pi\alpha/4)} \tag{46}$$

The specific dissipation function is given by Q^{-1} where Q is defined (in analogy with the quality factor 'Q' of harmonic oscillators) as the phase change in radians experienced as the wave travels the distance x in equation. (45) and is given by

$$Q = x\left(\frac{\omega}{c_0}\right)^{\alpha/2} \sin(\pi\alpha / 4) = \frac{1}{2}\tan(\pi\alpha / 4) \tag{47}$$

An alternate definition of 'Q^{-1}' has been given [6] by Caputo and Mainardi, who introduced in analogy with the optical case, a complex 'refractive index, $n = n_r - in_i$. The real and imaginary parts of the refractive index for the wave in eq. (40) are given by

$$n_r = \frac{c_0}{v} = \frac{\omega^{\alpha/2}\sin(\pi\alpha / 4)}{\omega c_0^{\alpha/2-1}} \tag{48}$$

and

$$n_i = \frac{c_{0\delta}}{\omega} = \frac{c_0}{\omega}\left(\frac{\omega}{c_0}\right)^{\alpha/2} \cos(\pi\alpha / 4) \tag{49}$$

According to Caputo and Mainardi [6], the specific dissipation function is given by

$$Q^{-1} = \frac{2n_i}{n_r} = 2\cot(\pi\alpha / 4) \tag{50}$$

which is equivalent to the expression in equation (47).

Kolsky's definition [26, equation 5.22] of Q^{-1}, $(= 2v\delta / \omega$, in our notation), also yields the same expression as in equation (50).

6.1. Further discussion

As explained earlier, the classical theory of viscoelasticity is based on the so called Standard Linear Solid Model characterized by the constitutive relation given in equation (6). This model gives a good qualitative description of the deformation behavior of viscoelastic solids, but has been found to be inadequate to give quantitative account as has been discussed in detail by Gorenflo and Mainardi [8]. More complex mechanical models lead to difficulties in formulation and solution of the resulting complicated differential equations and no satisfactory resolution has been possible [8]. Quantitative agreement is secured only when recourse is made to the so called fractional viscoelastic models. The stress-strain relations are then expressed in terms of fractional order integrals in both creep- and relaxation- representations mentioned earlier and further details can be obtained in reference [9]. In the space-time domain, the approach based on fractional calculus leads to the propagation of viscoelastic waves and the details can be found in Mainardi[9]. The approach developed in the present work is based on lattice dynamics and hence is a dynamical model. It is microscopic in approach and hence macroscopic properties are described in the so called long wave limit. Further deformation properties which are 'static' in nature are described in the zero frequency limit. As has been shown above the microscopic approach leads to the macroscopic properties in the appropriate limit. It has been noted [26] that the presence of spatial in-homogeneity in polymeric materials gives rise to fractional differential operators in time in the relevant evolution equations, while temporal in-homogeneity leads to fractional differential operators in space. Since only fractional order operation in time has been considered in the present work, there is an implicit in-homogeneity in space which results in a tendency of the particles to cluster and move in a collective fashion. Such collective motion can be considered as elastic waves of very large wavelength, much larger than the scale of the in-homogeneity. This provides a sort of justification for the limiting long wavelength approximation employed.

7. Summary

A microscopic approach based in lattice dynamics, but from the fractional calculus point of view has been developed for the theory of viscoelasticity. The system considered is a linear chain of fractional oscillators subject to a sinusoidal forcing at one end. For the system starting from a quiescent state, the response to sinusoidal forcing consists of a transient and a steady state part. The decay of the transient is characterized by a distribution of relaxation times and the expression for the relaxation spectrum has been obtained. The steady state is essentially a attenuated sinusoidal wave, whose phase velocity and attenuation have been studied and an expression for the specific dissipation function has been obtained. The results obtained are consistent with the results obtained by Mainardi and others from a macroscopic point of view.

Author details

B.N. Narahari Achar* and John W. Hanneken
Physics Department, University of Memphis, Memphis, USA

8. References

[1] Gross B (1953) Mathematical Structure of the Theories of Viscoelasticity Paris: Hermann.
[2] Bland D R (1960) The Linear Theory of Viscoelasticity Oxford: Pergamon.
[3] Christensen R M Theory of Viscoelasticity New York: Academic Press.

---

* Corresponding Author

[4] Zener C M (1948) Elasicity and Anelasticity of Metals Chicago: Chicago University Press.
[5] Caputo M and Mainardi F (1971) Linear Models of dissipation in Anelastic solids Riv. Nuovo Cimento (Ser. II) 1: 161-198.
[6] Caputo M and Mainardi F (1971) A new dissipation model based on memory mechanism Pure and Appl. Geophys. 91: 134-147.
[7] Glockle W G and Nonenmacher T F (1991) Macromolecules 24: 6426-6434
[8] Gorenflo R and Mainardi F (1997) Fractional Calculus: integral and differential equations of fractional order. In: Carpinteri A and Mainardi F , editors. Fractals and Fractional Calculus in Continuum Mechanics. Wien: Springer Verlag. pp 223-276.
[9] Mainardi F (1997) Fractional Calculus: Some basic problems of continuum and statistical Mechanics. In: Carpinteri A and Mainardi F , editors. Fractals and Fractional Calculus in Continuum Mechanics. Wien: Springer Verlag. pp 291-348.
[10] Mainardi, F (2010) Fractional calculus and Waves in Linear Viscoelasticity. London: Imperial College Press
[11] Born M and Von Karman T (1912) Phyzik, Z. 13: 297.
[12] Born M and Huang K (1954) Dynamical Theory of Crystal Lattices. Oxford,: Clarendon Press
[13] Maradudin A A, Montroll E M, Weiss G H and Ipotova I P (1971) Theory of Lattice Dynamics in the Harmonic Approximation (Solid State Physics Suppl. 3 2nd edition .New York: Academic Press.
[14] Bottger H (1983) Principles of the Theory of Lattice Dynamics. Weiheim: Physik Verlag
[15] Askar A (1985) Lattice Dynamical Foundations of Continuum Theories. Singapore: World Scientific.
[16] Narahari Achar B N, Hanneken J W, Enck T and Clarke T (2001) Dynamics of the Fractional Oscillator. Physica A 297: 361-367.
[17] Narahari Achar B N, Hanneken J W, and Clarke T (2002) Response Characteristics of the Fractional Oscillator *Physica A* 309 275-288 There is a factor r^{α} missing in equation.(27) of this reference. The corrected form of the kernel is given in equation (21a) in reference [16] cited below
[18] Narahari Achar B N, Hanneken J W and Clarle T 2004 Damping Characteristics of the Fractional Oscillator *Physica A* 339 311-319. A typographical error in equation. (13) of this reference has been corrected and the corrected equation appears as equation (37) in Ref. [18]
[19] Narahari Achar B N, Hanneken J W 2005 Dynamic Response of the Fractional Relaxor-Oscillator to a Harmonic Driving Force *Proc. IDETC/CIE 2005 ASME 2005 Int. Design Engineering Technical Conf. & Computers and Information in Engineering Conf.* (Long Beach, CA , 22-24 September 2005) DETC2005- 84345 pp 1-7
[20] Narahari Achar B N, Prozny T and Hanneken J W 2007 Linear chain of coupled oscillators: Response dynamics and its continuum limit *Proc. IDET/CIE 2007 ASME 2007 Int. Design Engineering Technical Conf. & Computers and Information in Engineering Conf.* (Las Vegas, Nevada, 4-7 September 2007) DETC2007-35403 pp 1-7
[21] Podlubny L 1999 *Fractional Differential Equations*,(San Diego, CA:Academic Press)

[22] Erdelyi A (ed.) 1955 *Table of Integral Transforms* Vol . I (New York: McGraw Hill)

[23] Schneider W R and Wyss W 1989 Fractional diffusion and wave equations. *J. Math Phys.* 30 134-144

[24] McLachlan N W 1963 *Complex Variable Theory and Transform Calculus* 2nd ed (Cambridge: Cambridge University Press)

[25] Sneddon I. N 1972 *The Use of Integral Transform* ,(New York: McGraw-Hill)

[26] Douglas J. F (2000)Polymer Science Applications of Path-Integration,Integral Equations, and Fractional Calculus. In: Hilfer R, editor.Applications of Fractional calculus in Physics. Singapore: World Scientific. Pp.241-330.

[27] Kolsky H 1963 *Stress Waves in Solids*(New York: Dover Publications) p106

Chapter 4

Die Swell of Complex Polymeric Systems

Kejian Wang

Additional information is available at the end of the chapter

http://dx.doi.org/10.5772/50137

1. Introduction

Die swell is a common phenomenon in polymer extrusion. When a viscoelastic fluid flows out of a die, the extrudate diameter is usually greater than the channel size. This is called die-swell, extrudate swell or the Barus effect. The degree of extrudate swell is usually expressed by the die-swell ratio (B) of extrudate diameter versus die diameter. A better understanding of such flow behavior will be beneficial for the optimization of processing parameters and the design of extrusion equipment, both of which affect product quality and production cost.

Innumerable valuable experimental data of die swell have been published, each focusing on different aspects affecting extrudate swell. There have been various interpretations of the extrusion swell of viscoelastic fluids from the macroscopic view of polymer rheology, such as a normal stress effect, an elastic energy effect, an entropy enlargement effect, an orientation effect and a memory effect. In fact, these interpretations are all related to each other[1,2]. It is generally believed that die swell is an important characteristic of the fluid elasticity during flow. The most common technique used to study rheological properties of polymer melts is capillary rheometry. In a capillary, polymer flows from the reservoir through a die and finally swells out of the exit. Under the action of extension, shear and compression, some molecular chains become disentangled, uncoiled or oriented in the convergent region resulting in an entry effect. During die flow, the resultant stress and strain cannot be relaxed completely. Simultaneously, some chains continue to be sheared and elongated during extrusion. On emerging from the die exit, the molecules are relaxed in elastic deformation by reentanglement and recoiling. The extrudate tends to contract in the flow direction and to grow in the normal direction, leading to extrudate swell[3]. As it does inside the capillary, the swelling evolves with time outside the capillary to reach a maximum at a certain extrusion distance. Graessley et al.[4] have reported that die swell occurs in two steps: (1) a very rapid swelling with a relatively large swell ratio very close to the die exit, which is known as running die swell; (2) a subsequent slow expansion to give

INTECH open science | open minds

© 2012 The Author(s). Licensee InTech. This chapter is distributed under the terms of the Creative Commons Attribution License http://creativecommons.org/licenses/by/3.0), which permits unrestricted use, distribution, and reproduction in any medium, provided the original work is properly cited.

an equilibrium die swell. Thus, the extent of the swelling will depend on both external factors as well as the intrinsic characteristics of the polymer[5]. The former include the geometry (contraction ratio and angle, die length and diameter as well as their ratio L/D) of the extrusion system (reservior, entry and die) [6,7] and the capillary operating conditions (the applied shear rate, stress and temperature) [8,9] and the external environment medium[10]. Extrudate swell can be used to assess the polymer viscoelasticity upon melt extrusion. Through rheological characterization, extrudate swell can also be correlated with the molecular structure of the polymer (molecular weight, and extents of crosslinking and long chain branching)[11].

Precise polymer processing requires an adequate quantitative description of flow. The viscous behavior determines the throughput, while the elastic properties are important for dimensional stability. Many theoretical studies have been conducted to improve our understanding and prediction of extrudate swell[12], in which a very critical aspect is the choice of an appropriate constitutive model[13,14]. In simulations, specific values of the model parameters—especially material parameters—must be defined. Some of these parameters can be measured by rheological experiments. Thus, quantitative analyses are able to relate extrudate swell to the viscoelasticity of polymer melts.

Extrudate swell was initially exploited to study the effects of die swell on processability and its correlation with the relevant rheological properties of polymer melts and the early work has been reviewed by Kontos and White[15] and Graessley[4]. In these early studies[16–18], a wide range of experiments on polymer melt swelling were performed. Some workers have reported that die swell varies with temperature[19], pressure during processing[20], molecular weight and its distribution[21–23], as well as molecular structure[24–27] and compounding ingredients[28]. Studies of die swell as a function of length to diameter ratio of die (L/D), entry speed and shape of the die have been reported by Han [29], Cotten[30], Lenk [31] and Vinogradov and Malkin[32]. Recently, the die swell of particle-filled polymers has attracted more attention because of their wider engineering applications[33,34]. These experiments have generated various valuable data of extrudate swell which need to be rationalized by theoretical studies.

Previous examinations of the existing data have identified some semi-empirical correlations relating the swell ratio to rheological parameters. In the 1970s, Bagley and Duffey[35], Graessley et al.[4], Han[36] and Tanner[37] proposed expressions for the relationship between the swell ratio B and the first normal stress difference or shear stress, on the supposition that polymer melt shear flow obeyed a simple law. One of the most famous approaches is that of Tanner[38] based on elastic recovery theory. In this model, the maximum diameter of the extrudate is related to the recoverable shear strain at the capillary wall.

Such continuum mechanics studies have been regularly reported from the early 1980s onwards. Investigations of the swelling phenomenon have been carried out using numerical simulations, especially by using the finite element method when considering complicated boundary conditions[39,40]. Most of the analyses have been conducted for laminar flow or at low Reynolds number, in which the swell ratio is greater than one. However, at higher Reynolds numbers, the dimensions of the extrudate are possibly smaller than those of the die. Few data have been published concerning the effect of Reynolds number on the die swell[41,42]. Han[29] suggested that the equilibrium die swell measurements are independent of rheometer geometry for common extrusion at relatively low flow rate through a capillary where the ratio of reservoir diameter to capillary die diameter is less than 10, and length to diameter ratio of the capillary die is greater than 20. Such a capillary is termed a long capillary, in which the effects of the entry flow from the reservoir on the die exit region flow can be neglected. Most theoretical models have been derived for a long capillary[43–47]. However, there have been few quantitative predictions of die swell in actual polymer processing operations in which the ratio of width to length in the die flow channel is relatively short[48-58]. The die swell behavior of polymeric materials should be very important in short capillary flow. However, entrance effects in the short tube flow of polymer melts are relatively complicated. Thus, it is necessary to extend the theoretical models of the die swell in a long capillary to that in a short capillary. This is one purpose of this paper.

On the other hand, particulate-filled polymer composites are widely used in engineering to improve polymer properties. Their flow properties are important when molding various products. Some swell experiments have been conducted on filled polymers in which it was noticed that the die swell was related to the filler shape, size, dispersion, concentration and interfacial modification of the inclusions in the matrix[59,60]. However, there are few quantitative theories of the swelling in composites, in contrast to the situation for pure polymer melts[37,61]. Thus, it is necessary to explore how the swell behaviors of composites differ from those of a pure polymer. A second aim of this paper is to develop a single unified quantitative extrudate swell model for both pure polymers and their composites.

The following sections first present a unified theoretical equation to describe extrusion swell in both long capillaries and short capillaries. The evolution of the finite distance of extrudate swelling and the particular features of the swelling of filled composites are also detailed. Finally, comparisons are made between the predictions of the model and experimental data.

2. Theoretical Model

2.1. The extrudate swell theory for a long capillary

Up to now, most of the extrudate swell models have been established on the basis of the analysis of long capillary flow. The swell ratio B has been related to the recoverable shear strain (S_R) or elastic strain energy[62]. One most pertinent systematic theories of extrudate swell of entangled polymeric liquids is that of Tanner, who first based his model on the K-BKZ constitutive equation and the free recovery from Poiseuille flow with the aspect ratio of length L to diameter D being very large[37]. The correlation was later confirmed for a wide class of constitutive equations, including PTT, pom–pom and general network type models for fully-developed tube flow[63]. It was found that the extrudate swelling ratio (B) may be expressed as a function of the first normal difference N_1 and the shear stress τ_{12}. Regardless of the fact that this clearly shows how the swell is related to the elasticity of viscoelastic polymeric fluids, the viscous heating and the time-dependent nature of swell are not considered.

More recently, Wang et al.[64,65] extensively exploited the variations of extrudate swell with both polymer characteristics ($\bar{M}_n$, $\bar{M}_n / \bar{M}_W$ and $\bar{M}_e$) and the operational processing conditions. A double molecular mechanism of disentanglement-reentanglement and decoil-recoil was proposed to rationalize variations of polymer elasticity during flow with chain conformation. The die swell behavior essentially results from the molecular dynamics of the system. Thus, it is desirable to establish a single primary theoretical framework to relate the extrudate swell to the intrinsic viscoelasticity and external conditions.

There are three kinds of polymer segments and chains in the extrudates, which have been defined by Song[66] as extending chain, coil chain and entangled polymeric chain. Besides the change in chain conformation and the degree of reorganization of the constituent chains, the extension and flow can also induce the dynamic and reversible disentanglement and reentanglement between polymeric chains, such that the polymeric melts then undergo a partial stress relaxation leading to extrudate swelling at the die exit. The swell ratio is affected by the length to diameter ratio and the residence time. Based on such a viewpoint of dynamics, Song developed a novel molecular theory of such multiple entangled constituent chains in order to analyze non-linear viscoelasticity for the polymeric melts. His derived constitutive equations and material functions in a multiple-flow field were subsequently quantitatively verified[67]. Based on the O-W-F constitutive equation and the multiple transient-network model as well as the double relaxation dynamics of the reentanglement-disentanglement transition (RE-DT) and recoil-uncoil transition (RC-UCT) from the Poiseuille flow, Song proposed that swell evolves in three stages (instantaneous swelling, delayed swelling and ultimate extrudate swelling)[68]. A new set of swelling equations incorporating molecular parameters, operational parameters and growth time in both the steady state and dynamic state were developed. Song's model successfully described the die swell through a long capillary of both linear polyethylene (HDPE) and linear polybutadiene (PBD) with the different molecular weights and different processing variables[69]. In this paper, Song's extrudate swell theory will be extended in order to analyze both the die swell out of a short capillary and the swell of polymer composites.

In the steady shear flow, the shear viscosity can be written as

$$\eta(\dot{\gamma}) = \eta_0 / [1 + (\frac{\eta_0 \dot{\gamma}}{G_0})^a]^n \quad (1)$$

The above expression is verified by experimental data, and can also be deduced from the O-W-F constitutive equation together with molecular dynamics[69]. The coefficient of the first normal-stress difference Ψ_1 in the steady shear flow is

$$\psi_1(\dot{\gamma}) = 2n\eta_0^{1+1/n} / (G_0)^{1/n} / \left[1 + \left(\frac{\eta_0 \dot{\gamma}}{G_0}\right)^a\right]^{n+1} \quad (2)$$

From experimental data of η and $\psi_1(\dot{\gamma})$, the molecular parameters of the zero shear viscosity η_0, the equilibrium modulus G_0, and the exponents n and a can be determined. The approximate value of the ultimate extrudate swell B was obtained as Eq. (3) by Song[69]:

$$B = \frac{1}{2}((n(\frac{\eta_0}{G_0})^{1/n}\dot{\gamma}/(1+(L/D)^{\alpha}))^{(1-W)} + 5.098)^{1/2} \tag{3}$$

B depends on both the molecular parameters and the operational variables [the capillary length to diameter ratio L/D and the shear rate $\dot{\gamma}$ or shear stress τ]. $(1-W)$ is the fraction of the recoverable conformation of the entangled polymeric chain in the flow, which can be calculated from the experimental swell ratio data by Eq. (4).

$$(1-W) = \frac{\ln(4B^2 + \frac{2}{B^4} - 5.098)}{\ln(n(\frac{\eta_0}{G_0})^{1/n}\dot{\gamma}/(1+(L/D)^{a}))} \tag{4}$$

Details of the derivation of the above equations (3) and (4) and their application to the extrudate swell of HDPE and PDB are given by Song[69]. Eq.3 may also be expressed in the form as shown in Eq. (5), where B_{equ} is the die swell ratio at $L/D=\infty$, and m_{γ} is the shear rate-dependent constant.

$$B = B_{equ} + \frac{b\dot{\gamma}^{m_r}}{(1+(L/D)^{a})} \tag{5}$$

2.2. Finite extrudate swell distance

As discussed above, the swelling evolves with time after the polymer is extruded from the capillary, during which time the molecules continue to exhibit similar disentanglement-reentanglement transitions and uncoil-recoil transitions to those occurring whilst it is still inside the capillary. To describe this phase, Eq. (3) can be reformulated as Eq. (6) by introducing two parameters k_t and f_w:

$$B = \frac{1}{2}(k_t(n(\frac{\eta_0}{G_0})^{1/n}\dot{\gamma}/(1+(L/D)^{\alpha}))^{(1-W)} + 5.098)^{f_w} \tag{6}$$

Here f_w is a variable which replaces the fixed value of 0.5 used by Song[69] in Eq. (3) and the coefficient k_t describes the evolution of the extrudate swell with time. k_t may be written as

$$k_t = 1 + A_0 \ln(t/t_{\infty})/(1+(\eta_0\dot{\gamma}/G_0)^{\alpha}) \tag{7}$$

As the swell approaches a maximum at $t = t_{\infty} = \infty$, k_t approaches unity corresponding to the model used by Song[69]. However in contrast to Song's model, in practice t_{∞} is not infinite and actually has a finite value which can be determined experimentally[70], i.e., the maximum ultimate swelling will be realized at time t_{∞} along the extrusion distance. If the extrusion distance is expressed as $Z \times D$, where D is the capillary diameter and Z is a numerical factor, t_{∞} and Z can be correlated as follows from the shear rate,

$$t_{\infty} = 8k_n Z / \dot{\gamma} \tag{8}$$

where k_n is a constant for a given material, whose approximate value is

$$k_n = (1+3n)/4n \tag{9}$$

where n is the constitute exponent of the fluid. From Eqs. (7)–(9), it can be shown that

$$\ln t_{\infty} = \ln k_n + \ln Z - \ln \dot{\gamma} - \frac{k_t}{1+(\tau\dot{\gamma})^{\alpha}} \tag{10}$$

Eq. (10) can be rewritten as Eq. (11).

$$\ln(t_{\infty} * \dot{\gamma}) - \ln Z = \ln k_n - \frac{k_t}{1+(\tau\dot{\gamma})^{\alpha}} \tag{11}$$

Thus, either Eq. (10) or Eq. (11) can be used to predict the time of maximum swelling and the corresponding swell distance.

2.3. Extrudate swell theory for a short capillary

The above model is based on the assumption that the chain elongation incurred on reservoir entry is fully relaxed in the capillary. This is only approximately true for extrusion in a long capillary and a very poor assumption for a short capillary where the the entry flow is more complicated and the entry effect is more prominent[71]. Liang[72] reported the results of many swelling experiments using a short capillary and some quantitative empirical relations between the swell ratio and the material characteristics and operational parameters have been obtained[37,70,73]. In this paper, we attempt to modify Song's model in order to describe the swelling behavior in short capillary extrusion.

Liang[74] described the die swell ratio as follows,

$$B = (1+\lambda_1 S_R)^{1/2} \tag{12}$$

$$S_R = (N_1 / 2\tau)^{1/2} \tag{13}$$

$$\lambda_1 = 0.5 \tan \alpha_0 \tag{14}$$

where S_R is the recoverable shear strain; λ_l shows the elastic strain induced by the stored energy in the capillary reservoir, which is related to the capillary geometry and the fluid viscoelasticity; λ_l is a function of the entry converging flow parameter; α_0 is the half-entry convergence angle of a viscoelastic fluid, which is given by a function of the constitute exponent n of the fluid, the entry pressure drop ΔP and the ratio of the capillary diameter D to its reservoir diameter D_r. The pressure loss in the entry region can be approximately expressed in terms of the Bagley entrance correction factor (e).

$$\tan\alpha_0 = \frac{1}{e}\left[\frac{4(1-(D/D_r)^{1.5(n+1)})}{3(n+1)} + \frac{2\xi((D_r/D)^{1.5(n+1)}-1)}{3(1-n)}\right] \tag{15}$$

$$\frac{1}{e} = \frac{\Delta P}{2\tau} \tag{16}$$

The total die swell ratio in Eq. (3) can be approximated as Eq. (17), which has the same form as Eq. (12).

$$B = (5.098/4 + S_R^{1-w}/4)^{1/2} \tag{17}$$

Eq. (17) can be modified to describe the extrudate swell out of a short capillary, as shown in Eq. (18),

$$B = \frac{1}{2}(k_t k_{L\dot{\gamma}} \frac{M}{4}(S_R)^{(1-W)} + 5.098/4)^{f_w} \tag{18}$$

where k_t is the time-dependent coefficient reflecting the swelling evolution as defined by Eq. (7) and $k_{L\dot{\gamma}} = [\frac{1+(\tau_0\dot{\gamma})^a}{1+(L/D)^a}]^{(1-w)}$ shows the effects of capillary length, i.e., the degree of relaxation. $M/4$ is the recoverable effect from the stored energy in the capillary reservoir. It may be *ca.* 1 for a sufficiently long capillary while it may be λ_l for a very short capillary as defined by Eq. (14). Thus, Eq. (18) may be used for both long and short capillaries since it includes sufficient variables, whereas Eq. (3) is only appropriate for long capillaries.

2.4. Extrudate swell theory of filled composites

In particle-filled composites, the filler is distributed in an entangled matrix network and the filling affects the network relaxation[74]. Regardless of the variation in viscoelastic properties, it has been found experimentally that the values of η and $\psi_1(\dot{\gamma})$ for filled composites are similar to those of the pure polymer matrix, i.e., the above extrudate swell theory may be modified for use with filled composites.

However, experiments have shown that the die swell is usually weakened with increasing filler concentration [75]. This suggests that the die swell of a composite can be expressed by Eq. (19),

$$B(\dot{\gamma},\phi) = B(\dot{\gamma},\phi = 0)F(\phi) \tag{19}$$

where $B(\dot{\gamma},\phi)$ and $B(\dot{\gamma},\phi = 0) = B_m$ are the die swell of the composite with a filler content of ϕ and the die swell of the pure matrix, respectively, and $F(\phi)$ is the filling effect. Eq. (19) implies that the viscoelastic behavior of the filled composites is dominated by the elasticity of the composite matrix in the high shear rate range. Liang[75] found that $B = \lambda\tau_w(1-\phi\varsigma)$, where τ_w was the wall shear stress, and λ and ς were constants related to the elasticity of the matrix melt and the geometry of the filler particles, respectively. The function $F(\phi)$ is analogous to those used for the viscosity of a suspension of spheres, and may be called a 'concentration shift factor'[76]. There are several forms for $F(\phi)$. Here we use the form shown in Eq. (20),

$$F(\phi) = [1-(\phi/\phi_c)^k]^p \tag{20}$$

$$\phi_c = (\dot{\gamma}/\dot{\gamma}_c)^q \tag{21}$$

where k, p and q are constants which depend on the filler-polymer system and the flow field; ϕ_c is the percentage decrease of the network elasticity caused by filling with the particles, which is directly related to the shear rate $\dot{\gamma}$ as described by Eq. (21); $\dot{\gamma}_c$ is the limiting shear rate when the network is almost completely destroyed, in accordance with percolation theory. An approximation to Eq. (21) may be rewritten in stress form as Eq. (22).

$$\phi_c = (\tau/\tau_c)^{q'} \tag{22}$$

3. Comparison of die swell given by the model with experiment

3.1. Experimental data

For comparison of the predicted swell ratios using the model described above with experimental data, several sets of raw swell data were taken from the literature and replotted in appropriate formats.

In testing the validity of the model for a system with large L/D, the rheometric and swell data of a pure IUPAC-LDPE standard was used as shown in Figures 1 and 2 [77,78]. The capillary had die diameter $D = 3.00mm$ and experiments were carried out at 150°C.

To show whether the model is valid in analyzing the maximum extrusion distance during swelling, we used the data for a semi-dilute polymer solution of a partially hydrolyzed polyacrylamide manufactured by Rhone Poulenc (Rhodoflood AD37, M_w= 6 x 10^6, degree of hydrolysis = 24%)[70]. Purified water containing 20 g/l of NaCl was employed as a solvent for the polymer. The polymer concentration was 3000 ppm. The swell tests were conducted in a stainless steel capillary with length of 51 mm and inner diameter of 0.60 mm.

The swell data for a rubber compound[73] were used to verify the effectiveness of the suggested equations for polymer extrusion in a short capillary. Sample SI was a calcium carbonate ($CaCO_3$) filled natural rubber (NR) compound, in which the content of $CaCO_3$ was 20 phr. Sample SII was a carbon-black (CB)-filled NR/styrene-butadiene rubber SBR) /cis-1,4-butadiene rubber (CBR) compound. The blending ratio of NR/SBR/CBR was 45/10/45, and the content of CB was 56 phr. Both the rubber compounds included some additives. An Instron capillary rheometer (model 3211) was used in the tests. Two capillary dies with different lengths were selected in order to measure the rheological properties of the sample materials. The length of the long die was 40 mm, the length of the short die was 0.2 mm, and the diameter of both dies was 1 mm.

Another set of data was for a polypropylene/glass bead composite[75]. The matrix resin was a common polypropylene (Pro-fax 6331). The melt flow rate (2.16 kg, 230 °C) and density were 12 g/10 min and 0.9 g/cm^3, respectively. The glass bead filler had a mean diameter of 219 nm and density of 2.5 g/cm^3. Rheological experiments with these samples were carried out using a Rosand capillary rheometer with twin cylinders. Two dies of different length with the same diameter (1 mm) were selected in order to measure the entry pressure losses. The length/diameter of the short die was << 1 and the other was 16. The test temperature was 190 °C, and apparent shear rates were varied from 50 to10^{-4}s^{-1}.

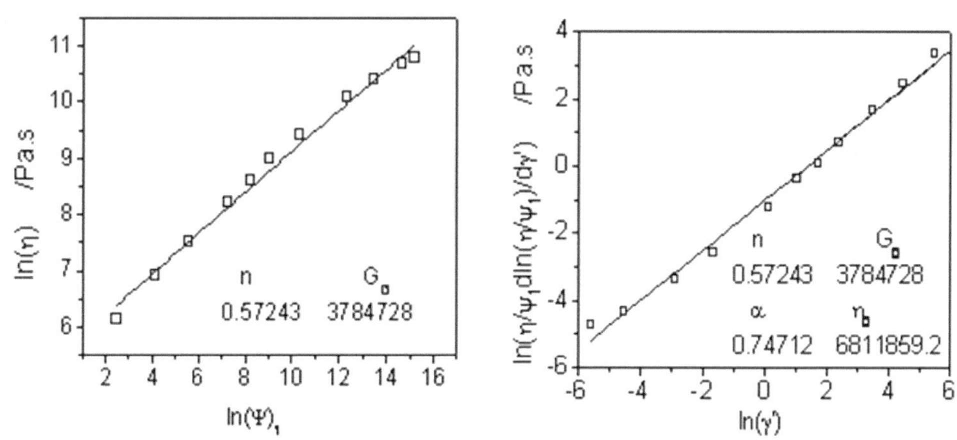

Figure 1. Plots of $\ln\eta$ versus $\ln\varphi_1$ and $\ln(\eta/\varphi_1 d\ln((\eta/\varphi_1))/d\dot{\gamma})$ versus $\ln\dot{\gamma}$ from the experimental data for LDPE[77] of $\eta \sim \dot{\gamma}$ and $N_1 \sim \dot{\gamma}$.

3.2. Comparison of the predicted values of swell ratio with the experimental data

3.2.1. Swelling simulations for pure LDPE extruded from capillaries with different L/D ratios

The dashed curves in Figure 2 present the experimental data[77] of extrudate swell ratio for LDPE for different capillary L/D ratio and shear rates from which $(1-W)$ can be calculated assuming $k_t = 1$ using Eq. (4), as listed in Table 1. f_w is determined by fitting B against L/D under different shear rates.

In addition, the swell ratio was also fitted by $B = B_0 - k_B(L/D)$ as suggested by Liang[75] (as shown by the dotted curves in Figure 2) and by Eq. (5) (as shown by the solid curves in Figure 2). The resulting values of key parameters for the three cases are compared in Table 1.

n			G_0 /Pa		a	η_0 /Pa.s
0.57243			3784728		0.74712	6811860
IUPAC-LDPE						
Dashed curves By Eq. (3) ($k_t = 1$)			Dotted curves by $B = B_0 - k_B(L/D)$		Solid curves by $B = B_{equ} + b\dot{\gamma}/(1+(L/D)^a)$	
$\dot{\gamma}$	$1-W$	f_w	B_0	k_B	B_{equ}	b
10	0.74	0.36	1.83	0.0021	1.573	2.4906
1	0.64	0.39	1.70	0.00185	1.561	1.9614
0.63	0.58	0.43	1.69	0.00125	1.557	1.6841
0.1	0.45	0.47	1.38	0.00030	1.316	0.7806

Table 1. Values of parameters for LDPE obtained from the plots in Figures 1 and 2

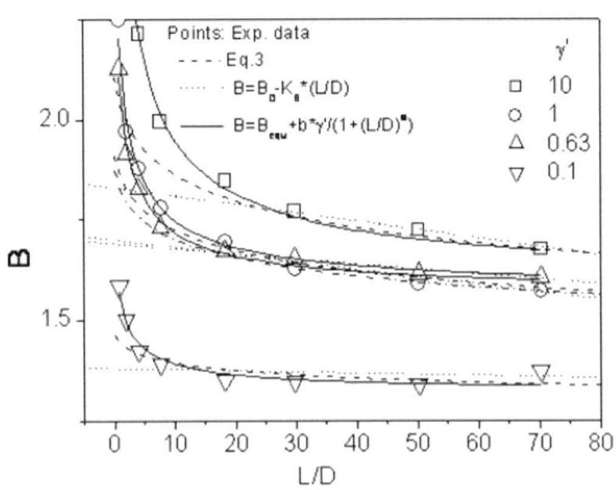

Figure 2. Predicted curves and experimental data of extrudate swell ratio for LDPE with different capillary L/D ratio and shear rates.

The comparisons of swell ratio predicted by Eq. (3), $B = B_0 - k_B(L/D)$ and Eq. (5) are also compared with experimental data (points) in Figure 2. All three models predict that the swell ratio decreases with increasing capillary L/D ratio and increases with increasing shear rate. It is also found that all three equations describe well the approximate linear relationship between the experimental swell ratio and L/D at larger values of L/D, i.e., in long capillaries. However, below a certain critical value of L/D, B increases rapidly with decreasing L/D. This trend is well predicted by Eq. 5 only. Liang's model[77] $B = B_0 - k_B(L/D)$ gives a completely different picture to experiment. Song's model Eq. (3) underestimates the swell ratio compared with experiment, showing that Eqs. (3) and (4) are only appropriate for swelling on extrusion from a long capillary, because of the assumption in Song's theory[69] that the entry effect is almost relaxed in capillary flow. In a short capillary, the chain elongation on entry cannot be neglected. The critical L/D ratio, which determines whether a capillary behaves as 'long' or 'short', increases with shear rate. This may be explained by the more incomplete relaxation of the elongated chains in the entry region at higher shear rate. The degree of relaxation depends on both the capillary length and the strength of the flow field as described by Eq. (5).

3.2.2. *Extrusion distance giving the maximum swell ratio for semi-dilute polymer solutions*

For a capillary with a L/D ratio of 85, the swelling of polyacrylamide solution was observed to reach a maximum at a given time t_∞ at a given distance (expressed as a multiple Z of capillary diameter D) from the capillary[70], i.e., the swelling is finite. In the simulations, n, a and τ_0 were obtained by fitting the experimental $\eta \sim \dot{\gamma}$ curve[70] using Eqs. (1) and (2). K_t and $(1-W)$ can be fitted using Eq. (3). t_∞ and Z were subsequently calculated from Eq. (11) and compared with the experimental data as shown in Figure 3. The calculated swell ratio B is close to the experimental data for $K_t = 3.5468$ and $(1-W) = 0.5964$. The simulated value of t_∞ decreases as a power law function of shear rate as observed from the experimental data in Figure 4a, as does Z as shown in Figure 4b.

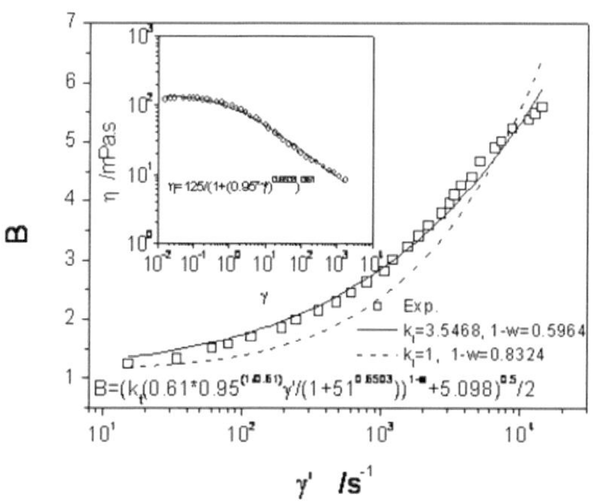

Figure 3. Calculated plots of B versus $\dot{\gamma}$ from the experimental data for a semi-dilute polymer solution

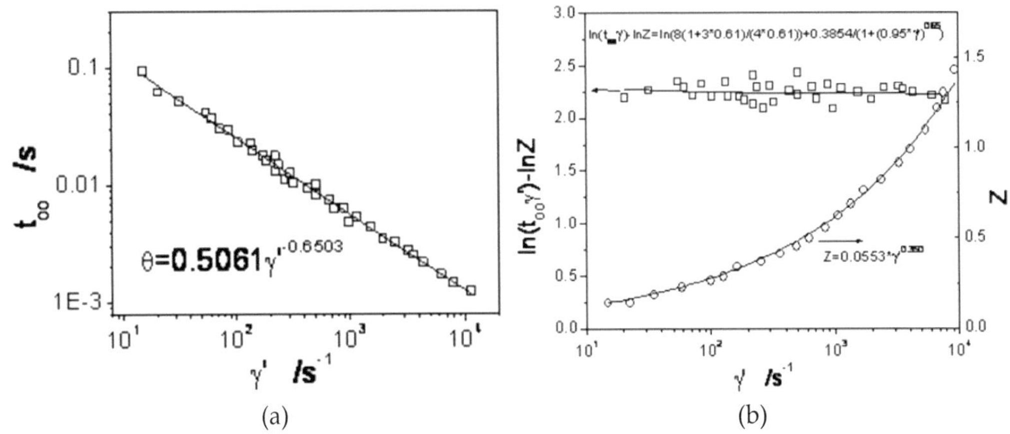

Figure 4. Curves of $\ln(t_{\infty}\dot{\gamma}/Z)$ and Z versus $\dot{\gamma}$ calculated from the experimental data.

3.2.3. *Verification of the swelling equation for rubber in a short capillary*

The recoverable strain can be calculated by Eq. (23) due to Tanner[37].

$$S_R = [2(B^6 - 1) - 2]^{0.5} \tag{23}$$

Eq. (24) can be approximately fitted to give values of $Mk_t / 4$, a and $1-W$ when $f_w = 0.5$.

$$B = \frac{1}{2}\left(\frac{Mk_t}{4}\left(\frac{1+(\tau\dot{\gamma})^a}{1+(L/D)^a} S_R\right)^{(1-W)} + \frac{5.098}{4}\right)^{0.5} \tag{24}$$

The experimental swell ratios[73] are shown in Figure 5, from which the values of the parameters in Eq. (24) can be obtained by fitting and they are listed in Table 2. Figure 5 also illustrates that the swell ratios predicted by Eq. (24) are in good agreement with the experimental data. It should be noted that the swell ratios reported in the literature[73] were measured when the extrudate was naturally cooled to ambient temperature. The swell ratio is considerably larger for extrusion from a very short capillary than from a long capillary. A larger fraction of the relaxation occurs in the melt state at higher temperature in the longer capillary than in the short capillary. The unrelaxed energy is not easily released from the capillary at low temperature. The fitted value of $(1-W)$ is larger for swelling on extrusion from the short capillary than from the long capillary, which reflects a greater amount of retained elasticity in the former case.

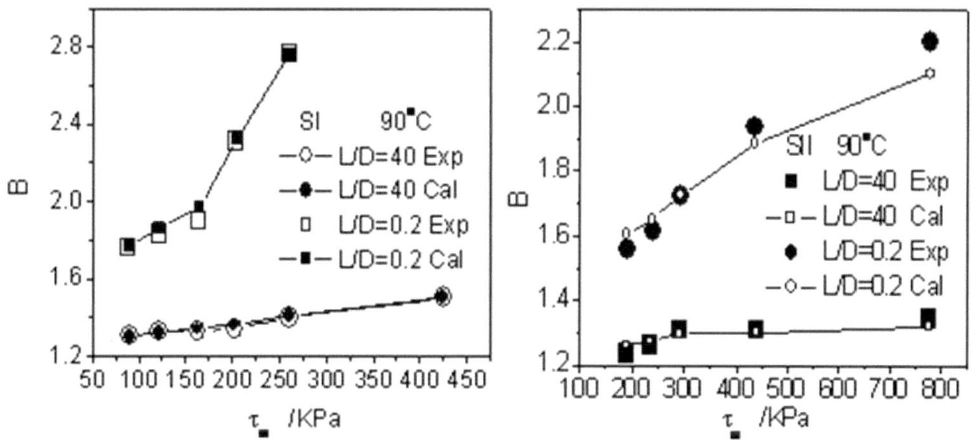

Figure 5. Plots of fitted swell ratio versus shear stress compared with experimental data.

The filler content of sample SI with a matrix of NR rubber is lower than that of sample SII with a matrix of NR/SBR/CBR blend[73]. As seen from Table 2, the average relaxation time τ is shorter, and the values of a and $(1-W)$ are lower for sample SI than for sample SII containing a higher filler content.

Samples	L/D	$k_t M / 4$	τ	a	$(1-W)$
SI	0.2	0.32	0.12	0.26	0.64
	40	0.18			0.56
SII	0.2	0.22	0.72	0.53	0.72
	40	0.48			0.70

Table 2. Values of the parameters in Eq. (24) for rubber compounds

3.2.4. *Application of the swelling equations for a PP/glass bead composite*

As shown in Figure 6, the viscosity of the PP/glass bead composite can be calculated from the experimental data of stress versus shear rate in the literature[75]. Eq. (1) accurately predicts the variation of viscosity with shear rate for composites with different filler percentages. The fitted values of η_0, τ_0, a and n are listed in Table 3; all of these values increase with increasing filler content in the composite while shear thinning becomes more marked when increasing the filler content.

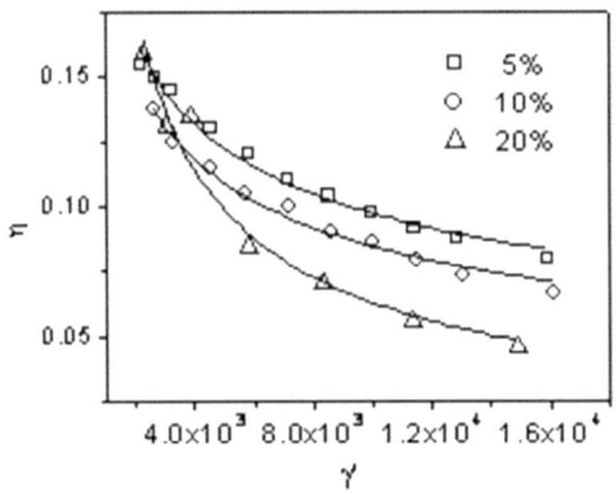

Figure 6. Experimental data for variation of viscosity with shear rate and the simulated values for the composites.

Filler content	n	τ_0 /s	a	η_0 /kPa.s
5%	0.2071	1.31785	1.6011	2.2607
10%	0.2175	1.47162	1.6849	2.8496
15%	0.3812	1.85625	1.7210	3.5809

Table 3. Structural parameters for PP/glass bead composites

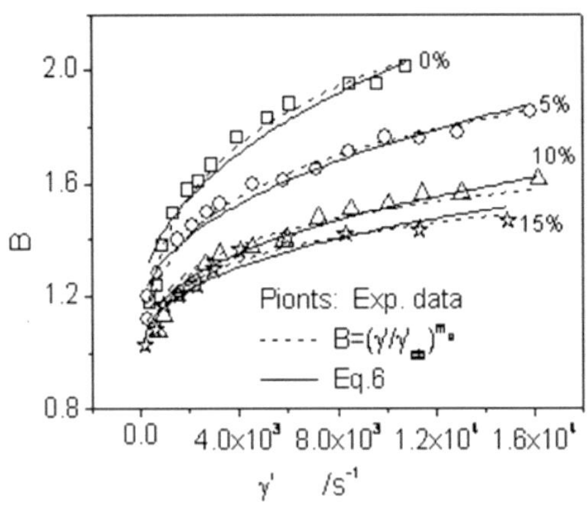

Figure 7. Comparison of predicted swell ratio with experimental data for PP/glass bead composites.

Eq. (6) can be used to predict the extrudate swell ratio of the composites with the fitted structural parameters listed in Table 3. Comparison of the predicted data with the experimental values are shown in Figure 7, assuming $k_t = 1$. The modified theory shows better agreement at higher shear rate which indicates its validity for particulate-filled composites. Table 4 shows that $(1-W)$, which is the residual fraction of the recoverable entanglement in polymeric chains after flow shear, decreases with increasing filler content. Smaller values of $(1-W)$ correspond to lower elastic recovery, i.e., swell ratios are lower for composites with higher filler content. Furthermore, it can be seen that f_w decreases (from

$f_w = 0.5137 \approx 0.5$ for the pure matrix melt) to lower values with increasing filler concentration. This shows that recovery is harder for an entangled network with higher filler content than for the corresponding matrix network.

Dashed curves given by Eq. (3)			Dotted curves given by $B = (\dot{\gamma} / \dot{\gamma}_{cb})^{m_b}$	
Filler content	$1-W$	f_w	$\dot{\gamma}_{cb}$	m_B
0	0.56165	0.5137	127.51	0.16
5%	0.55176	0.5003	138.73	0.1307
10%	0.53287	0.4400	153.45	0.09821
15%	0.49195	0.4217	169.62	0.08876

Table 4. Values of model parameters in the equations for the composite in Figure 7

In addition, it is interesting that the swell is well depicted by one simple equation $B = (\dot{\gamma} / \dot{\gamma}_{cb})^{m_b}$. $\dot{\gamma}_{cb}$ may be related to the composite rigidity or the density while m_B is sensitive to the shear, reflecting the recovery capability of the composites. When increasing the filler content in the composite, $\dot{\gamma}_{cb}$ becomes larger while m_B becomes smaller, showing that the elasticity is weakened while the rigidity becomes stronger, which corresponds to the observed decrease in swell ratio.

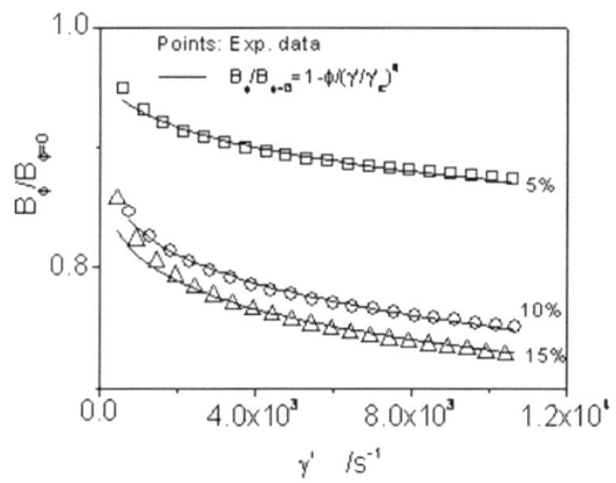

Figure 8. Comparison of plots of $B(\dot{\gamma},\phi) / B(\dot{\gamma},\phi = 0)$ versus shear rate with experimental data.

Plots of $B(\dot{\gamma},\phi) / B(\dot{\gamma},\phi = 0)$ against shear rate based on the experimental data in Figures 6 and 7 are shown in Figure 8. The concentration shift factor $F(\phi) = [1 - \phi / \phi_c]$ (here k and p in Eq. (20) are both unity) illustrates the reinforcement effect while reflecting the weaker swelling capability. In our opinion, there are two levels of network structure. The microscopic one involves molecular entanglement, whilst the other involves mesoscopic particles dispersed in the compound melt. The concentration shift factor shows that the swelling functions of the two levels may be separated. The mesoscopic network probably shrinks, counteracting part of the microscopic melt swelling. The offset effect can be described by the concentration shift factor. However, the filler is far larger than the polymer chains and does not vary the basic molecular relaxation dynamics of the melt matrix except at the interfacial region around the particles. The dispersed particles perturb the flow of the melt. It is interesting that this is consistent with previous publications showing that the dimensions of the extrudate were possibly smaller than those of the die at higher Reynolds numbers[41,42].

In addition, Eq. (21) is also successful in fitting the plots of $B(\dot{\gamma},\phi)/B(\dot{\gamma},\phi=0)$ against $\dot{\gamma}$ as Figure 8. $\dot{\gamma}_c$ increases while q decreases when adding more glass beads to the polypropylene. To a certain degree, $\dot{\gamma}_c$ represents a critical shear rate when the melt swelling is completely offset by the shrinkage of the mesoscopic network.

Filler content	$\dot{\gamma}_c$	q
5%	1.249E10	0.2616
10%	2.945E13	0.1694
15%	6.188E15	0.1482

Table 5. Parameters in Eq. (19) (p=1) for the composite in Figure 5

Plots of die swell versus filler concentration[75] are shown in Figure 9. Eq. (20) well demonstrates this correlation. In fitting the data with Eq. (20), it is found that with increasing shear rate the critical content ϕ_c decreases, while the die swell ratio at $\phi=0$ increases. Larger die swell ratios are observed for pure PP at higher shear rates, since more elastic energy is stored.

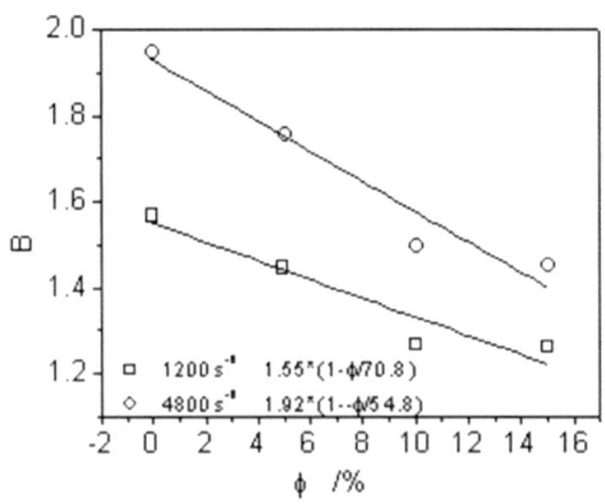

Figure 9. Extrudate swell ratio of PP/glass bead composites with different filler contents.

$B(\tau,\phi=0)=B(\tau,\phi)/F(\phi)$ is also appropriate for correlating shear stress and filler content for the composite as shown in Figure 10. In this case, in the expression $\phi_c=(\tau/\tau_c)^{q'}$ τ_c increases while q' decreases with increasing filler content. More importantly, $B(\tau,\phi)/F(\phi)$, i.e., $B(\tau,\phi=0)$ is almost a linear function of shear stress. It can be written as $B(\tau,\phi)/F(\phi)=(B/F)_{\tau=0}+k_\tau \tau$ where $B_{\tau=0}$ and k_τ decrease with increasing amount of filler in the PP melt.

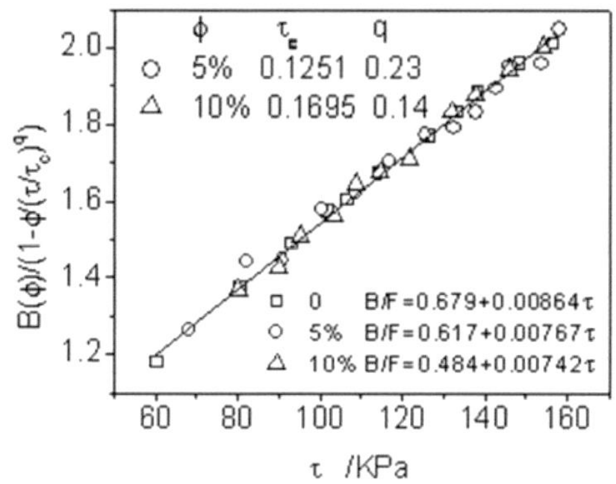

Figure 10. Plots of $B(\tau,\phi) / F(\phi)$ versus shear stress from experimental data for PP/glass bead composites.

4. Conclusions

Two limitations of Song's polymer extrudate swell theory have been identified for the first time. Song's model has been modified in order to predict the finite distance at which the swelling reaches a maximum. Furthermore, the model was extended to describe the die swell on extrusion from a short capillary by considering the entry effect in Song's molecular dynamics model and incorporating Liang's expression. The resulting modified model can be applied to extrusion swelling for both long and short capillaries, whereas Song's model is only appropriate for long capillaries. More importantly, the modified model is also suitable for analysis of the swelling of particle-filled composites which cannot be treated by Song's model. The composite swell ratio can be separated into the product of the matrix swell ratio and the concentration shift factor. The excellent agreements between the values predicted by the modified model and experimental data reported in the literature for a variety of different systems demonstrate its viability for a wide range of materials and experimental conditions.

Author details

Kejian Wang*

Institute of Plastics Machinery and Engineering,
Beijing University of Chemical and Technology, Beijing, China

5. References

[1] Liang J.Z. Plast Rubber Compos Process Appl 1991; 15, 75-79.

[2] Liang J. Z. J Mater Process Technol 1995; 52, 207-215.

[3] Lodge A.S. Elastic Liquids, Academic Press, New York, 1964.

* Corresponding Author

[4] Graessley W.W., Glasscock S.D. and Crawley R.L. Trans. Soc. Rheol. 1970; 14: 519–544.
[5] Wong A.C.Y. J. Mater. Proc. Tech. 1998;79:163–169.
[6] Koopmans R.J. J. Appl. Polym. Sci. 1992; 32:1755-1762.
[7] Liang J.Z. Plast. Rubber Proc. Appl. 1995;23: 93-96..
[8] Seo Y. J. Appl. Polym. Sci. 1990;30:235-240.
[9] Richardson S. Rheol. Acta 1969;9 :193-199.
[10] Koopmans R.J. J. Appl. Polym. Sci. 1992;32:1750-1767.
[11] Koopmans R.J. J. Appl. Polym. Sci. 1992;32:1741-1750.
[12] Swan P.L., Dealy J.M. Polym. Eng. Sci. 1991;31:705–710.
[13] Larson R.G., Constitutive Equations for Polymer Melts and Solutions, Butterworths, Boston, 1988.
[14] Giesekus H. J. Non-Newtonian Fluid Mech. 1982;11:69–109.
[15] White J.L. Rubber Chem. Technol. 1977; 50: 163–185.
[16] Spencer R. S., Dillon, R. E. J Colloid Sci 1948; 3, 163-171.
[17] Schreiber H.P., Rudin A., Bagley E.B. J. Appl. Polym. Sci. 1965; 9:887–892.
[18] Metzner A.B. Trans. Soc. Rheol. 1969;13:467–470.
[19] Brydson J.A. Flow Properties of Polymer melts. London: The Plastic Institute. 1981.
[20] Cogswell F.N. Plast. Polym. 1973; 41: 39–43.
[21] Shaw M.T. Polym. Eng. Sci. 1977; 17: 266–268.
[22] Rogers M.G. J. Appl. Polym. Sci. 1970;14 :1679–1689.
[23] Racin R., D.C. Bogue, J. Rheol. 1979;23:263–280.
[24] Mendelson R.A., F.L. Finger, J. Appl. Polym. Sci. 1975;19:1061–1078.
[25] Rokudai M. J. Appl. Polym. Sci. 1981;26:1427–1429.
[26] Liang J. Z., Nes J. N. s, Polymer Testing 1999;18:37–46,
[27] Dufrancatel-Veiller L., Lacrampe M. F., Pababiot J. J. Appl. Polym. Sci. 2001;80:1710–1724.
[28] Minagawa N. and White J.L. J. Appl. Poly. Sci. 1976; 20: 501–523.
[29] Han C. D. Rheology in Polymer Processing; Academic: New York, 1976.
[30] Cotten G.R. Rubber. Chem. Technol. 1979; 52: 187–198.
[31] Lenk R.S. Polymer Rheology. London: Applied Science Publishers Ltd. Ch.10, p.101. 1978
[32] Vinogradov G.V. and Malkin A.Y. Rheology of Polymers: Viscoelasticity and Flow of Polymers, Moscow: Mir Publishers. Ch. 5, pp. 355–379,1980.
[33] Rapeephun D., Jimmy Y., Pitt S. Polymer Testing 2005;24:2–11.
[34] Nattaya M., Manit N., Brian P. Gradyb, Rathanawan M. Polymer Testing 2008;27:470–479
[35] Bagley E.B., Duffey H.J., Trans. Soc. Rheol. 1970; 14:454-459.
[36] Han C.D., Charles M. Trans. Soc. Rheol. 1970;14:213–218.
[37] Tanner R. I. J. Polym. Sci. Part A-2: Polym. Phys. 1970; 8:2067-2071.
[38] Tanner R.I. J. Non-Newtonian Fluid Mech. 1980;6:289-302.
[39] Crochet M.J., Keunings R. J. Non-Newtonian Fluid Mech. 1980; 7:199–212.

[40] Barakos G., Mitsoulis E. J. Rheol. 1995;39:193–209.
[41] Middleman S. Fundamental of Polymer Processing. New York: Magraw-Hill. 1977.
[42] Georgiou G.C., Papanastasiou T.C. and Wilkes, J.O. AIChE. J. 1988; 34:1559–1562.
[43] Utracki, L.A., Bakerdjian, Z. and Kamal, M.R. J. Appl. Polym. Sci. 1975;19: 481–501.
[44] Meissner J. Pure Appl. Chem. 1975; 42: 551–612.
[45] Macosko C.W. Rheology, Principles, Measurements, and Application. New York: Wiley-Vch. Ch. 6, pp. 236–260, 1990.
[46] Treloar L.R.G. The Physics of Rubber Elasticity. 3rd Edn., Oxford: Clarendon Press. 1975.
[47] Barone J.R., Plucktaveesak N. and Wang S.Q. J. Rheology1998; 42: 813–832.
[48] Huang D., White J. L. Polym Eng Sci 1980; 20, 182-186.
[49] J.den Doelder C.F., Koopmans R.J. J. Non-Newt. Fluid Mec. 2008; 152(1-3): 195-202.
[50] Muksing N., Nithitanakul M., Grady B.P. and Magaraphan R. Polym. Testing 2008; 27(4): 470-479.
[51] Kumari K., Kundu P.P., Bulletin of Mater. Sci. 2008;31(2):159-167.
[52] Ayutthaya S., Isarankura N., Wootthikanokkhan J., J. Appl. Polym. Sci. 2008;107(6):3853-3863.
[53] Mitra S., Chattopadhyay S., Bhowmick A.K., J. Appl.Polym.Sci. 2008; 107(5): 2755-2767.
[54] Rowell R.M., J. Polym. Environ., 2007; 15(4):229-235.
[55] de Paulo G.S., Tome M.F., McKee S. J. Non-Newtonian Fluid Mech. 2007; 147(3):149-174.
[56] Mitsoulis E., J. Fluids Eng. Trans.of the ASME2007; 129(11):1384-1393.
[57] Mohanty S., Nayak S.K., Polym. Eng. Sci. 2007; 47(10):1634-1642.
[58] Peng B., Wu H., Guo S.Y., Lai S.Y., Jow J., J. Appl. Polym.Sci. 2007;106(3):1725-1732.
[59] Liang J.Z., Li R.K.Y., Tang C.Y. and Cheung S.W. J. Appl. Polym. Sci. 2000; 76:419–424.
[60] Faulkner D.L., Schmidt L.R. Polym. Eng. Sci. 1977;17,657–667.
[61] Bush M.B., Polym. Eng. Sci. 1993; 33, 950-958.
[62] Kamal K.K. and Joshua U.O. J. Elast. Plas. 2001;33:297-336
[63] Tanner R. I. J. Non-Newtonian Fluid Mech. 2005;129:85–87.
[64] Tapadia P., Wang S.Q. Macromolecules 2004;37:9083-9099.
[65] Zhu Z Y, Wang S Q , J Rheol. 2004;48(3):571-578.
[66] Song M S, Hu G X, Yang Z H, Xu Q , Wu S Z. J Mater Sci Tech. 2006; 22(1), 93-115.
[67] Song M S, Xu Q, Hu G X and Wu S Z, J Mater Sci Tech. 2006;22(5):664-686.
[68] Zhu C W, Song M S, Hu G X, Zhao J and Wu D M. Chin J Chem Phys. 2007;20(5): 563-581.
[69] Zhao J., Song M.S., Zhu C.W., Hu G.X., Wang K.J., Wu D. M. Chin. J. Chem. Phys. 2008; 21(1):55-88.
[70] Allain C., Cloitre M. and Perrot P. J. Non-Newtonian Fluid Mech. 1997; 73,51-66.
[71] Zheng R, Tang G J, Polymer communication 1986;3:161-172
[72] Liang, J. Z. J Mater Process Technol 1996; 59: 268-276.
[73] Liang J.Z., J. Appl. Polym. Sci. 2007; 104:70–74.
[74] Stabik J., Inter Polym Process 2004; 19(4):350-355.
[75] Liang J.Z. Polymer Testing 2002; 21:927–931.
[76] Gleissle W., Hochstein B. J. Rheol. 2003;47:897–910.

[77] Huang S.X., Lu C.J. J. Non-Newtonian Fluid Mech. 2006; 136:147–156.
[78] Meissner J. Pure Appl. Chem. 1975; 42,551–612.

Section 2

Biological Materials

Chapter 5

Viscoelastic Properties of Biological Materials

Naoki Sasaki

Additional information is available at the end of the chapter

http://dx.doi.org/10.5772/49979

1. Introduction

Almost all of biological tissues are viscoelastic and their viscoelastic mechanical properties are important in their characteristic functions. This is because constituents of tissues – cells, extracellular matrices, structural proteins, and so on – are viscoelastic. Even hard tissues have been shown to be viscoelastic. For long, main sample material for investigating the viscoelasticity has been amorphous polymeric materials, which brought about the remarkable development in phenomenological theories. Constructing mechanistic images, on the other hand, of the viscoelasticity of polymeric materials had been difficult because of dearth of the materials structural information. Contrary to these materials, for biological tissues, many structural investigations have been made and as a result detailed structural information is available. Making use of the information, it will be possible to investigate the viscoelastic properties of biological materials on the basis of their structure. Such a study will contribute to the construction of molecular theory for the viscoelasticity in amorphous materials.

In this chapter, some examples of viscoelastic nature of biological materials and then their relevance to the structure would be presented. In some cases, a mechanistic model for the viscoelasticity will be presented. As measuring method varies depending on the modulus value of the specimens, the various methods used in studying viscoelastic properties of biological materials will be illustrated.

2. A Brief introduction to viscoelasticity

2.1. Introduction to elasticity and viscosity

Elasticity is a material property that generates recovering force at an application of an external force to deform the material. When an external force is applied to a material and the material is in an equilibrium deformation, the external force is balanced by an inner force. The inner force is the recovering force. The recovering force divided by the cross sectional

INTECH
open science | open minds

© 2012 The Author(s). Licensee InTech. This chapter is distributed under the terms of the Creative Commons Attribution License http://creativecommons.org/licenses/by/3.0), which permits unrestricted use, distribution, and reproduction in any medium, provided the original work is properly cited.

area that the external force is working on is defined as stress, σ. Suppose the material is initially in a shape of rod of length L_0 and cross-sectional area, A_0. The force is applied to the length direction and the length after the deformation is L. The deformation is generally normalized as,

$$\varepsilon = \frac{L - L_0}{L_0}, \tag{1}$$

where ε is called the engineering strain or the Caushy strain. This is used for small deformations. For small strains, a Hookean relation,

$$\sigma \infty \varepsilon \tag{2}$$

has been known to hold. The proportionality factor is defined as a modulus and the modulus is a material's constant. If the material was deformed by a tensile force, the modulus is defined as Young's modulus (Fig. 1 (a)). If the deformation was caused by a shear

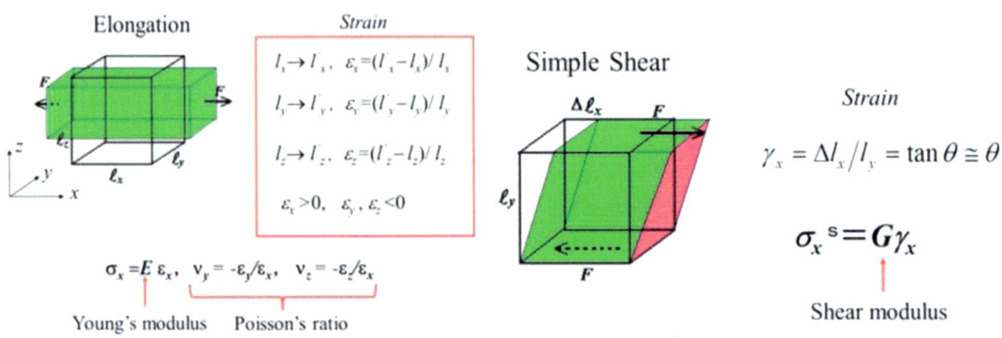

Figure 1. Deformation modes and corresponding moduli. (a) Elongation and Young's modulus, and (b) shear deformation and the shear modulus.

shear force, the modulus is the shear modulus (Fig. 1(b)). In each case, the Hookean constitutive equations are written as,

$$\sigma = E \cdot \varepsilon \quad (E\text{: Young's modulus}) \tag{3}$$

and

$$\sigma = G \cdot \gamma \quad (G\text{: shear modulus}) \tag{4}$$

where γ is the shear strain. After eliminating the external force, the recovering force and the deformation are completely diminished. Elasticity is a property of a material to resist the deformation by the external force.

Viscosity is a characteristic nature of a fluid. When a fluid is constantly flowing on an infinitively wide plain, the flow velocity is largest on the fluid surface. On the plain bottom the flow velocity is zero. Every part of the fluid is sheared (Fig. 2). The shear is generated by an applied force causing the flow and an internal force against the former. The internal force is also transformed into a stress generated in the flow. Newton postulated the quantitative relationship between the stress and the shear rate as,

$$\sigma \propto \frac{d\gamma}{dt}. \tag{5}$$

The proportionality factor is defined as a viscosity coefficient, η, also a materials constant of a fluid. Then the constitutive relation for the viscosity is

$$\sigma = \eta \frac{d\gamma}{dt}. \tag{6}$$

Viscosity is a property of a fluid to resist the force for flow. The fluid described by the equation is classified as Newtonian. After stopping the flow, the fluid maintains its deformation strain at the time of stopping. There is a saying that water conforms to the shape of the vessel that contains it.

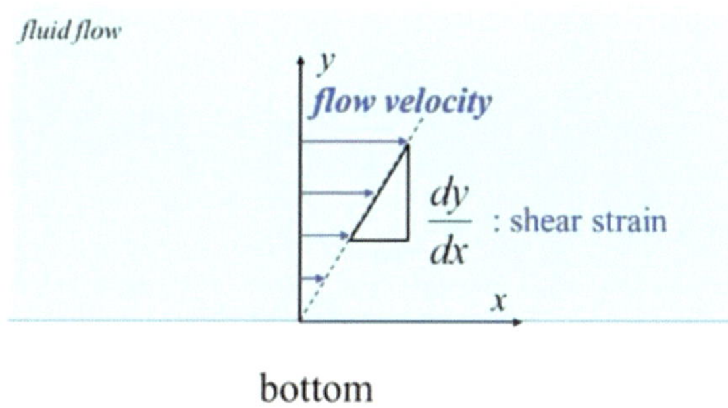

Figure 2. Shear strain in a fluid. A velocity gradient perpendicular to the bottom emerged.

2.2. Introduction to viscoelasticity

When a step strain excitation is applied to an idealistically elastic material, the response of the material will be that shown in Fig. 3(a). This response can be easily expected from the Hookean constitutive equation. If the same strain excitation is applied to an idealistically viscous material, the response will be those as shown in the Fig. 3(b) and (c), which is also expectable from the Newtonian constitutive equation. Materials that exhibit viscoelastic properties can be considered as having both elastic and viscous components. Then, in the case of viscoelastic materials, the response is expected to have both characteristics of elasticity and viscosity. Schematic drawing of the stress response according to many experimental results is shown in Fig. 4. Here we define the stress relaxation as follows. Fig. 5 shows a step strain excitation and a stress response to it for a viscoelastic material. The instantaneous and equilibrium moduli, m_g and m_e, respectively, are material's constants. The relaxation function $\phi(t)$ is a material's function. Phenomenological investigation of materials from the stand point of viscoelastic properties is to determine these material's constants and function and, then, deduce the molecular mechanism corresponding to the viscoelastic behavior. For this purpose, more generalized tool shall be introduced. Any given strain excitation can be constructed by a linear combination of many small step strains applied to the material at every Δt seconds. According to the linear response theory, stress response to the excitation also can be a linear combination of responses to small step strains applied to the material (Fig. 6). From any given strain excitation $\varepsilon(t)$, stress response $\sigma(t)$ would be obtained through the Boltzmann's superposition equation,

$$\sigma(t) = \int_{-\infty}^{t} \frac{d\varepsilon(u)}{du} m(t-u)du. \tag{7}$$

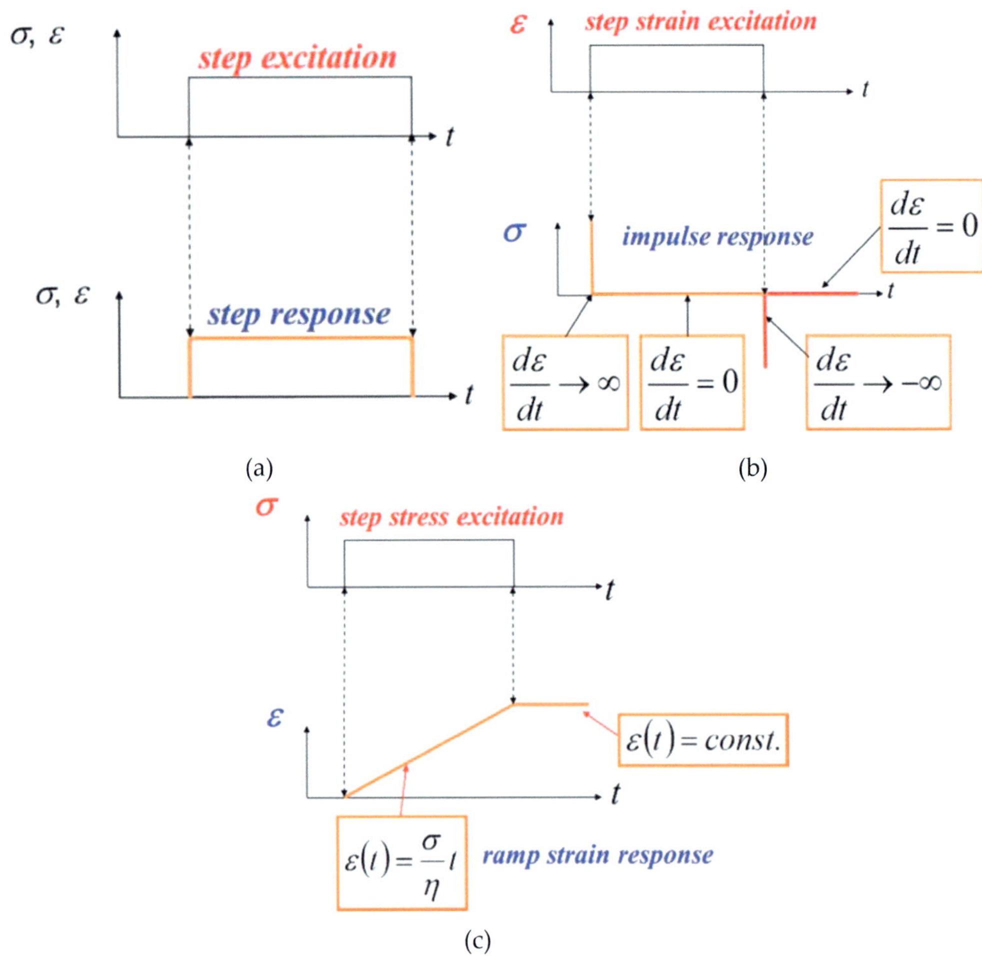

Figure 3. Responses of idealistically elastic materials (a) and that of idealistic viscous materials (b) and (c) to step strain and stress excitations, respectively.

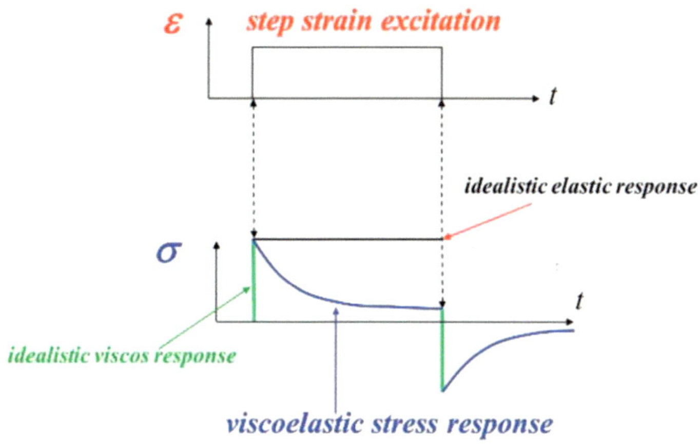

Figure 4. Response of viscoelastic materials to a box-shaped strain excitation.

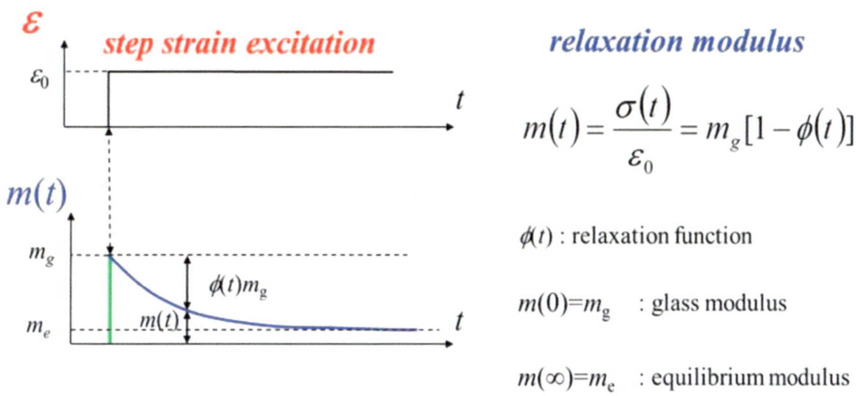

Figure 5. Definition of the response of a viscoelastic material to a step strain excitation.

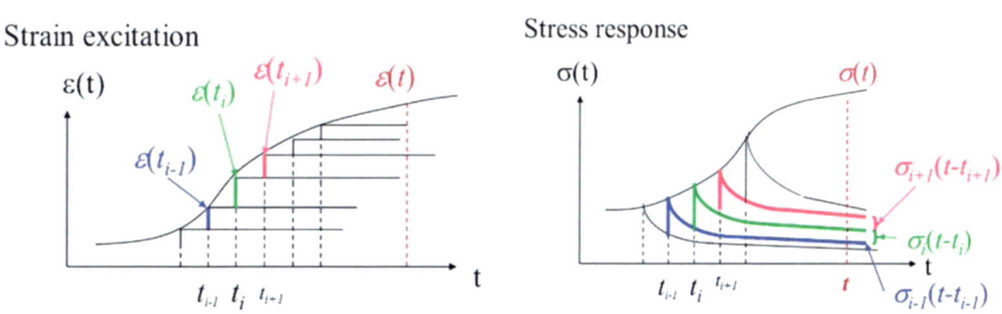

Figure 6. Generalized strain excitation and the stress response of viscoelastic materials to the former. From the relationship between the excitation and the response, Boltzmann's equation, eq. 7, was derived.

For stress relaxation experiments, a step strain excitation is needed. But, actually, it is difficult to realize such an excitation strain. The realistic one should be a lamp followed by a plateau strain. An acceptable lamp width can be determined from the Boltzmann equation above. If the relaxation function is described as

$$\phi(t) = 1 - \exp(-t/\tau) \tag{8}$$

where τ is a characteristic relaxation time of a material, the stress response will be

$$\sigma(t) = A\tau e^{-t/\tau}\left(e^{t_1/\tau} - 1\right). \tag{9}$$

In this equation, A is a constant and t_1 is the lamp width. If $t_1 \ll \tau$, it is rewritten as

$$\sigma(t) = At_1 \exp(-t/\tau) \tag{10}$$

which is a usual relaxation modulus form of materials with single relaxation time τ as a response to the step strain excitation.

2.3. Viscoelastic investigation

As indicated above, the objective of the viscoelastic studies of materials is to understand the molecular mechanism corresponding to the viscoelastic behavior. For this purpose, to know the relaxation function of the material is needed. Almost all actual relaxation data, however, cannot be described by a relaxation function with single relaxation time as shown by eq. 8. In order to describe the relaxation modulus, then the relaxation function, the following methods have been traditionally employed: (A) multi relaxation time analysis, (B) relaxation time distribution analysis (Ferry, 1980), and (C) analysis using specific functions.

(A) and (B) are substantially similar analyses. For the analysis (A), experimental relaxation modulus, $E(t)$, is fitted by

$$E(t) = E_e + \sum_{i=1}^{N} a_i \exp(-t/\tau_i) \tag{11}$$

where E_e is an equilibrium modulus. If N is large and a_i is a continual function of τ, $a(\tau)$, eq. 11 can be described as

$$E(t) = E_e + \int_{-\infty}^{t} a(\tau)\exp(-t/\tau)d\tau. \tag{12}$$

This description represents the relaxation time distribution analysis and $a(t)$ is a relaxation time spectrum. By a numerical transformation of the experimental data, the relaxation spectrum will be obtained. At this point, relaxation spectrum does not have any more information than the original relaxation data. It is still difficult to deduce molecular events underlying the relaxation.

For (C), it has been established that the non-single relaxation timed relaxation process can be well described by a power law relation or stretched exponential function, depending on the materials and their state;

$$E(t) = E_0 t^{-n} + E_e \quad (0 < n \leq 1) \tag{13}$$

or

$$E(t) = E_0 \exp[-(t/\tau)^{\beta}] \quad (0 < b \leq 1). \tag{14}$$

One of the advantages of using these functions for describing the relaxation process is easily to establish the mechanistic model for the relaxation process. As these two functions have been found to describe well the mechanical or electric relaxation phenomena in many amorphous glassy materials, they are regarded as universal functions.

When there is a dynamic structural inhomogeneity in glassy materials, Jonscher proposed that the power in the power law relation, eq. 13, indicates the correlation of molecular motions among motional groups and the variation in the number of motional unit in each molecular motional group (Jonscher, 1983). On the other hand, for explanation of the exponent value in the stretched exponential function, eq. 14, a diffusion-trap model is proposed (Klafter et al., 1986; Philips, 1996). When a step strain is applied to a material, many excited sites will emerge inside the material. In amorphous glassy materials, molecular motion of constituents could be regarded as a diffusion of a free volume of the constituents. When the free volume comes to an excited site, the site will be relaxed. This would be an elementary process of the relaxation. If the excited sites in the material is distributing on a fractal lattice, total relaxation of the material is expected to be described by the stretched exponential function, eq. 14. The exponent value, β, is strongly related the geometry of the exciting site distribution (Potuzak et al., 2011).

3. Stress relaxation experiments

In the introductory section, only the stress response of materials to strain excitation was explained. This is because our viscoelastic study of biological materials mainly has been employing the stress relaxation experiments. Therefore, in this chapter, the relaxation modulus of biological materials is discussed in conjunction with their structural information. In this section, the empirical determination of the relaxation modulus of materials would be explained.

3.1. Apparatus and equipments

Figure 7 shows a schema for the stress relaxation experiments. By moving the load cell (LC), Kyowa Electric Works LTS series, strain is applied to the specimen through an equipped attachment mounted on a probe and the stress response to the excitation is detected by LC. LC is mounted on an auto micro-stage, Sigma Koki CTS-50X, derived by a stage controller Sigma Koki MARK-12. The sensitivity of LC and the attachments are chosen depending on the specimens. Soft materials are examined by a tension or indentation methods depending on the specimen. When the specimen can be shaped in a sheet, relaxation modulus was measured by the tensile strain application.

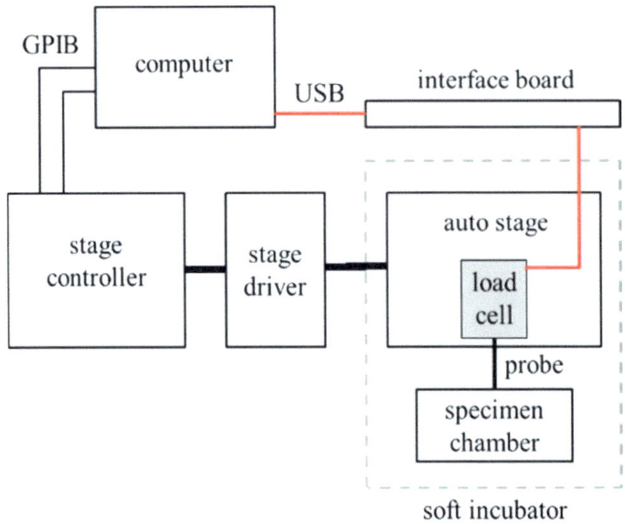

Figure 7. Block diagram for stress relaxation measuring apparatus.

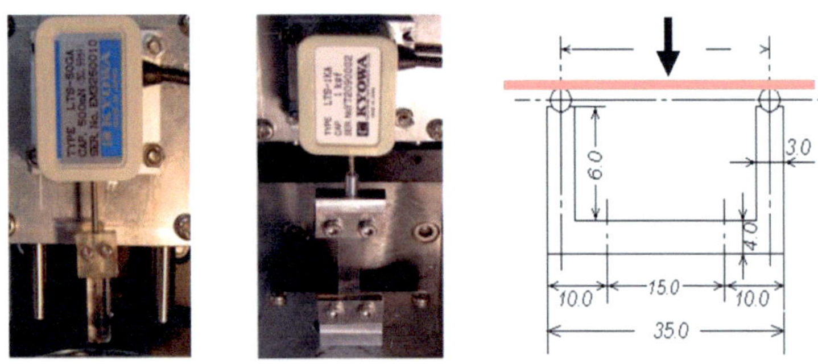

(a) (b) (c)

Figure 8. (a) shows the attachments for tension measurements. Cell-seeded gels are examined by the indentation. Fig. 8(b) shows the indenter for this measurement. Hard materials are shaped in bean examined by a bending method. Fig. 8(c) shows three point bending equipment used for stress relaxation of bone specimens. In both measurements, specimens are soaked in Ringer's solution at 37°C during the relaxation experiments. Detected LC signals were passed to adata logger, Kyowa Electric Works PCD-300B, and finally stored into a PC.

3.2. Measurements

At each measuring mode, prior to the start of the stress relaxation experiment, an upper clump in the case of tensile strain application and an indenter head in the cases of indentation and beam bending must be placed at a zero strain point. For this procedure, the clump or indenter head were derived in 5 μm step. The judgment for an initial touch of the probe head on the specimen was made on the indication of signals from LC starting to change to more than a few με. The stress relaxation experiments were started one hour after the zero strain state detection, waiting for stress relaxations brought about by the zero point detecting procedures. For the stress relaxation measurement, strain values less than 1/2 of yield strain value were applied to the specimen at the deformation rate of 3.3 mm sec^{-1}.

4. Viscoelastic properties of agarose gel and cell-seeded agarose gel

Agarose gel is a material of multi-purpose use. For example, it has been used widely as a cell culture matrix. In the following sections, how cultured cells in agarose gel change the mechanical properties of matrix agarose gel by precipitating extracellular matrix would be shown. For the precise estimation of the change in mechanical properties of the matrix, those of agarose gel itself must be quantitatively estimated. Here results obtained for pure agarose gel at first and then cell-seeded ones would be presented. Agarose used in this experiment and those in the following sections is type VII agarose for cell culture use purchased from Sigma.

4.1. Agarose gel

Recent investigation on the agarose gelation revealed that a phase separation takes place slightly below the gelation temperature, where the gelation process was expected to proceed competitively with the phase separation of the solution (Morita et al., 2008). It is expected that the mechanical properties of so prepared agarose gel depend remarkably on the thermal history during the preparation. For the reproducible data, gel preparation must be carefully conducted. Agarose type VII powder (Sigma Co. Ltd) was dissolved in distilled and deionized water and stirred for 12 hrs at room temperature. The aqueous solution was then incubated at 90°C for 5 hrs. The incubated solution was quenched from 90°C to 4°C, where the gelation temperature was determined to be 32°C. Before mechanical test, the gel specimens were swollen with the distilled and deionized water. Quench temperature 4°C is far below the spinodal point curve in the phase diagram. Fig. 9 shows the relaxation modulus of 2% (w/v) of Type VII agarose gel plotted against time. The modulus value was estimated by using the Hayes equation,

$$E = \frac{F\left(1-\nu^2\right)}{2da\kappa}, \tag{15}$$

where F is the recovering force of gel applied to the indenter surface, a is the radius of the indenter and d is the deformation depth (Hayes et al., 1972). For a specimen of thickness h, κ is a function of the ratio, a/h, and Poisson's ratio, ν, of the gel. Following Watase *et al.*, Poisson's ratio of our agarose gel was considered to be very close to 0.5 because of its near incompressibility (Watase et al., 1983). κ comes from the solution of the integral equation, and its value is numerically calculated and tabulated (Hayes et al., 1972). The radius of the indenter head used in this work was a = 3 mm, and the average thickness of specimens was h = 3 mm. Since Poisson's ratio was assumed to be 0.5, κ of 3.609 was employed. There is a large relaxation up to 10^4 sec followed by a plateau which is characteristic of gels. As both axes are scaled in logarithm, it is clear that the relaxation modulus is not described by a power law relation. After examining the data, the empirical equation

$$E(t) = E_0 \exp\left[-(t/\tau)^{\beta}\right] + E_e \quad (0 < \le 1) \tag{16}$$

has been found suitable. In the Fig. 9, determined parameter values are listed. Recent investigation on the stretched exponential function revealed that β value suggests the relaxation mechanism at molecular level for homogenous systems: β=3/5 for relaxation arising from short-range forces and β=3/7 for relaxation dominated by long-range forces (Philips, 1996). According to β~0.4 obtained here for agarose gel, the driving force for the relaxation would be long-range ones.

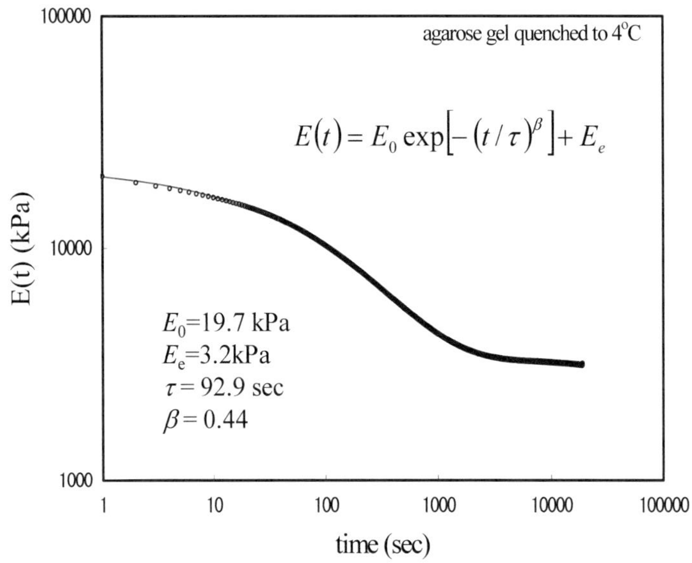

Figure 9. Relaxation modulus curve for agarose gel fabricated by quenching to 4°C.

4.2. Osteoblast seeded agarose gel

On the basis of the relaxation of agarose gel indicated above, how cells change their matrix by precipitating organic and inorganic materials has been investigated (Hanazaki et al., 2011). As seeded cells were osteoblast-like cells, MC3T3-E1, the system can be regarded as a model system of osteogenesis. MC3T3-E1 cells precipitate collagenous proteins and hydroxyl apatite (HAp)-like minerals. Fig.10 shows stress relaxation curves for osteoblasts seeded agarose gel cultured up to 18 days. Relaxation curve for agarose gel is also plotted as a control. All specimens showed the characteristic stress relaxation curve of agarose gel; a large relaxation up to 10^4 sec followed by a gel plateau where the former was attributed to molecular motions of polymer chains between two adjacent cross-links of the gel and the latter to the elasticity of the gel network. Each relaxation modulus curve was able to be described well by eq.16, and parameter determined by the fitting was listed in Table 1. Change in modulus value both E_0 and E_e as a function of culturing period was not so simple. Up to day 15, the change was not significant and at day 18, both modulus values significantly increased. The relaxation time τ and its distribution β did not change significantly through the cultivation period. Fig. 11 shows X-ray diffraction profiles of the cell-agarose gel composite specimens cultured for the indicated periods. The profile for a sintered HAp is also shown as a reference (Okazaki et al., 1997). There is a large peak at about $2\theta \sim 20^\circ$, which was attributed to the diffraction peaks related to the agarose gel matrix (Foord & Atkins, 1989). The diffraction profile of the specimen after 9 days of culture showed diffraction peaks attributable to HAp crystal, where the peak at $2\theta \sim 25^\circ$ was indexed as (002) and those at $\sim 32^\circ$ were indexed as (211), (112), and (300) (Okazaki et al., 1997). With further extension of culturing time, the intensities of peaks corresponding to HAp crystal gradually increased. In Figs. 12(a) and 12(b), E_0 and Ee values are respectively plotted against culture days. Relative integrated diffraction intensity in Fig. 11 is also plotted against days, defined as the ratio of intensity from 002 plan at each culture day against that at the day 18, $I_{002}/I_{002}(18)$. The diffraction intensity is generally proportional to the amount of crystals, provided that the state of crystals in each specimen is not greatly different (Kakudo & Kasai, 1972). In Fig. 11, the half width of each (002) peak was almost the same throughout the culture time up to 18 days. In this case, the intensity can be considered to be proportional to the amount of HAp crystals in the specimen. The $I_{002}/I_{002}(18)$ vs. culture time plot indicates that the amount of HAp crystals increases monotonically with culture period. The result indicates that even though HAp content proportionally increases with culture time, modulus values increment started only from the day 18. To explain this discrepancy between HAp increment and modulus values increment, the efficiency of the reinforcement of agarose gel matrix by precipitated mineral particles was appreciated. Fig. 13 shows a schema of the precipitation process of mineral particles into agarose gel matrix. In the early stage of HAp particle precipitation by MC3T3-E1 cells, the number of mineral particles increases inside a gel network [schema (a) in Fig. 13]. In this state of a cell-seeded agarose gel system, precipitated mineral particles do not contribute to the composite modulus. When a mineral particle cluster percolates inside the gel network, the modulus of the composite will increase [schema (b) in Fig. 13]. The percolation could occur almost simultaneously in every network around each cell. The effect of this percolation is considered to be similar to the increase in the crosslink density, which will lead to the increase in E_e. At the same time, the stiffness of network chains around the mineral percolation would be increased and their mobility would be decreased by the precipitated mineral particles. The empirical equation employed in this study, eq. 16, basically assumes that a relaxing entity can be described by a serial combination of a spring and a dashpot, where the former represents an elasticity, ε, and the latter a viscosity, η, in a material. The relaxation time, τ, is expressed as

$$\tau = \frac{\eta}{\varepsilon}. \tag{17}$$

An increase in the stiffness of the network chain implies an increase in ε and a decrease in their mobility, that is, an increase in η, which would result in an increased E_0 with relaxation time almost unchanged. Above the percolation threshold, the increase in the modulus of the system, (both E_0 and E_e values) with culture time would become remarkable. In our case, the percolation was considered to take place between days 15 and 18 of culture.

culture period (day)	sample size	E_0 (kPa)	τ (sec)	β	E_e (kPa)
agarose	4	16.7 ± 0.6	162 ± 17	0.49 ± 0.01	2.19 ± 0.32
1	4	15.1 ± 2.4	178 ± 61	0.43 ± 0.03	2.36 ± 0.38
9	4	13.9 ± 3.0	144 ± 54	0.44 ± 0.04	2.56 ± 0.40
12	4	12.4 ± 2.0	168 ± 59	0.48 ± 0.03	2.21 ± 0.41
15	4	15.0 ± 2.0	136 ± 27	0.46 ± 0.01	2.22 ± 0.13
18	4	21.3 ± 3.6	173 ± 41	0.49 ± 0.02	3.38 ± 0.16

(From Hanazaki et al., J. Biorheology. In press. DOI 10.1007/s12573-011-0043-2. With permission.)

*All the parameter values are presented as the average value ± the standard error.

Table 1. Parameter values determined by fitting equation (16) to experimental data.

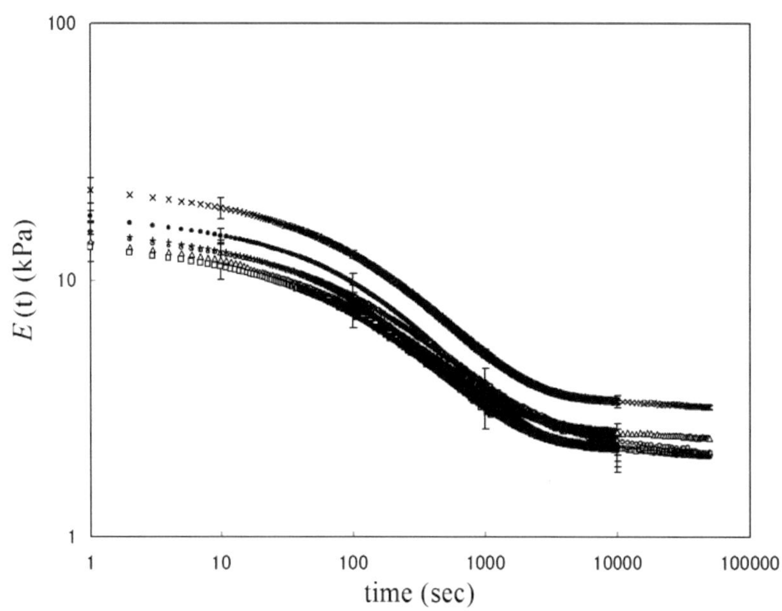

Figure 10. Relaxation modulus curves for MC3T3-E1 seeded agarose gels cultured for 1 day (○), 9 days (△), 12 days (□), 15 days (+), and 18 days (x). Relaxation modulus for agarose gel (•) is also shown as a control. (From Hanazaki et al., J. Biorheology. In press. DOI 10.1007/s 12573 -011- 0043-2. With permission.)

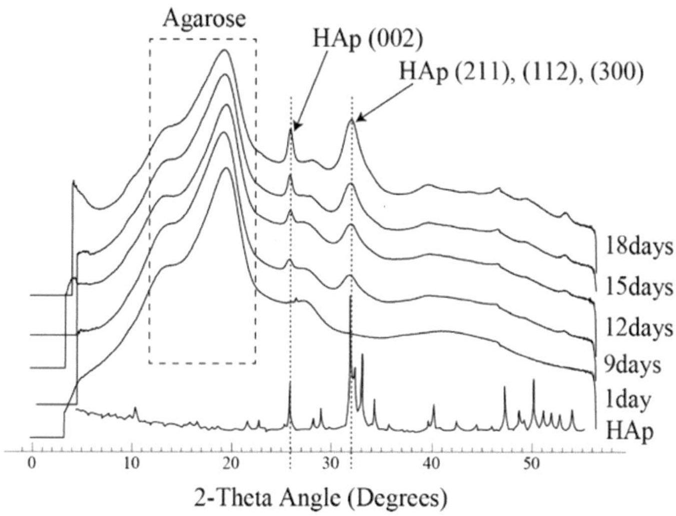

Figure 11. X-ray diffraction profiles for cell-seeded agarose gel cultured for indicated days. That for HAp powder is also shown for reference (Okasaki at al., 1997). (From Hanazaki et al., J. Biorheology. In press. DOI 10.1007/s12573 -011- 0043-2. With permission.)

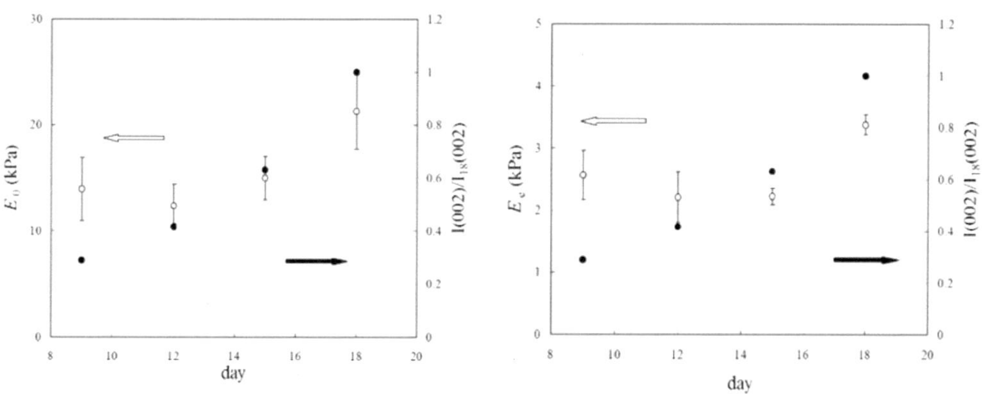

Figure 12. Comparison of E_0 (a) and E_e (b) values with diffraction intensity for HAp 002 as a function of culture period. (From Hanazaki et al., J. Biorheology. In press. DOI 10.1007/s12573-011-0043-2. With permission.)

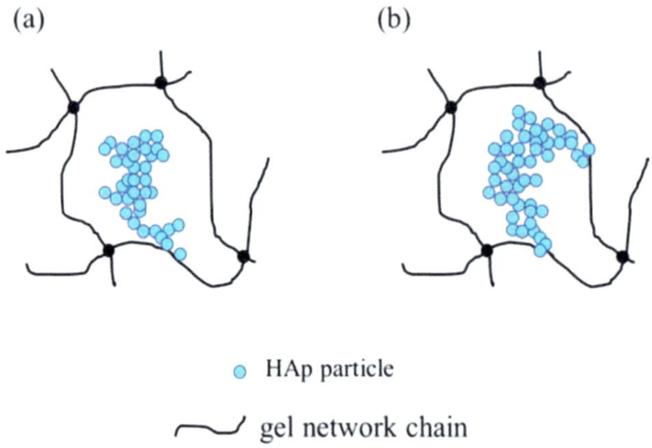

Figure 13. A model explaining the discrepancy of the mechanical behavior from the X-ray diffraction profiles as a function of culture period. (a) Precipitation of HAp in agarose gel up to day 15. (b) Partial percolation of HAp precipitates in a agarose net work area after day 18. (From Hanazaki et al., J. Biorheology. In press. DOI 10.1007/s12573-011-0043-2. With permission.)

4.3. Chondrocyte seeded agarose gel

Chondrocytes are responsible for the synthesis, maintenance, and gradual turnover of an extracellular matrix (ECM) composed principally of a hydrated collagen fibril network enmeshed in a gel of highly charged proteoglycan molecules. Each chondrocyte is surrounded by a narrow tissue region of pericellular matrix (PCM), the elastic modulus of which has been regarded to be larger than that of a chondrocyte and smaller than those of territorial matrix (TM) and inter territorial matrix (ITM). The unit consisting of a chondrocyte with PCM is generally termed a chondron. Since the volume fraction of chondrocytes in articular cartilage has been reported to be about 5-10%, chondrons will occupy more than this volume fraction. Even though the modulus of a chondron is much smaller than those of TM and ITM, the contribution of chondrons as mechanical elements to the mechanical function of articular cartilage is not negligible. To clarify the contribution of the viscoelastic nature of chondrons to that of articular cartilage tissue, relaxation modulus for chondrocyte-seeded agarose gel and that cultured for three weeks were measured as a model system of articular cartilage (Sasaki et al., 2009). The relaxation modulus curves for chondrocyte-seeded agarose gel were compared with that for agarose gel (AG). Fig. 14(a) shows a microscopic image of chondrocytes-seeded in agarose gel (AGC0) that was obtained immediately before mechanical measurements. The density of chondrocytes was 30×10^6 cell/m*l* and the average size of a chondrocyte was about 10 μm in diameter. Chondrocytes were shown to be dispersed almost homogeneously. Fig. 14(b) shows a microscopic image of chondrocyte-seeded agarose gel after 21 days of cultivation (AGC3). The magnification is the same as that in Fig. 14(a). Toluidine Blue staining was performed. Material was synthesized around each cell after 21 days of cultivation. Fig. 15 shows the relaxation modulus of AGC3 compared with those of AGC0 and AG. The relaxation modulus of AGC0 was increased by cultivation to be that of AGC3. In a short time region, up to 10^2 sec, the increment was not statistically significant. After10^3 sec, the relaxation modulus value of AGC0 significantly increased (*p<0.05). In order to analyse the change in the viscoelastic properties of chondrocyte-agarose composite originated from the PCM-like material production by chondrocytes, the differences in the relaxation modulus values among three samples were calculated. Fig. 16 shows the difference in the relaxation modulus values between AG and AGC0 (○),

$$\Delta E_{\mathrm{AGC0-AG}} = E_{\mathrm{AGC0}} - E_{\mathrm{AG}},$$

AGC3 and AGC0 (▲),

$$\Delta E_{\mathrm{AGC3-AGC0}} = E_{\mathrm{AGC3}} - E_{\mathrm{AGC0}},$$

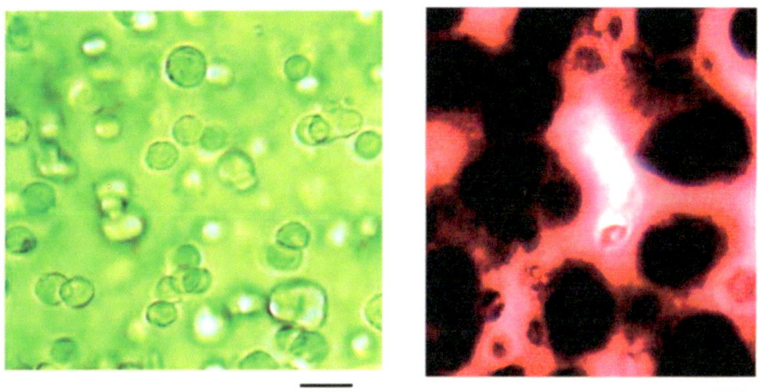

Figure 14. Microscopic images of chondrocyte-seeded agarose gel (a) before culture and (b) cultured for 21 days.(From Sasaki et al., J. Biorheology 23, 95-101 (2009). With permission.)

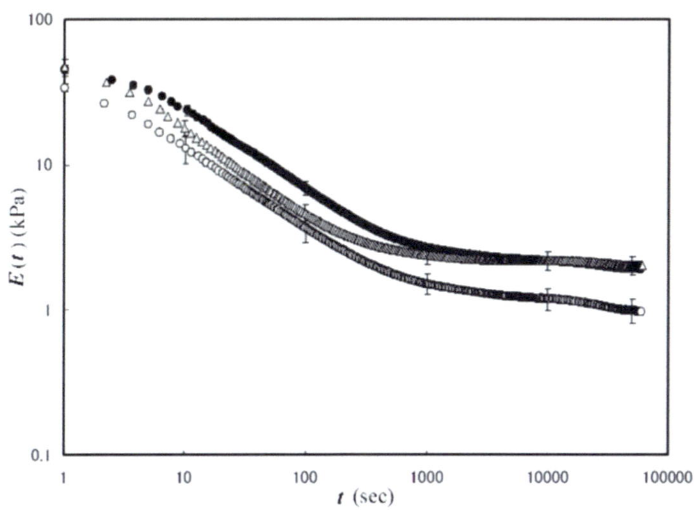

Figure 15. Relaxation modulus curves for chondrocyte seeded agarose gel immediately after seeded (○) and that cultured for 3 weeks (△). Relaxation modulus curve for agarose gel is also shown as a reference (●). (From Sasaki et al., J. Biorheology 23, 95-101 (2009). With permission.)

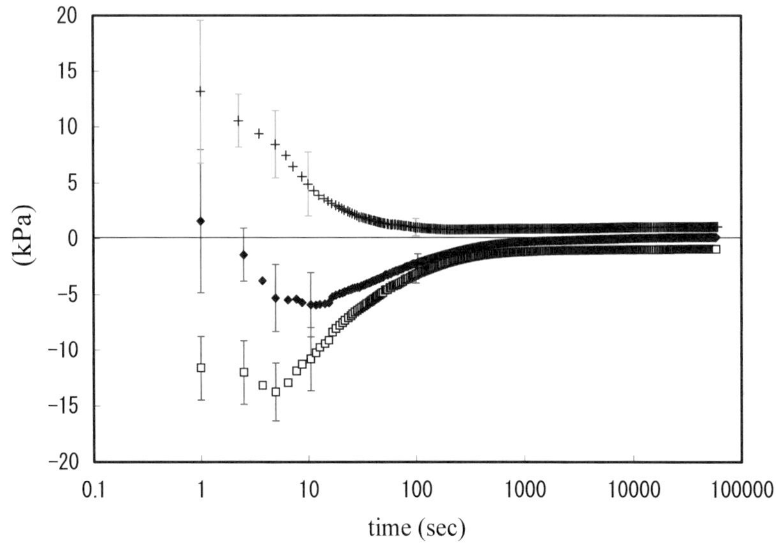

Figure 16. Differences of relaxation modulus curves. $E_{AGC0}(t) - E_{AG}(t)$ (○), $E_{AGC3}(t) - E_{AG}(t)$ (●), and $E_{AGC3}(t) - E_{AGC0}(t)$ (+). (From Sasaki et al., J. Biorheology 23, 95-101 (2009). With permission.)

and AGC3 and AG (◆),

$$\Delta E_{AGC3-AG} = E_{AGC3} - E_{AG}$$

where subscript letters AG, AGC0, and AGC3, indicate the relaxation modulus observed for AG, AGC0, and AGC3 specimens, respectively. The difference was remarkable in the gel viscoelasticity region. In $\Delta E_{AGC0\text{-}AG}$, as the modulus of AG is larger than that of AGC0, the difference was negative, showing the reduction of relaxation modulus by the seeding of chondrocytes. In the Figure, the modulus value difference looks to become levelled off in the short time region from 1 sec to more than 10 sec. The characteristic relaxation time of agarose gel has been reported to be around 10^2 sec (Watase et al., 1980), while that of a chondrocyte was determined to be 1 ~30 sec (Leipzig et al., 2005; Jones et al., 1999; Shieh et al., 2006). The behaviour of $\Delta E_{AGC0\text{-}AG}$ in the short time region may be a reflection of the viscoelastic chondrocyte contribution as an elastic component in short time region of the relaxation process. $\Delta E_{AGC3\text{-}AGC0}$ represents the modulus increase by the PCM-like material production by chondrocytes. The modulus generally increased throughout the relaxation process. However, the increment also seems to be levelled off in the short time region as the reduction in $\Delta E_{AGC0\text{-}AG}$. $\Delta E_{AGC3\text{-}AG}$ (◆) shows the re-increase in the relaxation modulus from AGC0 by the production of PCM-like material by chondrocytes. The plateau modulus value of AGC3 at times larger than 10^3 sec almost completely recovered the value of AG (the horizontal axis of ΔE=0), indicating that the gel modulus value once reduced by the chondrocyte seeding was increased with the PCM-like material production by chondrocytes after 21 days of cultivation. On the other hand, as $\Delta E_{AGC3\text{-}AG} = \Delta E_{AGC0\text{-}AG} + \Delta E_{AGC3\text{-}AGC0}$, the increment from$\Delta E_{AGC0\text{-}AG}$ was also shown to be significant in both the plateau (t larger than 10^3 sec) and the relaxation region (less than 10 sec) (*$p<0.05$ and **$p<0.01$, respectively). The change in the relaxation modulus in $\Delta E_{AGC3\text{-}AGC0}$ can be attributed to the material produced by chondrocytes after 21 days of cultivation. For the increase in gel modulus value in general, two possibilities would be responsible; (1) the increase in the modulus of gel network conforming polymer chain and (2) the increase in the cross-link density. The case (1) would result in a shortened relaxation time and the case (2) would provide a larger equilibrium modulus. In AGC3, the specimen seems to obtain a larger value of equilibrium modulus by the production of PCM-like material. According to Bushmann *et al.* (1995), the PCM-like material produced by chondrocytes was firstly deposited in the intra-network space of agarose gel in the vicinity of each chondrocyte surface in early cultures and with the cultivation period the material deposition area extended. From the stress relaxation results above, it is deduced that the PCM-like material in AGC3 could provide apparent cross-link points, which would contribute to the increase in the apparent cross-link density as compared with that in AGC0.

5. Stress relaxation of bones

In the previous section, viscoelastic properties of model systems of cartilaginous and osteogenesis tissues were shown. In this section, viscoelastic mechanical properties of actual tissues, in particular bones, would be described.

5.1. Relaxation modulus of cortical bone

Bone has been often regarded mechanically as a composite material of hydrated organic matrix mainly composed of collagen and hydroxyapatite (HAp)-like mineral phase. It is thought that the pliant collagen is reinforced by stiff mineral particles, and, as a composite, the brittleness of the mineral is compensated for by the viscoelasticity of the collagen. Recently, the existence of non-collagenous glue proteins that connect mineralized collagen fibres has been revealed (Fantner *et al.*, 2005). These organic phases have been suggested to be responsible for the toughness of bone. Because of this viscoelasticity of collagen fibres and non-fibrous proteins in the bone matrix, bone itself has noticeable viscoelasticity (Currey, 1965; Sasaki, 2000). Detailed experimental studies on the viscoelasticity of bone have been carried out only recently, despite the fact that it has been known for a long time that bone is viscoelastic (Currey, 1964; Lakes *et al.*, 1979). According to the results of those experimental studies, the relaxation modulus of bone could not be adequately described by a single relaxation time and was found, unlike synthetic polymeric materials, to be thermo-rheologically complex (Vincent, 1982).

In our previous papers, as a new empirical equation for the description of stress relaxation of cortical bone, we proposed that stress relaxation of cortical bone could generally be described by a linear combination of two Kohlraush-Williams-Watts (KWW) functions (Iyo *et al.*, 2004; Iyo *et al.*, 2006),

$$E(t) = E_0\left\{A\exp\left[-(t/\tau_1)^{\beta}\right] + (1-A)\exp\left[-(t/\tau_2)^{\gamma}\right]\right\},\ \left[0 < A, \beta, \gamma < 1\right], \tag{18}$$

where E_0 is the initial modulus value, $E(0)$.τ_1 and τ_2 (>> τ_1) are characteristic times of the relaxation processes, A is the fractional contribution of the fast relaxation to the whole relaxation process, and β and γ are parameters describing the shape of the relaxation modulus. It has been revealed that the first term represents the relaxation in the collagen matrix in bone and the second term is related to the change in a higher-order structure of bone that is responsible for the anisotropic mechanical properties (Iyo *et al.*, 2004). It seems to be possible to relate the viscoelastic properties and the hierarchical structure of bone by investigating these mechanical parameters. In this section, the effect of the structural anisotropy on the relaxation modulus of cortical bone would be presented, where relaxation parameters in eq. 18 will be analyzed in conjunction with the structural information of bone.

The bone samples used in this study were obtained from the mid-diaphysis of a 36-month-old bovine femur. Optical microscopic examination showed that all of the samples were generally plexiform but partly transformed into Haversian bone. The samples were cut using a band saw under tap water. In order to examine Young's moduli parallel and normal to the BA, we cut out specimen plates whose longer axes were parallel and normal to the BA, respectively. The cut sections were shaped into rectangular plates with approximate dimension of 0.5 cm (width), 3.2 cm (length), and 0.1 cm (thickness) by using emery paper under tap water. A plate with the longer edge parallel to the BA was coded specimen P, and a plate with the longer edge normal to the BA was coded specimen N. Fig. 17 shows the geometry of specimens.

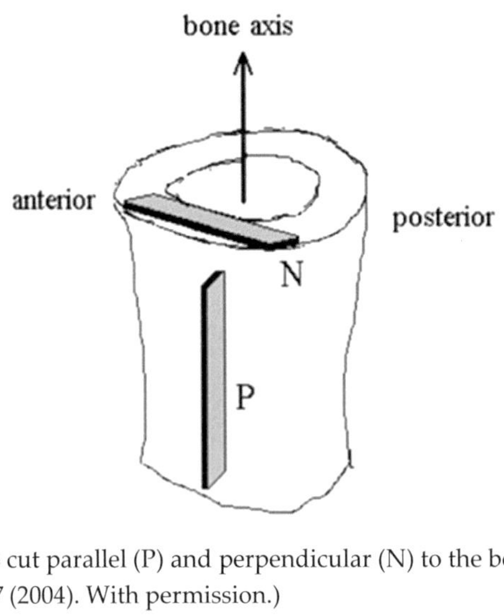

Figure 17. Bone specimens cut parallel (P) and perpendicular (N) to the bone axis.(Iyo et al., J. Biomechanics 37, 1433-1437 (2004). With permission.)

Fig. 18 shows the average values of relaxation Young's modulus, $E(t)$, plotted against time for specimens of P and N. At several points, standard errors are shown by vertical bars and are listed in Table 2. As mentioned above, stress relaxation of cortical bone has been revealed to be expressed by a combination of two relaxation processes according to eq. 18: a fast process (KWW1 process) with a relaxation time, τ_1, no more than 100 sec and a slow process (KWW2 process) with a relaxation time, τ_2, in the order of 10^6 sec. In this experiment, fitting of the average data to eq. 18 was performed. The relaxation modulus results obtained were described well by eq. 18. The relaxation parameters determined by the fitting, as well as the coefficient of determination, R^2, and the mean square error, s, are listed in Table 2. The average initial value of the relaxation Young's modulus, $E(0)$, of P was significantly larger than that of N ($p<0.05$, ANOVA).

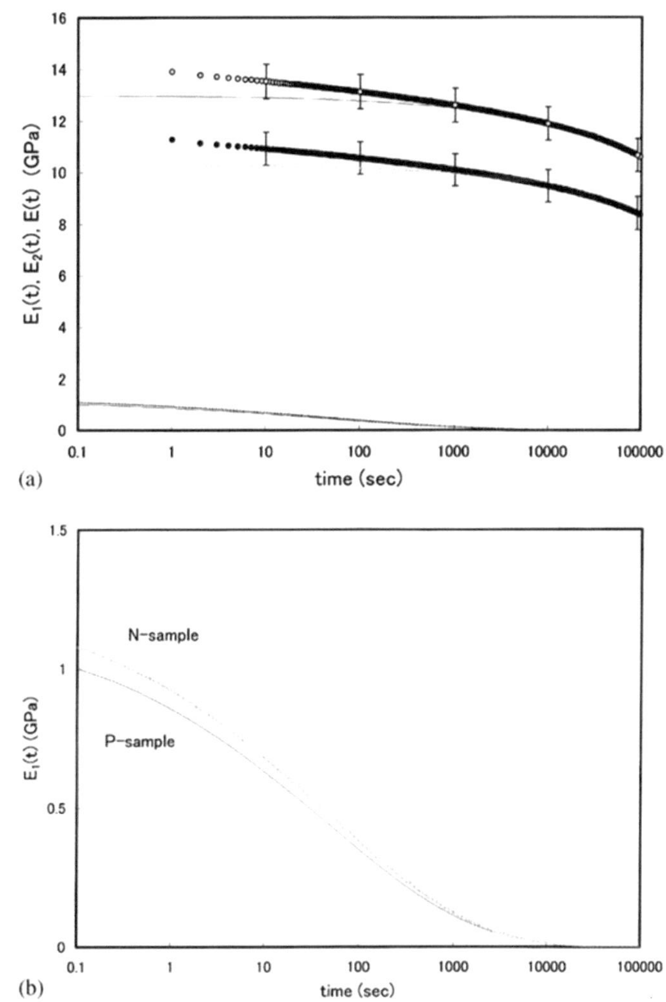

Figure 18. Relaxation modulus curves for P (○) and N (•) bone specimens. (a) Decomposition into the fast relaxation and slow relaxation processes are shown in lines for each relaxation curve. (b) Magnification of decomposed fast relaxation modulus curves. (Iyo et al., J. Biomechanics 37, 1433-1437 (2004). With permission.)

In the figure, lines represent the relaxation modulus of the KWW1 process, $E_1(t)$, and that of the KWW2 process, $E_2(t)$, for P and N specimens decomposed from the data according to eq. 18 using parameters of the best fit results listed in Table 2. Despite a difference in structural anisotropy of the specimens, the KWW1 relaxation process of a P specimen is indistinguishable from that of an N specimen at this magnification. In order to quantify the anisotropic mechanical properties of cortical bone, anisotropy ratio (AR) has been defined as the ratio of Young's modulus of bone in the direction parallel to the BA, E_P, against that normal to the BA, E_N, AR=E_P/E_N (Hasegawa *et al.*, 1994). AR values estimated from our results using average values are listed in Table 3, where the AR value for E_0 was listed as AR_0, and AR values for E_1, $AR_1= E_{P1}/E_{N1}$, and E_2, $AR_2=E_{P2}/E_{N2}$, were also estimated.

sample code	sample size	E_0 (GPa)	A_1	τ_1 (sec)	β	τ_2 (X10^6 sec)	γ	Standard Error (GPa) 10 sec	10^2 sec	10^3 sec	10^4 sec	R^2	s
P	7	14.2	0.08	49	0.28	9.3	0.35	0.67	0.66	0.66	0.65	0.99989	0.0086
N	5	11.6	0.11	50	0.26	6.4	0.37	0.64	0.63	0.62	0.63	0.99986	0.0195

(Iyo et al., J. Biomechanics 37, 1433-1437 (2004). With permission.)

Table 2. Relaxation parameters according to the empirical equation (18) determined for the average relaxation modulus curve.

sample code	E_0 (GPa)	E_1(GPa)	E_2(GPa)	AR_0	AR_1	AR_2
P	14.2	1.19	13.0	1.22	0.93	1.26
N	11.6	1.28	10.3			

(Iyo et al., J. Biomechanics 37, 1433-1437 (2004). With permission.)

Table 3. Decomposition of initial Young's modulus value into those of the KWW1 and KWW2 processes and anisotropic parameters.

The E_1 value for a P specimen was almost equal to that for an N specimen ($p>0.6$, ANOVA), and AR_1 (=0.93) was close to 1, indicating that relaxation Young's modulus in the KWW1 process was insensitive to anisotropic morphology of bone. An elementary process of KWW1 relaxation processes was thought to be attributed to a component of bone that was mechanically isotropic.

AR_2 (=1.26) for the KWW2 process was similar to that of AR_0 (=1.22) for the whole bone, indicating that an elementary process of the KWW2 relaxation process originates from a component causing the anisotropy of the whole bone. The difference between the whole relaxation Young's modulus value of a P sample from that of an N sample is represented by the difference in the respective KWW2 relaxation modulus values. The relaxation time for the KWW2 process, τ_2, for P-specimen was larger than that for N-specimen. Values of γ for P and N specimens were similar but larger than β values. This indicates that the KWW2 process is attributable to a mode that is governed by the structural anisotropy in bone.

5.2. Change in the relaxation modulus of cortical bone by the change in the mineral fraction

It has been regarded that the stiffness of bone is originated from minerals because modulus value of HAp minerals is almost 100 times larger than that of collagen. In the application as artificial bone materials, materials are required to have bio-compatibility, resistance to corrosion, adequate fracture toughness and fatigue strength. As for the bio-mechanical-compatibility, in order to obtain the matching in modulus, it is possible to fabricate a composite material of stiff materials with pliant matrix. Changing the stiff component, we will be able to have materials with similar modulus as bone. At the same time, with the change in stiff component, reinforcement state of the matrix can be changed. This means the viscoelastic properties of the matrix changes with the stiff component. We aimed to investigate the viscoelastic properties of bone with changing mineral content (Sasaki & Yoshikawa, 1993). Demineralization of bone specimens was performed in 0.5 M EDTA, pH8.0 at 4°C. Mineral fraction was determined by weighing EDTA treated bone specimens. Fig. 19 shows the relaxation moduli for bovine femoral cortical bone specimens of five different mineral contents. The set of relaxation modulus curves appear to be different parts

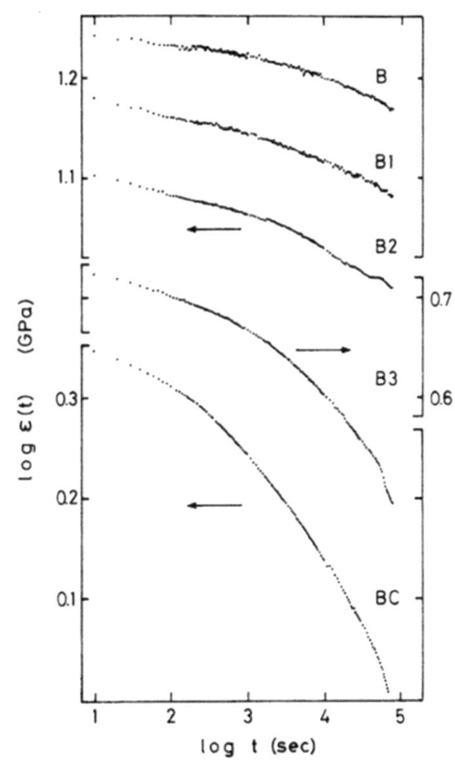

Figure 19. Relaxation modulus curves for bone specimens with various mineral contents; B ϕ_M =0.41 (volume fraction), B1 ϕ_M =0.35, B2 ϕ_M =0.33, B3 ϕ_M =0.24, and BC =0 (bone collagen). (Sasaki et al., J. Biomechanics 26, 77-83 (1993). With permission.)

of a single large master curve, suggesting the reduced variable method, that is, the time-mineral content superposition principle could be applicable. Fig. 20 shows the synthetic master curve constructed for bone specimens with different mineral contents. The synthetic curve looks smooth and the scatter of the data points is small. Fig. 21 shows the vertical shift factor, b_m, plotted against the mineral content. The filled circles are taken from the mineral content dependence of the elastic modulus of bone after Katz (1971). Mineral content dependence of b_m accords well with that of elastic modulus itself. This fact indicates that the superposition procedure was carried out correctly. Then, the result indicates the time-mineral content superposition principle. A polymer-filler system has been considered to have the same reinforcing mechanism as the model discussed. But in the usual polymer-filler system, the time-filler-fraction superposition principle does not hold. The size of the commercially available filler is at least of the order of a few μm. By the analysis of the horizontal shift factor, the reinforcing effect depends on the filler-matrix surface area, not on the filler size. The mineral particle in bone, where the time-mineral content reduction was concluded to be applicable, has been recognized to be of the size of a few hundred Å at most. The reason why the time-filler fraction superposition principle does not hold in the polymer-filler system is deduced to be related to the very large filler size compared with the mineral particles in bone, as well as an adhesive weakness between filler and matrix. This fact leads to the suggestion that, in order to improve the relaxation properties of mineral-resin composite as artificial bone, the mineral size should be reduced to, say, submicron level.

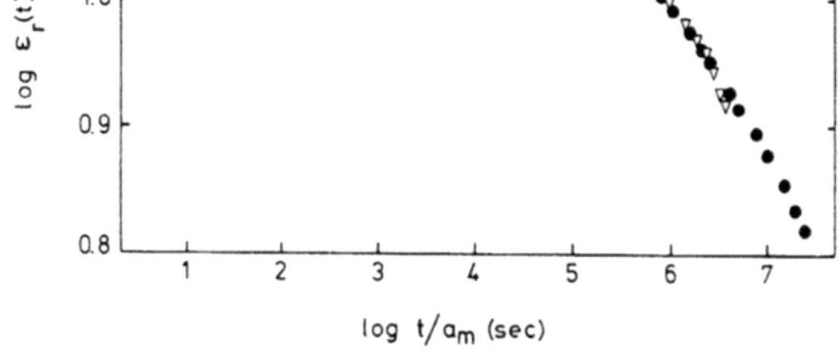

Figure 20. A master curve constructed by superimposing the relaxation modulus curves in Fig. 19. For the successful superposition, both the vertical and horizontal shifts were needed. (Sasaki et al., J. Biomechanics 26, 77-83 (1993). With permission.)

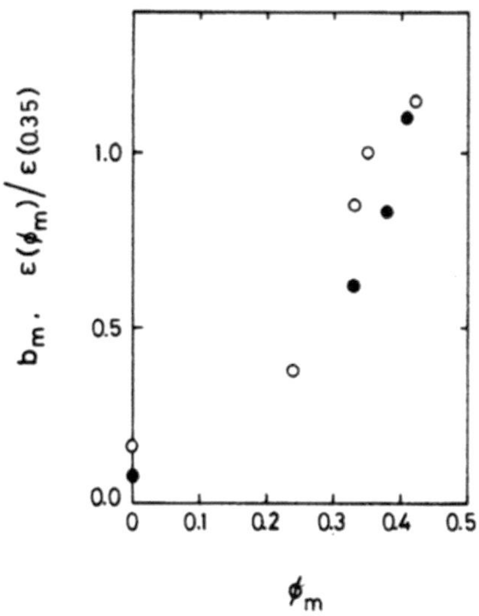

Figure 21. Vertical shift factor as a function of mineral content ϕ_M (volume fraction) (○). Mineral content dependence of elastic modulus is also plotted (•). (Sasaki et al., J. Biomechanics 26, 77-83 (1993). With permission.)

Author details

Naoki Sasaki
Faculty of Advanced Life Science,
Department of Interdisciplinary Sciences, Hokkaido University, Japan

6. References

Buschmann, M. D., Gluzband, Y. A., Grodzinsky, A. J., Hunziker, E. B. (1995). Mechanical compression modulates matrix biosynthesis in chondrocyte/agarose culture. *Journal of Cell Science* 108, 1497-1508.

Currey, JD. (1965) Anelasticity in bone and echinoderm skeletons. *Journal of Experimental Biology* 43, 279-292.

Currey, J. D. (1964) Three analogies to explain the mechanical properties of bone. *Biorheology* 2, 1-10.

Fanter, G., Hassenkam, T., Kindt, J., Weaver, J. C., Birkedal, H., Pechenik, L., Cutroni, J. A., Cidade, G. A. G., Stucky, G. D., Morse, D. E., Hansma, P. K. (2005) Sacrificial bonds and hidden length dissipate energy as mineralized fibrils separate during bone fracture. *Nature Materials* 4, 612-615.

Ferry, J. D. (1980) *Viscoelastic Properties of Polymers.* Wiley & Sons, New York.

Foord, S. A., Atkins, E. D. T. (1989). New X-ray diffraction results from agarose: extended single helix structure and implications for gelation mechanism. *Biopolymers* 28(8), 1345-1365.

Hanazaki, Y., Ito, D., Furusawa, K., Fukui, A., Sasaki, N. (2012) Change in the viscoelastic properties of agarose gel by Hap precipitation by osteoblasts cultured in agarose gel matrix. *Journal of Biorhelogy*, in press, DOI 10.1007/s12573-011-0043-2.

Hasegawa, K., Turner, C. H., Burr, D. B. (1994) Contribution of collagen and mineral to the elastic anisotropy of bone. *Calcified Tissue International* 55, 381-386.

Hayes, W. C., Keer, L. M., Herrmann, G., Mockros, L. F. (1972). A mathematical analysis for indentationtests of articular cartilage. *Journal of Biomechanics* 5(5), 541-551.

Iyo, T., Sasaki, N., Maki, Y., Nakata, M. (2006) Mathematical description of stress relaxation of bovine femoral cortical bone. *Biorheology* 43, 117-132.

Iyo, T., Maki, Y., Sasaki, N., Nakata, M. (2004). Anisotropic viscoelastic properties of cortical bone. *Journal of Biomechanics* 37, 1433-1437.

Jones, W. R., Ting-Beall, H. P., Lee, G. M., Kelly, S. S., Hochmuth, R. M., Guilak, F. (1999). Alterations in the Young's modulus and volumetric properties of chondrocytes isolated from normal and osteoarthritic human cartilage. *Journal of Biomechanics* 32, 119-127.

Jonscher, A. K. (1983) *Dielectric Relaxation in Solids*. Chelsea Dielectrics, London.

Kakudo, M., Kasai, N. (1972) *X-ray diffraction by polymers*. Kodansha, Tokyo.

Katz, J. L. (1971). Hard tissue as a composite material—1. Bounds on the elastic behavior. *Journal of Biomechanics* 4, 455-473.

Klafter, J., Schlesinger, M. F. (1986) On the relationship among three theories of relaxation in disordered systems. *Proceedings of National Academy of Science* U. S. A. 83, 848-851.

Lakes, R. S., Katz, J. L., Sternstein, S. S. (1979) Viscoelastic properties of wet cortical bone-I. Tortional and biaxial studies. *Journal of Biomechanics* 12, 657-678.

Leipzig, N. D., Athanasiou, K. A. (2005). Unconfined creep compression of chondrocytes. *Journal of Biomechanics* 38, 77-85.

Morita, T., Narita, T., Tokita, M. (2008). Spinodal decomposition of agarose gel. *Foods & Food Ingredients Journal Japan.* 213(5), 452-460.

Okazaki, M., Taira, M., Takahashi, J. (1997). Rietveld analysis and Fourier maps of hydroxyapatite. *Biomaterials* 18(11), 795-799.

Phillips, J. C. (1996). Stretched exponential relaxation in molecular and electric glasses. *Reports on Progress in Physics*. 59, 1133-1207.

Potuzak, M., Welch, R. C., Mauro, J. C. (2011) Topological origin of stretched exponential relaxation in glass. *Journal of Chemical Physics* 135, 214502-1-7.

Sasaki, N., Imai, T., Hshimoto, A., Yasuda, H. (2009). Effect of pericellular matrix formation by chondrocytes cultured in agarose gel on the viscoelastic properties of agarose gel matrix. *Journal of Biorhelogy*, 23, 95-101.

Sasaki, N. (2000) Viscoelastic properties of bone and testing method. In *Mechanical testing of bone and the bone-implant interface*. (Edited by An, Y. H. and Draughn, R. A.) Boca Raton FL: CRC Press; p. 329-348.

Sasaki, N., Yoshikawa, M. (1993) Stress relaxation in native and EDTA-treated bone as a function of mineral content. *Journal of Biomechanics* 26, 77-83.

Shieh, A. C., Athanasiou, K. A. (2006). Biomechanics of single zonal chondrocytes. *Journal of Biomechanics* 39, 1595-1602.

Vincent, J. F. V. (1982) *Structural Biomaterials.* Macmillan Press, London.

Watase, M., Nishinari, K. (1983) Rheological properties of agarose gels with different molecular weights. *Rheologica Acta* 22(6), 580-587.

Watase, M., Nishinari, K. (1980) Rheological properties of agarose-gelatin gels. *Rheologica Acta* 19, 220-225.

Chapter 6

Viscoelasticity in Biological Systems: A Special Focus on Microbes

Supriya Bhat, Dong Jun, Biplab C. Paul and Tanya E. S Dahms

Additional information is available at the end of the chapter

http://dx.doi.org/10.5772/49980

1. Introduction

Over billions of years of evolution, living organisms have developed into complex biosystems, of which the basic unit is the cell. Cells have a complex molecular structure with a certain level of rigidity. Living cells, whether isolated or part of a larger collective, live under constant mechanical stress from their external environments. Cells have developed adaptive mechanisms to maintain homeostasis and viability, which interestingly follow the basic principles of classical mechanics.

Cell mechanical properties have myriad biological significance and so there has been significant interest in the past decade to measure the response of cells to external mechanical signals. Cellular mechanics and rheological properties (*e.g.* stress-strain relationships) are known to play a role in biological processes such as cell growth, stem cell differentiation, cell crawling, wound healing, protein regulation, cell malignancy and even apoptosis (programmed cell death) [1,2].

A living cell is a complex dynamic system, far from static, which constantly undergoes remodeling to adapt to varying environmental conditions. The mechanical changes in cells under normal conditions and in response to external signals are highly complex and extremely difficult to measure *in vitro*. The interplay of cellular constituents enables adaptation to changing demands of mechanical strength and stability. The field of rheological science deals with the mechanical behavior of biological materials and over the past decade several rheological methods have been developed to quantify the mechanical behavior of cells in response to external conditions and forces.

To understand cell mechanics we first need an appreciation of how cells operate in a mechanical context. Firstly, how do cells maintain their shape and flexibility to accommodate cellular requirements? Cell surface layers are strong, playing a crucial

INTECH
open science | open minds

© 2012 The Author(s). Licensee InTech. This chapter is distributed under the terms of the Creative Commons Attribution License http://creativecommons.org/licenses/by/3.0), which permits unrestricted use, distribution, and reproduction in any medium, provided the original work is properly cited.

mechanical role in maintaining cellular shape and resisting turgor pressure, yet at the same time flexible enough to allow cell growth and division. Multi-cellular eukaryotes are arranged into specialized structures of varied composition (*e.g.* tissues) which protect them from harsh environments. Single-celled prokaryotes or eukaryotes can also form sophisticated structures, such as biofilms or mycelia, but even as single cells they are able to bear mechanical stress and maintain integrity which is quite astonishing. Cells range from soft mammalian tissue cells to those with very firm plant cellulosic fibers, so understanding how different cellular components provide diverse mechanical properties is of great interest. Further, how do mechanical properties support locomotion? Active movements depend on cell type and ultimate function within tissues or the greater environment, and their associated mechanical properties change according to environmental demands. Another question relates to how external factors, such as temperature, pH, aridity and xenobiotics affect cell mechanical properties. Further, what is the role of cell mechanics in processes such as active and passive diffusion, adhesion, community formation and evolution of complex tissues in higher eukaryotes, and how do the mechanics of isolated components work in concert within dynamic live systems that are far from equilibrium?

Several of these questions remain unanswered, however early success studying the rheological properties of non-living materials has provided a strong foundation for studying biological systems. Over the past decade, several obstacles have been overcome and thus a new perspective of cellular mechanics is emerging [2].

1.1. Viscoelasticity in biology

If cellular metabolism could be frozen at any given moment, disrupting the production of metabolites and supra-molecular structures, the cell would simply become fluid-like with the associated mechanical properties. Living cells behave both as an elastic solid and as a viscous fluid, and so are considered viscoelastic. Such materials, including biological molecules and cells, cannot be fit using classical models of either elasticity or viscosity. Cellular viscoelasticity arises from the combination of high water content conflated with a polymerized structural matrix. On the one hand, the biopolymers which support cell shape provide strong enough mechanical properties to resist environmental pressures, but on the other hand their organization is highly dynamic and linked to metabolic conditions.

Cellular mechanical properties can be characterized using viscosity, elasticity and creep compliance. Herein we focus on viscoelasticity studies measuring either biological sample deformation induced by an external force or the force resulting from sample deformation. The cell cytosol, which contains the majority of cell fluid, can be treated as the coexistence of liquid and solid phases. The latter is composed of proteins, DNA, RNA and cytoskeleton filaments as well as organelles suspended in the viscous buffered saline, saturated with metabolites and proteins. The liquid component has a high compression modulus, meaning that without the structural components and macromolecules it would be very fluid (less viscous). In contrast, the solid phase has a lower compression modulus, exhibiting more elastic properties. In cells as a whole, the viscosity of the cytoplasm dominates the transport and movement of subcellular macromolecules (such as DNA and proteins), elasticity controls the response of the cytoplasm to mechanical stresses at shorter timescales (seconds), organelles and cytoskeleton elements over long timescales (minutes), and the contractile mechanical regime governs responses at even longer time scales (minutes to hours).

The quantitative characterization of elastic material is the elastic modulus, described as the ratio of stress to strain, but the deformation (strain) of a fluid under an external stress changes as a function of time and is referred to as strain rate. Thus fluid viscosity is the ratio of stress to strain rate. The mechanical responses of biopolymers fall into a category between that of an elastic solid and a fluid, defined as viscoelasticity.

In this chapter we will describe the methods to measure viscoelasticity in biological systems and their significance. We generalize the mechanical properties of prokaryotic and eukaryotic cells since it has been argued that viscoelastic properties are universal across cell types and species, however the mechanistic basis of this phenomenon is not well understood [4]. This chapter is organized into four main sections: (i) a brief summary of various rheological methods used for cell micromanipulation and the novel application of atomic force microscopy to measure cell mechanical properties, (ii) the main mechanical load bearing components of cells and associated studies that have helped to establish a consensus understanding of cell mechanical properties, (iii) a survey of the various factors that affect cell viscoelastic properties, but which currently lack clear interpretation, and finally (iv) a future perspective on the significance of cellular viscoelasticity.

2. Methods of measuring cellular viscoelasticity

In microrheological studies, several experimental techniques and theoretical models are combined to accurately quantify cellular mechanics. Several methods that are routinely used for cell rheology are briefly reviewed in the following section. Micropipette aspiration, microplate rheometry, and optical stretching are used to probe whole cell mechanics, while probe- or particle-based techniques such as magnetic probes, optical tweezers and particle tracking cell rheology only measure local mechanical properties. The probes or particles within a structure are subjected to an external force and the subsequent response measured, usually by tracking their displacement, to characterize associated mechanical properties. Atomic force microscopy uses a probe, but attached to the end of a cantilever with well-defined rigidity able to the mechanics of whole cells and their surface layers. A schematic diagram of various devices is shown in Figure 1.

The mechanical properties measured by a probe or particle technique are highly dependent on the strength of, and relationship between the particle/probe and the structure being studied. This relationship is highly complex, varies with cell type and can lead to significant misjudgement of the stiffness. Probes can be modified with different molecules such as antibodies, peptides and cadherins to target antigens, integrins and cytoskeletal components. However these measurements do not account for interactions between the probes and cells which can lead to confounding results [5]. Cells have a heterogeneous composition and their various compartments have diverse mechanical properties. If only one cell position is probed, as is common, the mechanical properties of the whole cell will not be well represented. Comprehensive methods have been developed for measuring the mechanical properties of mammalian cells and precise methods have not yet been adequately developed for prokaryotes. A summary of the advantages and disadvantages are listed in the Table 1.

Methods	Applications	Drawbacks	References
Micropipette aspiration	Measurements of non-linear deformations with high accuracy Both soft and rigid cells can be used	Quantitative measurements rely heavily on theoretical models Pipette geometry can limit measurements	[6,7]
Microplate rheometer	Several manipulations in the same instrument are possible A large range of forces are measurable (1nN - 1μN) Control of cellular pre-stress	No subcellular resolution	[8,9]
Optical stretcher	No physical contact required and non-destructive Less time consuming with simple setup	Causes heating of cells Extensive modeling is required to obtain force profiles	[10,11]
Magnetic probes	Large range of frequencies [0.01–1,000 Hz] Parallel measurements of large number of cells possible High timescale resolution	Can only probe microenvironment inside the cell	[12,13]
Particle tracking microrheology	Quantitative measurements of shear modulus possible Can be used under physiological conditions	Used only for soft tissue cells	[14-16]
Optical tweezers	High timescale resolution Force and position can be controlled more accurately	Local heating and prototoxicity can result Can only be used at low forces in the linear regime	[17,18]
Atomic force microscopy	High spatial resolution Measures large range of forces More precise and easy to control cantilever position	Slow scan rate	[19, 20]

Table 1. Comparison summary of different methods to measure cell rheology.

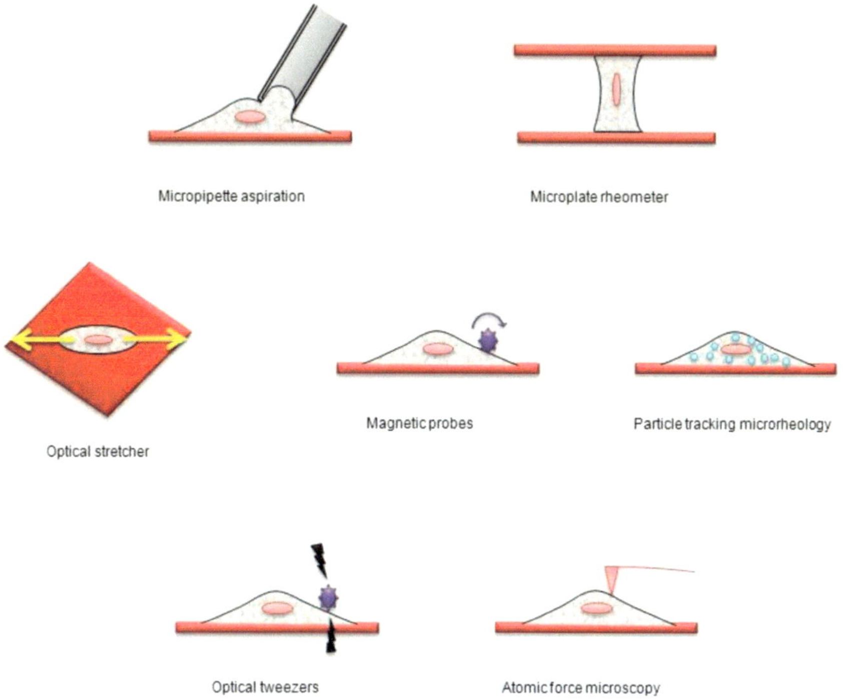

Figure 1. Schematics for the common methods used to measure cell rheology.

2.1. Micropipette aspiration

Micropipette aspiration is a widely used method and offers a versatile way of measuring the mechanical properties of living cells. A cell is aspired onto the end of a micropipette by a negative pressure gradient, and the aspiration length of the cell inside the micropipette is recorded as a function of time [21]. Cells experience large non-linear deformations in response to aspiration suction pressure (0.1 pN/μm^2), giving rise to forces (10 - 10^4 pN) that allow tracking of edges with high accuracy (± 25 nm). This method can measure the elastic and viscous properties of very soft materials like red and white blood cells, and stiffer cells such as endothelial (skin) cells and chondrocytes (cartilage) [7]. Measurements are interpreted using basic continuum models to solve for elasticity and viscosity parameters.

In most cases, cells suspended in buffer or saline solution that are aspirated appear round, but shape and degree of deformation depends on cell type. Many mammalian cells are naturally spherical (*e.g.* white blood cells) or spontaneously adopt this shape when detached from a surface. Recall that cells can behave as liquids and/or solids [7], the response of which is similar until a hemispherical projection is formed in the pipette. Beyond that point, a further increase in the suction pressure causes constant cortical tension, creating a liquid-like cell able to flow completely into the pipette [22]. On the other hand, the surface of a solid cell will extend into the pipette to a new equilibrium position. Under constant pressure, the mechanical properties of the cell determine how far it is pulled inside the micropipette. The measurement of suction pressure, location of the cell and movement of its edge in the micropipette can be used to calculate the viscous and elastic parameters [7].

There are several draw-backs associated with this method since quantitative measurement of cell stiffness relies heavily on theoretical models [6]. The most recent application of micropipette aspiration is to capture and hold cells for manipulation [7].

2.2. Microplate rheometry

Over the past decade there has been much progress in the study of rheological behaviour for single cells, including the development of several models which explain the mechanical properties of cytoskeletal networks. A microplate rheometer, invented by Thoumine and Ott in 1997 [8], consists of two parallel microplates that support cell adhesion and spreading. The plates can be made flexible and in some cases are coated with an adhesive material. A whole cell is stretched or compressed between the two plates while the flexible microplate is used as a nN-scale force sensor. Integration with an inverted microscope allows cells that are being manipulated to be directly observed. Several adaptations to the instrumentation allow measurement of creep function, area of expansion modulus, contraction forces of single cells, adhesive interactions and stress-strain responses. The latter can be used to measure viscoelasticity.

The force applied to the cell is directly proportional to the relative stiffness of the microplate, hence measurement of cell deformation offers valuable information on cell mechanical properties. Measureable forces range between 1 nN – 1 μN, and several manipulation modes are possible, including compression, traction, aspiration and adhesive rupture. Following their invention, microplate rheometers have been improved by a number of scientists. The behaviour of cells in response to compression, traction, aspiration or adhesion, for which elastic, viscous and contractile regimes can be distinguished based on time scale, can be used to calculate the viscoelastic modulus of living cells. The instrument has been modified for probing single layers of cells at once using the novel cell monolayer rheology (CMR) technique, making possible harmonic oscillation experiments and step shear or step stress experiments which reveal different viscoelastic regimes [9].

2.3. Optical cell stretcher

The optical stretcher is a non-destructive tool that can be used to quantify cell deformation. The optical stretcher was first developed by Guck et al. [23] based on the principle that a dielectric object, when placed between two opposed and non-focused laser beams, experiences a net force. The cell is suspended or "optically trapped" and is stable only if the total force is zero. Additive surface forces are capable of stretching an object, such as a cell, along the beam axes [11]. Both the exerted force and corresponding time-dependent deformations can be quantified. An optical stretcher allows the measurement of cell mechanics without physical contact, but the intense laser exposure tends to heat cells [24]. Recently, however, it has been demonstrated that heating does not affect cell cytoskeletal structure, proliferation, motility, or viability [25]. Another potential limitation is the complexity of the force profiles, which require extensive theoretical modeling. Nonetheless, the optical stretcher has proven to be an effective way of measuring the contribution of cytoskeletal filaments to cell viscoelasticity since single suspended cells are probed as well-defined viscoelastic objects [25].

2.4. Optical tweezers

Optical tweezers can be used to capture tiny dielectric particles with a highly focused laser beam [17]. The movement of dielectric particles can be controlled with two main optical forces. The scattering force acts along the direction of beam propagation, while the electric field induces a dipole in the dielectric particles, producing an electric field gradient that pulls particles towards the focus. When the gradient force dominates, the dielectric particles can be confined in a stable three-dimensional optical trap. The trap is then moved to manipulate the bead, so the applied force and resultant particle displacement are interpreted in terms of mechanical response [26]. To obtain viscoelastic information, an oscillatory force is applied to the dielectric bead by oscillating the laser position with a movable external mirror. The resultant amplitude of the bead motion and the phase shift are interpreted in terms of viscoelastic response. The experimental data from optical tweezers can only be used to study the viscoelastic responses at low force in the linear regime [18]. Local heating and phototoxicity ("opticution" as coined by Block) can result from the intense laser powers required to trap biological samples.

2.5. Magnetic probes

Another way to explore the viscoelasticity of cells involves manipulation of their movement through an externally applied magnetic field. The earliest application of this technique was pioneered by Freundlich and other researchers [12,27], in which magnetic particles were embedded in the material of interest (*e.g.* cells). The viscoelasticity can then be probed using magnetic tweezers or magnetic twisting. The ability to functionalize magnetic colloidal beads allows for their specific localization within the cell [28, 29].

The magnetic tweezer technique involves the manipulation of a super paramagnetic bead with an applied magnetic field generated by four pairs of soft ferromagnetic cores, each wound with a separate field coil arranged at special angles [30]. The movement of the bead is monitored by its induced magnetic dipole as it interacts with the field gradient of the strong magnet to which it is exposed. The corresponding displacement of the magnetic bead is used to measure cell properties. The movement of magnetic beads can also be controlled, albeit only in one direction, by a strong magnetic field gradient arising from electromagnets generated by axis-symmetrically arranged magnetic coils on a sharp iron tip [31]. Multiple pairs of electromagnetic tips are required for more complex movements of the magnetic bead, which for these experiments are smaller than the size of the cell, and thus are limited to probing the viscoelastic response of a microenvironment rather than the whole cell.

Twisting magnetometry [32] and the more recently developed magnetic twisting cytometry [33] can also be used to measure the movement of magnetic beads, which usually consist of colloidal metal or polycrystalline iron oxide. The cell is deformed under a twisting magnetic field that is applied perpendicularly to the initial magnetic field once it has been turned off [34]. The change in the magnetic field direction causes reorientation of the magnetic bead towards the twisting field, and once both are turned off, the rate of magnetic bead rotation and the amount of recoil are measured to interpret local viscoelasticity.

2.6. Particle-tracking microrheology

In particle-tracking microrheology, fluorescent microbeads are injected into live cells and diffused randomly in their cytoplasm. These beads are so small (< 1μm) that their inertial forces are negligible and they move according to Brownian motion. The movement of the fluorescent beads can be observed by fluorescence microscopy, and route distance can be converted to bead displacement which is used to calculate frequency-dependent viscoelastic moduli and/or the creep compliance of the cytoplasm [14]. For particle-tracking microrheology of living cells, the applied deformation and resultant stress is not oscillatory and is used to probe the mechanical properties of adherent cells on planar substrates, showing strong elastic responses over short timescales but with dominant viscous responses over longer time periods [15].

Particle-tracking microrheology has been used to study the viscoelastic responses of live cells and their cytoplasm under pharmacological treatment, serum starvation and at the edge of tissue wounds, as well as the mechanical responses of their nuclei [35-37]. For these studies, target cells can be deeply embedded in a 3D matrix, a condition more similar to cells in their physiological environment and difficult to probe by other methods.

2.7. Atomic force microscopy

The advent of atomic force microscopy (AFM) provided a valuable tool to image cell surface structure at sub-nm resolution and to probe the global and local nano-mechanical properties of cells. Such a non-invasive method makes it possible to investigate live cells under physiological conditions. The key component of AFM is a sharp tip mounted on a cantilever (usually silicon or silicon nitride), which is raster-scanned over the sample surface by piezoelectric micropositioners (Figure 2). Lateral or vertical displacement of the cantilever is detected by a position sensitive photodiode, which signals the fast feedback loop to maintain a constant relationship (*e.g.* force or distance) between tip and sample and the computer which is used to generate an image of the sample surface. AFM can be operated in many different modes, including force spectroscopy (FS) which is used to probe the mechanical properties of the cell surface layer or whole cell [38].

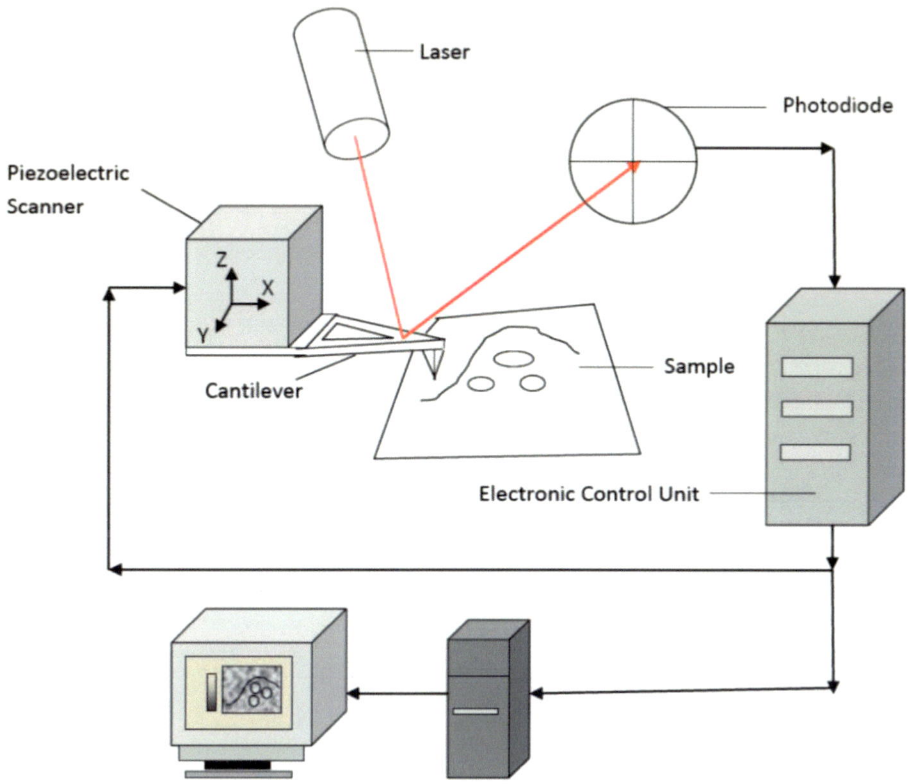

Figure 2. A schematic representation of the atomic force microscope. Printed with permission (Springer, USA).

AFM offers the further advantage of being able to correlate sample topography with mechanical properties across the sample surface using indentation forces as small as 10 pN. With force spectroscopy (single point) or force mapping (multiple points), the tip approaches the sample, indents the sample and then retracts at each point, generating a force versus distance curve at a specific point on the cell surface (Figure 3). Cantilever deflection as a function of distance of the tip from the cell surface is initially represented by photodiode voltage as function of piezo displacement. This voltage is then converted to cantilever deflection and finally a force or indentation distance. The extent to which the sample is deformed depends on its viscoelastic properties.

Cantilever deflection can be converted to force using Hook's law:

$$F = k \times d \tag{1}$$

where k is the cantilever spring constant, and d is cantilever deflection.

Force spectroscopy and mapping are used to quantify the mechanical behavior of the cell with the help of theoretical models. The indentation of the biological sample can be determined by subtracting the difference between cantilever deflection on hard surfaces and on soft biological surfaces. Based on the Hertz model, Sneddon [40] developed a theory describing the relationship between loading force and indentation. Most commercially available AFM tips are either conical or parabolic, and hence these two types of AFM tips are considered during modeling. The relationship between loading force and indentation are given by following equations [41]:

$$F_{parabolic} = \frac{4E\sqrt{R}}{3(1-v^2)}\delta^{3/2} \tag{2}$$

$$F_{con} = \frac{2E\tan\alpha}{\pi(1-v^2)}\delta^2 \tag{3}$$

where R is radius of curvature for a parabolic AFM tip, α is the half opening angle of conical tip, δ is the indentation of the cell as a result of loading force F, 'E' is the Young's modulus of the sample, which describes the magnitude of elasticity and v is Poisson ratio, which is assumed to be 0.5 for soft biological materials. The Young's modulus of microbial cells is determined from the non-linear portion of the force indentation curve with equations 2 or 3 [41].

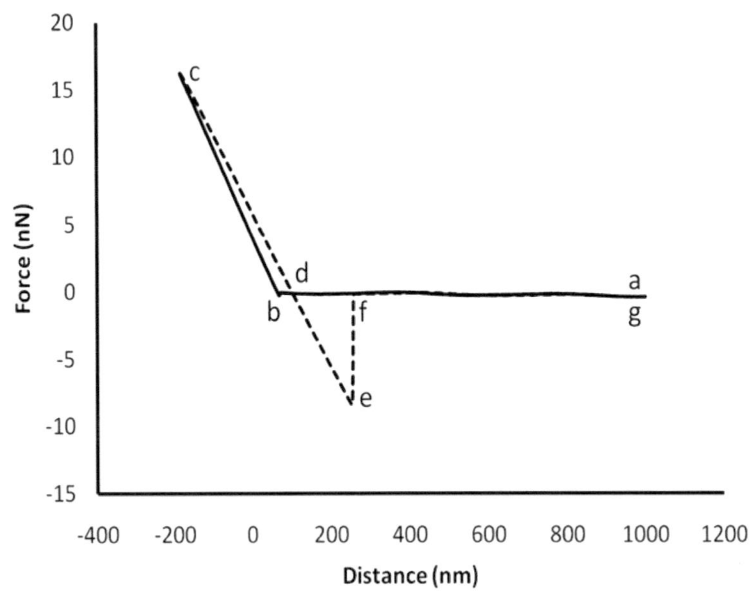

Figure 3. A representative force-distance curve taken on the surface of an *Aspergillus nidulans* cell wall. Solid and dashed lines represent approach and retract cycles respectively. Point b indicates jump into contact of the AFM tip to the sample. Section b-c represents the force required to indent the sample a given distance, and is used to measure cantilever deflection and to calculate sample indentation [39].

The spring constant of a fungal cell wall can be determined using the following equation:

$$k_w = \frac{K_c m}{1-m} \tag{4}$$

where k_w is the spring constant of the hyphal cell wall, also called relative rigidity, K_c is the spring constant of the cantilever and m is the slope of the approach curve, corrected for that of a hard surface. This equation can also be used to determine the spring constant of cylindrical bacterial cells [42]. Models developed by Zhao et al. [20] can be used directly to calculate the Young's modulus of fungal hyphal walls. Fungal cell wall elasticity depends not only on the spring constant, but also hyphal radius (R), and cell wall thickness (h):

$$E = 0.8\left(\frac{k_w}{h}\right)\left(\frac{R}{h}\right)^{1.5} \tag{5}$$

Since the slope of the approach portion of the force curve provides information on sample stiffness, the spring constant determined from the equation 4 can also be used to determine the elastic modulus of round-shaped fungal spores using the following equation [43]:

$$k_w = 2E\left(\frac{A}{\pi}\right)^{1/2} \tag{6}$$

where E is the elastic modulus of the spore and A is the contact area between the AFM tip and sample. The contact area between an AFM tip and spore sample can be determined from the following equation [44]:

$$A=\pi\left(2R\delta_p-\delta_p^{\,2}\right) \tag{7}$$

where δ_P is the indentation below the circle of contact calculated the from following equation:

$$\delta_P = \left(\frac{\delta_t-\delta_r}{2}\right) \tag{8}$$

where δ_t is the maximum indentation and δ_r is the residual depth of indentation. These values are determined experimentally from the force versus distance curve.

In the above section we have outlined how to quantify the elastic behavior of microbial cells, and most of the available literature describes bacteria elastic properties with Young's modulus. However, biological samples are not purely elastic but viscoelastic. Therefore, the microbial cell can be modeled as a combination of both properties.

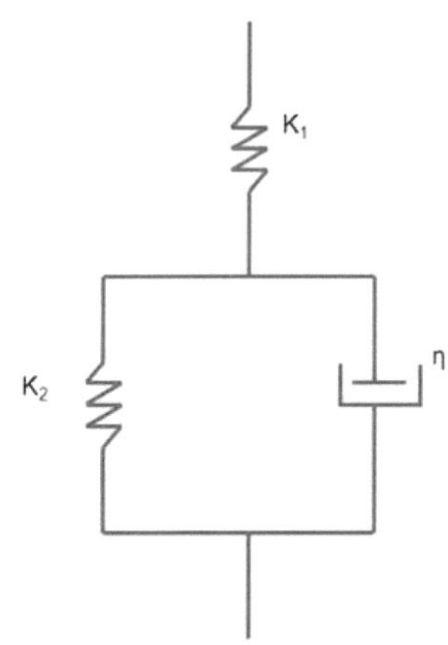

Figure 4. A schematic representation of standard solid model used to determine viscoelastic parameters. Adapted from [19] with permission.
k_1 is the instantaneous elastic response, k_2 is the delayed elastic response as a function of creep and η is the viscocity.

Vadillo-Rodrigue et al. (2009) [19] explained the viscoelastic properties of bacterial cell walls using a standard solid model which describes both an instantaneous and a delayed elastic deformation. Based on this model they have derived the following equation that describes the experimentally obtained creep response data:

$$Z(t) = \frac{F_0}{K_1} + \frac{F_0}{K_2}\left[1 - \exp\left(-t\frac{K_2}{\eta_2}\right)\right] \tag{9}$$

where, Z(t) is the position of the z piezoelectric transducer as a function of time t, K_1 is the spring constant that represents initial deformation, K_2 is the spring constant after creep response, and η_2 is viscosity.

The contribution of elastic and viscous components can be determined from the force-time curve taken at the center of cells when applying a constant force, F_0, for at least a 10 second period. Cantilever deflection is determined and using equation 1 is converted to force and then to an indentation-time curve, which is also called creep response. The indention of the cell over time at a constant force can be theoretically determined from equation 4 and fitted to the indentation-time curve shown in Figure 5. The experimentally determined data fit very well with the theoretical data obtained from the model. Microbial cells in particular exhibit two types of responses when a force is exerted on their surface. The first is the instantaneous linear relationship of the force versus distance curve, attributed to whole cell turgor pressure, while the non-linear region is thought to correspond to the response of the cell envelope.

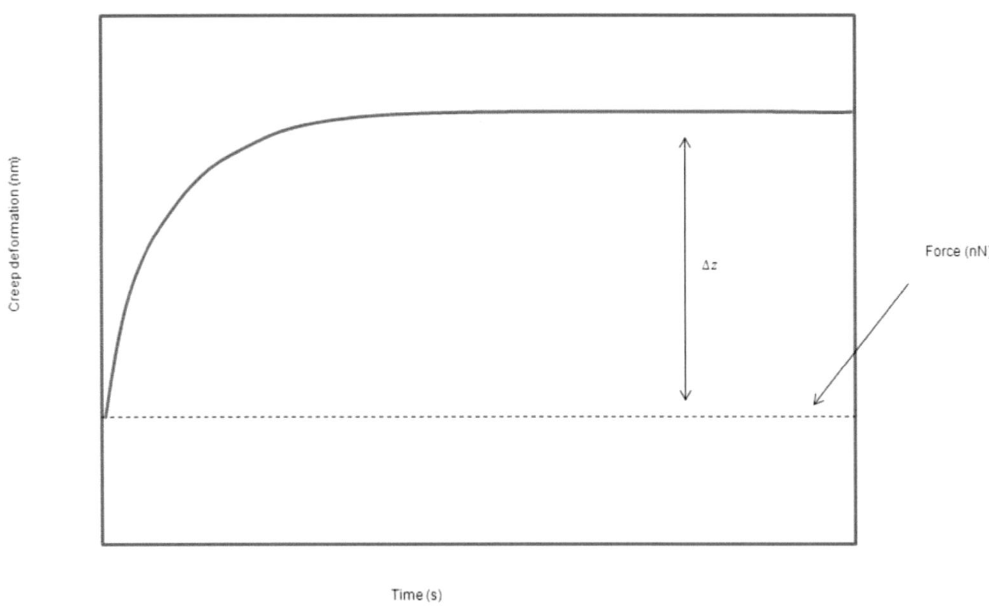

Figure 5. A typical creep deformation of an *Escherichia coli* cell at a constant force as a function of time. Adapted from [45] with permission.

AFM has been employed to measure the elasticity of a wide variety of cells ranging from bacteria, fungi, cancer cells, stem cells, osteoblasts, fibroblasts, leukocytes, cardiocytes developing embryos, cells at different cell cycle stages, and those treated with drugs. A broad spectrum of new measurements is possible by exploiting and manipulating the interaction between tip and sample in a quantitative way. Elasticity is most often measured with conical AFM tips. Spherical tips give rise to elasticity measurements 2-3 times that of conical tips, likely based on the large contact surface area. In comparison with other methods, AFM is more advantageous based on its ability to image the sample surface at high resolution while measuring an indentation map of the sample. The combination of imaging and force spectroscopy provides information about how cell surface structure affects elasticity and viscoelasticity. However, measurements depend on tip shape, which cannot be determined during sample scanning. Despite some limitations, AFM applications are rapidly developing. New instrumental designs and modification of the associated theoretical models will ensure an effective way to measure the elasticity and viscoelasticity for a wide variety of biological samples.

3. Viscoelastic cellular components and super structures

Although several methods have been developed to quantify cellular responses to deformation during locomotion, adhesion and mitosis, reliable tools are not available to quantify the distribution of mechanical forces between the various sub-cellular components [26]. Biological cells range in size between 1-100 μm and are comprised of constituents that provide mechanical strength, such as the cell envelope composed of multiple complex and distinct structures, cell walls composed primarily of polysaccharides interspersed with proteins, the cell membrane composed of phospholipid bilayers and membrane proteins, complex cell organelles of different sizes and shapes made of a variety of macromolecules, the cytoskeleton composed of microtubule networks, actin and intermediate filaments, other proteins and macromolecules such as DNA and RNA. The structure and function of each of these constituents may vary depending upon cell type. For instance, fungi are encased in cell walls, whereas bacteria have more elaborate cell envelopes with a peptidoglycan (polysaccharides cross-linked with peptides) layer and one or more cell membranes. Human cells, generally by virtue of being part of more elaborate structures, have only a cell membrane. It is not well understood how cells and their associated components sense mechanical forces or deformation, and convert such signals into biological responses [46].

The small size of prokaryotes, in comparison with larger eukaryotic cells, was a considerable obstacle in the development of methods for directly measuring their mechanical properties [47], solved largely by FS methods now routinely used. Cellular mechanical strength mainly relies on the outermost layers, such as the cell wall, envelope, or membrane, in addition to internal structural components such as the cytoskeleton. Extracellular components, such as those used to help form elaborate community structures (*e.g.* biofilms) also contribute to viscoelasticity and mechanical strength. There has been a major focus on the viscoelastic properties imbued to the cell by its cytoskeleton, which has been highly conserved throughout evolution and influences not only internal cell dynamics but overall function. Nonetheless, external cellular components also play a significant role in mechanics despite their limited study.

3.1. Cell envelopes, walls and membranes

The bacterial (prokaryotic) envelope is a structurally remarkable cell component that defines a cell from its external environment and serves a protective function. The envelope also helps maintain cell shape and mechanical integrity and is responsible for important biological functions, such as the selective transfer of material in and out of the cell, and necessary changes accommodating cell growth and division. The envelope has a multilayer geometry which withstands positive turgor pressure exerted by the cytoplasmic fluid from the cell interior, and so protects the cell from osmotic swelling and burst. Thus quantifying the cell envelope mechanical properties of prokaryotes that contribute to their mechanical strength is of interest.

The mechanical properties of the cell envelope relate to its structural composition, comprised of distinct layers made up of polysaccharides, lipids and proteins (Figure 6). Structural differences in the two different types of prokaryotic cell envelopes led to the classification of two distinct groups of bacteria namely, Gram-positive and Gram-negative. The former is named for the Gram stain retained by the thick peptidoglycan layer outside the cytoplasmic membrane (Figure 6A), while the latter having a relatively thin peptidoglycan layer sandwiched between inner cytoplasmic and outer membranes (Figure 6B) does not retain Gram stain. Many researchers have studied the flexibility of the bacterial cell envelope [48-50], for which the majority of its viscoelastic nature is attributed to the peptidoglycan layer.

Cell stiffness, required to maintain bacterial shape, is dependent on the cell envelope. When the peptidoglycan layer is removed from rod-shaped whole cells by chemical treatment, soft, highly deformable and osmotically sensitive spherical cells (spheroplasts) are the result, indicating that the elastic response of cells is largely dominated by the peptidoglycan matrix [51,52]. In isolation, peptidoglycan is very flexible, exhibiting purely elastic properties with a modulus of 25 MPa [53]. However, bacterial cell envelopes show a time dependent response to externally applied forces, meaning that their overall mechanical properties are more accurately described as viscoelastic [45].

Cell envelope composition makes a major contribution to viscoelasticity. The cell envelope of the Gram-positive bacterium *Bacillus subtilis* is significantly stiffer than the Gram-negative *E. coli*, attributed to the difference in their peptidoglycan layer thickness. FS measurements of local viscoelasticity for live bacterial cells show that the time required for *B. subtilis* to reach asymptotic creep deformation is higher than that of the Gram-negative *E. coli* and that covalent crosslinking increases cell envelope stiffness. The same study also showed a difference in the deformability of wild type *E. coli* (*lpp*+) and its Lpp mutant *lpp-* (Lpp a major peptidoglycan-associated lipoprotein and one of the most abundant outer membrane proteins in *E. coli* cells). Thus lipopolysaccharides, peptidoglycan thickness, the bound form of the peptidoglycan–lipoprotein complex and stabilizing cations all play an important role in maintaining viscoelasticity [19]. A micromanipulation study using optical fiber probes to test bursting strength also showed *E. coli* to have a weaker envelope than the Gram-positive *Staphylococcus epidermis*, more susceptible to mechanical stress [47]. Researchers thus attribute the elastic component of the cell envelope to the peptidoglycan layer and the viscous component to the liquid phase of the membranes [19,45]. The viscoelasticity of the bacterial cell envelope has also been shown to depend on its degree of hydration [19] and is thought to play an important role during cell division. During cell division, polymerization–depolymerization reactions in the FtsZ assembly cause softening and fluidization, reducing viscoelasticity and reflecting the more dynamic and active motion of individual FtsZ filaments in the lipid membrane [54].

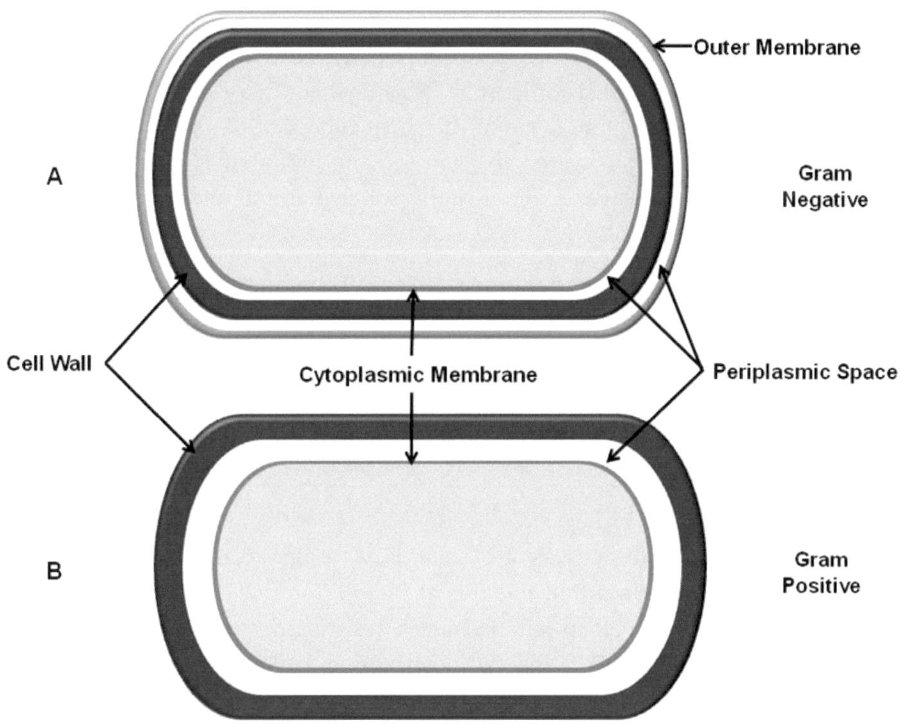

Figure 6. Simple models of (A) Gram-positive and (B) Gram-negative bacteria.

Earlier work on several fungi showed that the viscoelasticity of cell wall components allow for growth, cell division and spore germination [55-57]. The mechanical strength of fungal cells is largely attributed to their rigid but flexible cell walls which contain four major structural components: β-(1,3)-glucan, β-(1,6)-glucan, chitin (N-acetylglucosamine) and glycoproteins. The β-[1, 3]-glucan and β-[1,6]-glucan are more fibrous components whereas glycoproteins form a gel-like matrix, thought to impart viscoelasticity to the fungal cell wall. The mature cell wall has covalent cross-links formed between the chitin and glucan residues to provide the wall with sufficient strength and viscoelasticity. A study of the fungi *A. nidulans* showed that the immature hyphal wall region, having less well ordered wall components, and those devoid the cell wall component β-galactofuranose have lower viscoelasticity than wild type mature regions [58, 59] and this was conjectured to relate to reduced cross-linking. Determining the mechanical properties for cell walls of live fungal hyphae is challenging [60] and so there have been few studies measuring their mechanical properties.

Cell wall mechanical properties of the budding yeast *Saccharomyces cerevisiae* were determined through compression in a micromanipulator. Individual cells were compressed between two parallel surfaces and elastic parameters extracted from the force-deformation data [52]. A FS study of *Termitomyces clypeatus* showed an increase in cell wall rigidity and elasticity until the organism reached the stationary phase, followed by a decrease in these parameters at the onset of the death phase. The adsorption of chromium was also found to significantly affect the surface nano-mechanical properties of *T. clypeatus* [61].

Spores, another morphological form of fungi, appear in several studies describing their surface morphology [56,62,63] and adhesive properties [3,64] however, little information is available regarding their mechanical properties. Changes in the cell wall mechanical properties are a key factor in the emergence of the germ tube [55,56]. Both rodlet-covered and rodlet-free spores of *A. nidulans* were subjected to nano-indentation measurements by FS in air, showing that the rodlet layer is significantly softer than the underlying portion of the cell wall [43].

The nano-mechanical properties of whole human cells has been a topic of great interest for the past few decades and the viscoelastic properties of epithelial cells [4,65], stem cells [66], red cells [67,68] and cancer cells [69,70] are well-studied but beyond the scope of this chapter. The microbial cytoskeleton is less well studied, so the next section highlights significant data from isolated cytoskeletal components and those in the context of human cells.

3.2. Cell cytoskeleton

The filamentous network inside eukaryotic cells is a major contributor to the 3D morphology of a cell, acting as a scaffold to support the cell interior (Figure 7). This cross-linked biopolymer network has a role in cell mechanics, resisting deformation in response to external mechanical stresses. Besides being the determining factor for cellular shape, the cytoskeleton is involved in cell division, cell movement, adhesion and locomotion. Cellular viscoelastic responses can be largely dependent on the cytoskeleton, composed of three major groups of elements: microtubules, intermediate filaments and microfilaments. With the tremendous progress in biophysics, the structures of each cytoskeletal components are now well understood even at the molecular level, but we are just beginning to determine their contribution to cell mechanics.

3.2.1. Microtubules

Microtubules are the largest of the filamentous structures making up the cytoskeleton, for which the basic building block is tubulin heterodimers made up of α and β subunits. Structural analysis shows that the α and β tubulins alternately line up to form proto-filaments, which are further laterally arranged into a small lattice and closed to form a 25 nm wide cylindrical structure [71]. The α subunit forms the end of the microtubule localized to the centrosome, close to the center of the cell, and the microtubule grows out to peripheral regions of the cell ending with a β tubulin. The alternating α and β tubulins render the microtubule polar and its orientation guides the unidirectional movement of microtubule motor proteins from the kinesin and dynein families [72,73]. Microtubule aggregation is dependent on temperature and the relative amount of tubulin-GTP dimers [74]. The inherent dynamic instability caused by the hydrolysis of tubulin-GTP dimers puts microtubules in a state of continuous and rapid assembly and disassembly, depending on cell cycle phase. Given the limited generation of cellular microtubules, they are expected to play a minor role in cell mechanical responses and a major role in cell shape. Nevertheless, they do indirectly influence the mechanical properties of cells by regulating the actin network through myosin-II. In neural cells, axons, dendrites and microtubules play a more direct role in cell mechanics, because they form tight bundles in which microtubule-associated proteins (MAPs) bind and stabilize parallel arrayed microtubular filaments [75-77].

Experiments carried out on gels and cross-linked networks made of microtubules demonstrate that they exhibit the strongest bending stiffness among all the cytoskeletal components. Therefore, elasticity dominates over viscosity when microtubules are deformed under an external force. Under a low external load, microtubule network stiffness exhibits a linear elastic response as a function of microtubulin unit concentration [78], which closely relates to microtubule polymerization rate and the final microtubule length of tubulin dimers formed [79,80]. Fast growing microtubules are more likely to form short filaments, based on defects in the microtubule lattice that further influence the tubulin dimer bonds and make the microtubules more flexible. Microtubule viscoelasticity also depends on the intermolecular interactions between tubulin dimers, but their lateral and longitudinal interactions along the microtubule do not equally contribute to the total mechanical response. Both the shear and circumferential moduli of the longitudinal bonds in microtubule protofilaments are several orders of magnitude higher than those of lateral bonds. Since microtubules exhibit a mechanical response with enthalpic elasticity arising from the bending and stretching of microtubule filaments, factors influencing the tubulin spacings play a critical role in determining microtubule viscoelasticity [78]. Among these factors, the microtubule-associated proteins are an important contributor as they bind to the microtubule surface through electrostatic interactions. In the presence of MAPs, the spacing of adjacent microtubules is doubled creating a cushion against compression. There is also evidence to show that the elastic modulus increases with the addition of crosslinkers to the microtubule networks [81]. Factors affecting microtubule stiffness, such as taxol and GTP analogs, also influence their mechanical properties by stabilizing and preventing their depolymerization [82]. Therefore, microtubule viscoelasticity is a function of both subunit concentration and crosslinking.

3.2.2. Intermediate filaments

Intermediate filaments are the non-polarized cylindrical fibrils of the cytoskeleton named for their size with a diameter of around 10 nm, intermediate to microtubules and microfilaments [83,84]. Intermediate filaments are constructed by a group of related proteins, which have been divided into five subgroups according to their sequence identity. All of the proteins associated with type I, II, III, and IV immediate filaments are localized to the cytoplasm, while those of type V (*e.g.* lamins) localize to the cell nucleus and form a network underneath the nuclear membrane. Similar to microtubules, intermediate filaments are composed of protein dimers, but in this case elongated ones.

Intermediate filaments have been discovered in diverse cell types, with more than 60 associated genes coded in humans [85,86]. They are found to be highly extensible compared to filamentous actin (F-actin) and microtubules, which are able to retain enormous strains while sustaining the intact filament structure [87,88]. Their unique extensibility implies their it could play a special role in cell mechanics, which would set it apart from other cytoskeletal elements. Since most genes associated with intermediate filaments are responsible for coding keratins, keratin-based intermediate filaments are considered an important player in cellular mechanics [89,90], regulating viscoelastic properties and the motility of cancer cells. The architecture of the keratin filament is regulated by phosporylation, accounting for the viscoelastic responses of carcinoma cells during large deformations, and the actin network is also regulated by phosphorylation under the regulation of kinases [91-93], phosphatases and other regulatory proteins [89,94].

The mechanical properties of intermediate filament networks have been studied by applying classical models to homogeneous and isotropical samples, which link the elastic shear modulus to the mean mesh size of the cytoskeleton [95]. This means that the viscoelastic response of the network is dependent upon its subcellular organization, filament composition, and overall protein concentration. The nonlinear relationship between the mesh size of these networks and the elastic shear modulus [96] underscores whole cell mechanical properties, elasticity and viscoelasticity of the network, which can be significantly changed by small perturbations [97]. In the non-polar network, the intermediate filaments branch in an attempt to adapt to the cellular demands of the micro-environment [98].

The crosslinking interactions between proteins making up intermediate filaments are postulated to be vital for cell mechanics, inspiring the study of the associated protein tail. Mutation of the desmin proteins at the filament tail causes slight changes in network flexibility, but does not have a significant effect on the mesh size or shear modulus [99]. The crosslinking interaction may be expected to rely on electrostatic interactions between filament proteins, and by extension the salt concentration in their local micro-environment. Indeed, an increase in cations enhances the stiffness of the intermediate filament network. In the case of small external stresses, the elasticity of the network shifts from the linear to nonlinear state as a function of stress magnitude. On the other hand, under greater stress the network has a modified nonlinear elasticity [100]. Intermediate filament networks with a greater number of bundles have higher persistence lengths and flexural stiffness, in contrast to those with a lower number of bundles, demonstrating the importance of bundle number and thus protein type on cell elasticity.

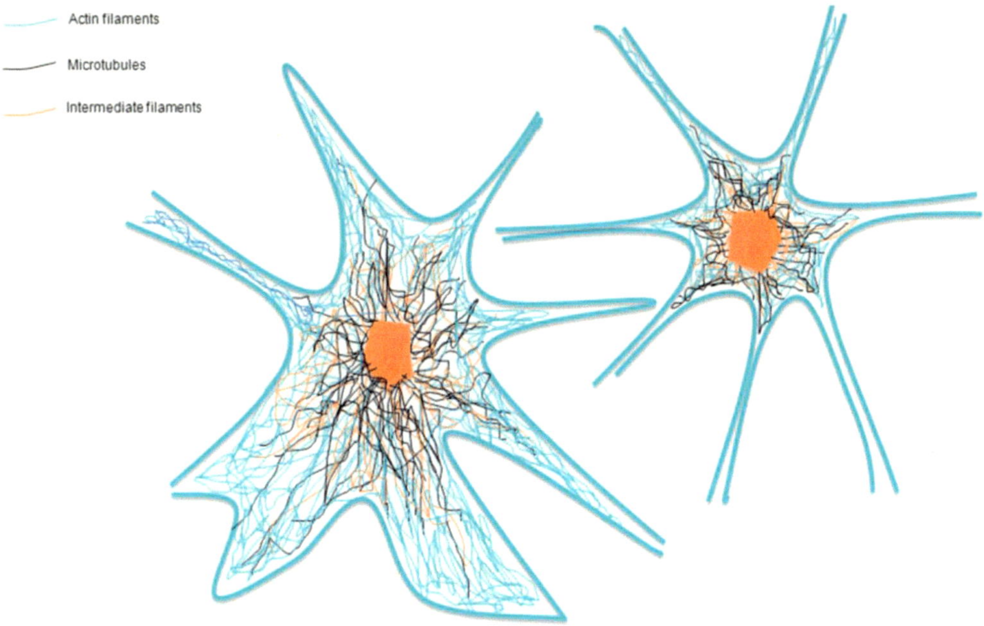

Figure 7. Schematic of a cytoskeleton as a dynamic 3-dimensional scaffold made up of actin filaments, intermediate filaments and microtubules.

3.2.3. Actin filaments

Actin filaments are the thinnest (≈ 8nm) and most plentiful among all the components of the eukaryotic cytoskeleton [101]. Under physiological conditions, actin filaments grow by polymerization of the ATP-bound actin monomers, addition of monomers to the end of the severed filament fragments, or by branching out of the existing filaments. Actin filaments are polar fibers, with helically arranged actin monomers all pointed in the same direction, which can form a cortex underneath the cytoplasmic membrane to support cell movement. More than 100 proteins promote the formation of a single actin filament, filament cross-linking, networks and bundle formation. Actin filaments are continuously assembled and disassembled in response to cell signaling.

F-actin exhibits viscoelasticity as a semi-flexible polymer with linear and nonlinear responses to external forces [102,103]. To quantitatively evaluate the contribution of F-actins to cytoskeleton mechanics, such networks have been reconstituted *in vitro* with purified protein. The reconstituted F-actin forms a gel-like network in which the mesh size is dependent on the F-actin concentration [104]. Under a low applied force, semiflexible networks only composed of F-actins deform in a manner proportional to the force load, but under a larger force which affects F-actin filament contour length, the deformation can be described as strain-stiffening [95]. The mechanical properties of F-actin filaments are affected by thermal fluctuations, which cause transverse bending, decreasing its end to end distance [105]. When an F-actin filament is under strain, fluctuations in the F-actin filament cause straightening with an associated consumption of energy. Thus, F-actin filaments exhibit an entropic elastic response. The magnitude of F-actin deformation reflects the reduction in the number of fluctuations in the actin filament [106].

Since the elastic response of F-actin filaments is length dependent, the concentration of actin and crosslinking proteins are closely related to the magnitude of the filament elastic modulus. The characterization of F-actin filament mechanical properties is also convoluted with actin concentration, the type and concentration of crosslinking protein, and the magnitude of the applied force. If the fluctuations are deconvoluted, the inherent elastic modulus of the filament dominates its mechanical response [106]. Crosslinking proteins covalently link individual subunits, allowing the actin filament to form a bundled structure with a larger diameter. Therefore, F-actin networks in combination with cross-linking proteins could behave differently from networks formed by only F-actins, the mechanical properties of which are more akin to the properties of cells. Once the actin filament becomes stiffer with the addition of crosslinkers, its elasticity dominates, and contributions from thermal fluctuations play a smaller role in the total mechanical response [106]. The entropic elasticity is reduced as actin and cross linker concentrations increase. In this case, the deformation of the actin filament depends on the bending and stretching of the filament itself [106]. The rigidly cross-linked actin filament has a more linear response and tends to break under small strains [107,108], and such filaments can sustain much larger stresses compared to pure F-actin networks, providing a reasonable explanation for their ability to resist a broad range of external stresses or internal tension.

3.3. Biofilms contribute to viscoelasticity

Biofilms have serious implications in industry, medicine and environmental systems making the study of their physical properties imperative to shed light on their growth mechanisms and adhesion properties, key factors in biofilm control. A biofilm is a surface-associated three dimensional consortia of microbial communities, surrounded by a matrix of protective biopolymers, macromolecular debris, sediment and precipitate, making it a highly complex mechanical, compliant, and viscoelastic structure. Biofilms can also be highly heterogeneous consisting of mixed populations of bacteria, fungi, protozoa [112] with interspersed pores and channels. Biofilms exhibit enormous resistance to external stress factors, with exopolysaccharides (EPS) contributing to their overall mechanical stability and enabling them to withstand external forces. Hydrodynamic forces can have a strong influence on biofilm formation, structure and thickness, EPS production, mass and metabolic activities [109-111]. Their dynamic structure ultimately affects how we view, model and study their mechanical behaviour. A schematic presentation of biofilm dynamic behavior is depicted in Figure 8.

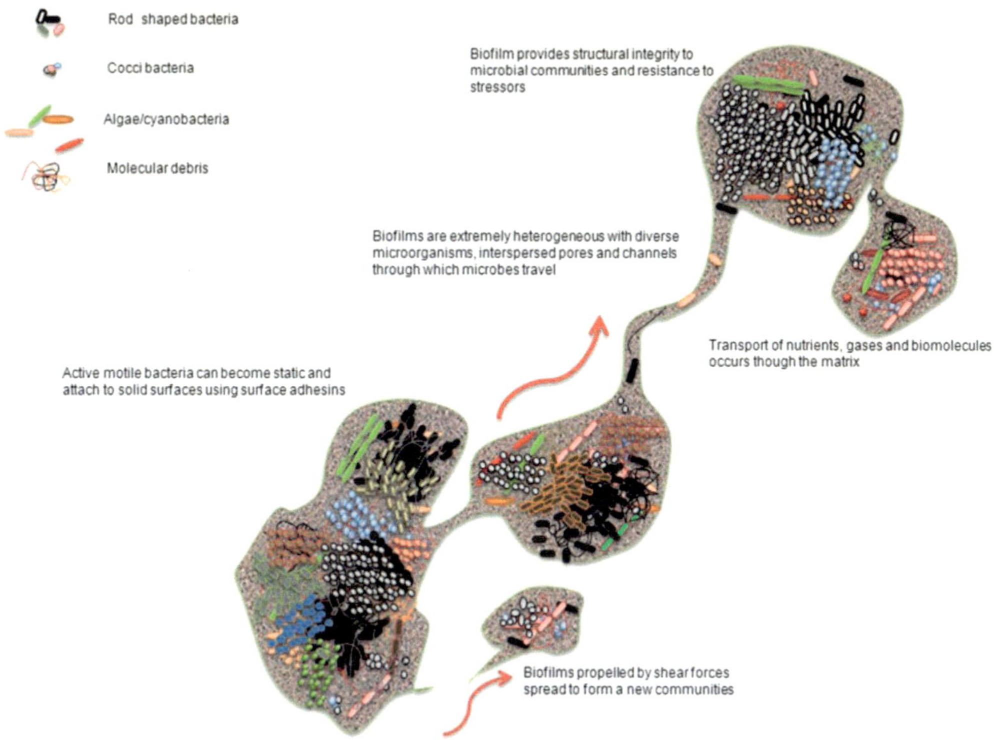

Figure 8. A model of biofilm growth and their dynamic behaviors.

Several attempts have been made to study the rheological properties of biofilms, yet they remain understudied, likely for the following reasons: (i) biofilms are extremely complex and heterogeneous with no defined geometry, making the use of conventional techniques difficult, (ii) their size and diverse nature make sample handling difficult and underscore the need to study biofilms in the natural environment to best represent their complexity (Figure 8). Studies have shown that biofilms formed under low shear conditions (laminar flow) are characterized by spherical micro-colonies divided by water channels [109]. It has also been shown that biofilms formed under higher detachment forces (*e.g.* shaking) produce more EPS to increase mechanical strength and to withstand shear stress [113].

A decade ago it was demonstrated that biofilms are effectively viscoelastic, a property that plays a major role in the various stages of biofilm growth [109,112,114-117]. Biofilms, all of which exhibit classic viscoelastic behaviours [109,115,116,118,119], can grow in a wide variety of environments including laboratory conditions, under flow and even in hot springs, demonstrating their ability to absorb elevated changes in shear stresses. Biofilms are elastic for shorter periods, can resist shear stresses and detachment, and over longer periods of time they flow as viscous fluids and become streamlined [109,112,114]. Viscoelasticity of the biofilm matrix has been shown to determine its structural integrity, resistance to stress, and ease of dispersion [120]. Simple stress-strain and creep tests of mixed culture and single species biofilms showed that the deformation of individual cell clusters was related to the fluid shear stress, and that both mixed species biofilms and those from pure cultures behaved like viscoelastic fluids [114]. It has been proposed that the viscoelasticity of biofilms allows them to resist detachment as demonstrated for *Staphylococcus aureus* biofilms [117]. Lieleg et al. [121] showed that high shear stress may transiently fragment the biofilm but does not cause it to detach from the surface. This resilience has been attributed to the viscoelastic behavior of the bacterial biofilm. Biofilms of various *Pseudomonas aeruginosa* strains subjected to physical deformations in fluid were shown to be viscoelastic fluids, which behave like elastic solids over periods of a few seconds but like linear viscous fluids over longer times. Therefore in several studies, bacterial biofilms have been reported to behave as viscoelastic materials [38,122,123], while in other studies they are described as elastic [124,125]. Chemical perturbations can reduce biofilm viscoelasticity and hence slow down recovery to their original state [126].

Several experimental models have been generated to help understand the rheology and mechanical behaviour of biofilms. Rheology of undisturbed dental plaque biofilms made up of *Streptococcus mutans* and those from pond water showed a linear viscoelastic behaviour for which the Burger model was successfully applied to study creep compliances [127]. There are a large number of studies in the literature that have used different techniques to measure various material properties of biofilms, each of which provide information about their mechanics. Some of the most common methods used to measure the tensile strength of biofilms include cone and plate rheometry [128] and later the centrifugation method [125]. Particle-tracking microrheology has been successfully used to measure the strength of single species biofilms of *S. aureus* and *P. aeruginosa* [129]. The strength and apparent viscoelastic modulus of *P. aeruginosa* biofilms grown on membrane filters has also been investigated using a uniaxial compression experimental device and a film rheometer [118]. Directly applied and controlled loading forces have been used to quantify various biofilm viscoelasticity parameters, usually without a hydrodynamic flow, for example microbead FS [120], the micro-cantilever technique [130,131], indenters [122] or T-shaped probes [75], which are used to pull (tensile testing under a normal load) or push (compression testing under a normal load) the biofilm (see review 38 and references therein). On the other hand, there are several methods that use hydrodynamic loading, where biofilms are subjected to a fluid flow in flow cells [109,112,132], or Couette–Taylor type reactors [133,134]. Real time monitoring of cell growth and proliferation corresponding to viscoelasticity changes within a biofilm have been investigated in *Streptococcus mutans* biofilms using a Quartz crystal microbalance with a dissipation monitoring device [135].

Although there have been a large number of techniques developed to measure various biofilm parameters, a fully effective method is pending. The diverse magnitude of viscoelasticity parameters might reflect diverse biofilm properties, growth environments and source organisms. The viscoelastic properties of biofilms are adaptations to stress factors and shear forces, and are achieved through modifications to the secreted EPS. A better understanding of biolfilm mechanical properties and viscoelastic behaviours may inform effective strategies for biofilm removal or control.

4. Factors affecting viscoelasticity of biological materials

Cell components imparting viscoelastic properties to the entire cell are well studied, but there are very few examples in the literature in which external factors are shown to affect cell viscoelasticity. The structural integrity of biological systems is partially dependent on the degree of hydration. Humidity affects the mechanical properties of biopolymers, but the intensity of this effect depends on the type of biopolymer. For example, effects of humidity on cellulose are much less severe than on peptidoglycan. Thwaites et al. [136] demonstrated that the viscoelastic behavior of *Bacillus subtilis* depends on humidity, which affects the viscoelasticity of the peptidoglycan layer. Increasing humidity gives rise to cell wall hydration, allowing water to form hydrogen bonds with peptides and other cell wall components, making the cell wall more pliable [136]. Bacterial and fungal spores are very rigid under dry conditions, but increased humidity leads to removal of the outer spore hydrophobic layer, making it softer than under dry conditions and leading to changes in viscoelasticity and eventually germination [20, 136].

Cells also require optimal temperature conditions for their survival. Temperature affects the proper function and conformation of biomolecules, and hence indirectly affects cell mechanical properties. The rigidity of *E. coli* has been found to increase as a function of temperature, attributed to the folding of lipoproteins in the outer membrane leading to an increase in turgor pressure [136]. An earlier study showed that a temperature sensitive mutant of *B. subtilis* (indole^{-} and thymine^{-}) was converted to a spherical shape from the wild type rod shape, attributed to loosening of the cell wall strength [137]. Hochmuth et al. [7] demonstrated that for red blood cells the time dependent viscoelastic behavior was reduced with increasing temperature.

The elastic behavior of the bacterial cell surface depends on the cell volume and ionic strength of the surrounding medium, which is related to its osmolarity. Abu-Lail and Camesano [138] observed that the elasticity of *E. coli* increased with reduced solvent polarity, and that bacteria in the least polar solvent have the highest Young's modulus [138]. Further, the spring constant of bacteria in a high ionic strength solvent of is higher than that in low ionic strength.

5. Summary

In summary, the viscoelastic properties of biological structures are responsible for their mechanical behavior which in turn is required for normal cell function. Viscoelasticity of whole biological cells is the combined contribution of cellular components, and several creative methods have been put forward to measure the associated parameters. Emerging experimental tools enable quantitative deformation studies of individual cells, biological polymers and macromolecules, which have led to understanding the relationship between mechanical properties and function. Nonetheless, studying the mechanical behavior of cellular components remains challenging. There are several theoretical models to determine the mechanical properties of cells and their components, but based on diverse findings and the different cell types used to determine viscoelastic parameters, it is difficult to compare cell rheology measurements. There are uncertainties associated with the methods developed to determine mechanical properties, and any single method cannot be used for all cell types. To date, atomic force microscopy appears to be the most effective method for measuring the viscoelasticity of biological materials.

It is certain that cellular viscoelasticity plays a great role in normal cell function such as cellular homeostasis, cell-cell communication, stress adaptive mechanisms, tissue formation, and locomotive functions. The most basic requirement of cells is their mechanical strength, which has potentially led to the evolution of complex multicellular organization in higher animals and even molecular evolution in the most primitive prokaryotes. Vital components of any given cell, such as its envelope, cytoskeleton and EPS, are devoted to maintaining a unique viscoelasticity, making the significance of this property in biological systems of great importance. Considering the diversity of living cells in nature, viscoelasticity remains universal, making its study exceptionally important, but nonetheless the study of cellular viscoelasticity remains in its infancy. The contribution of the cell cytoskeleton to viscoelasticity remains the most well studied, but there are still unresolved issues regarding its contribution, such as how mechanical force propagates through the cell cytoskeleton without a change in its composition. Although it is widely thought that the mechanical characteristics of cell components are significantly affected by external physical factors, our knowledge in this area is inadequate.

6. Conclusions and outlook

Through constant adaptation and survival, cells have acquired sophisticated structures made up of simple biomolecules which have remarkable mechanical integrity. Recent progress in the development of novel experimental techniques provides almost unlimited opportunities in the field of cell mechanics. By applying the basic rheology principles of non-living materials to live cells we can establish strong connections between cellular mechanics and function. More emphasis on the viscoelastic materials of cells such as the cell membrane, wall, envelope, and elaborate structures adopted by multiple cells including biofilms and tissues, will provide further insight into their contribution to cell mechanics. Combining the powerful experimental techniques discussed in this chapter, the wealth of knowledge from biochemistry with theoretical models (not discussed here, see [139] for a review) will allow us to further explore the importance of cellular viscoelasticity. In future, the resolution of several remaining gaps will lead to a fundamental and novel understanding of cellular function associated with cytomechanics.

Author details

Supriya Bhat, Dong Jun, Biplab C. Paul and Tanya E. S Dahms*
Department of Chemistry and Biochemistry, University of Regina, Regina, SK, Canada

* Corresponding Author

Acknowledgement

This work was supported by National Science and Engineering Research Council (NSERC) grants to TESD (228206-07). SB, JD and BCP were partially supported by the Faculty of Graduate Studies and Research (University of Regina). The authors wish to thank all those researchers, past and present, who have directly or peripherally helped pioneer our current understanding of cellular mechanics.

7. References

[1] Janmey PA. The cytoskeleton and cell signaling: component localization and mechanical coupling. Physiol Rev 1998 Jul;78(3):763-781.

[2] Hoffman BD, Crocker JC. Cell mechanics: dissecting the physical responses of cells to force. Annu Rev Biomed Eng 2009 Apr; 11:259-288.

[3] Bowen WR, Lovitt RW, Wright CJ. Direct Quantification of Aspergillus niger Spore Adhesion in Liquid Using an Atomic Force Microscope. J Colloid Interface Sci 2000 Aug;228(2):428-433.

[4] Trepat X, Lenormand G, Fredberg J, J. Universality in cell mechanics. Soft Matter 2008 Jul; 4:1750-8.

[5] Hoffman BD, Crocker JC. Cell mechanics: dissecting the physical responses of cells to force. Annu Rev Biomed Eng 2009 Apr; 11:259-288.

[6] Guilak F, Alexopoulos LG, Haider MA, Ting-Beall HP, Setton LA. Zonal uniformity in mechanical properties of the chondrocyte pericellular matrix: micropipette aspiration of canine chondrons isolated by cartilage homogenization. Ann Biomed Eng 2005 Oct;33(10):1312-1318.

[7] Hochmuth RM. Micropipette aspiration of living cells. J Biomech 2000 Jan;33[1]: 15-22.

[8] Thoumine O, Ott A. Time scale dependent viscoelastic and contractile regimes in fibroblasts probed by microplate manipulation. J Cell Sci 1997 Sep;110 (Pt 17)(Pt 17):2109-2116.

[9] Fernandez P, Ott A. Single cell mechanics: stress stiffening and kinematic hardening. Phys Rev Lett 2008 Jun;100(23):238102.

[10] Balland M, Desprat N, Icard D, Fereol S, Asnacios A, Browaeys J, et al. Power laws in microrheology experiments on living cells: Comparative analysis and modeling. Phys Rev E Stat Nonlin Soft Matter Phys 2006 Aug;74(2 Pt 1):021911.

[11] Guck J, Ananthakrishnan R, Mahmood H, Moon TJ, Cunningham CC, Kas J. The optical stretcher: a novel laser tool to micromanipulate cells. Biophys J 2001 Aug;81(2):767-784.

[12] Freundlich H, Seifriz W. Ober die Elastizitat von Solen und Gelen. Zritschr. f. physik. Chem., 1923 Jan; 104:233-28.

[13] Wang N, Butler JP, Ingber DE. Mechanotransduction across the cell surface and through the cytoskeleton. Science 1993 May;260(5111):1124-1127.

[14] Wirtz D. Particle-tracking microrheology of living cells: principles and applications. Annu Rev Biophys 2009 Jun;38:301-326.

[15] Tseng Y, Kole TP, Wirtz D. Micromechanical mapping of live cells by multiple-particle-tracking microrheology. Biophys J 2002 Dec;83(6):3162-3176.

[16] Crocker JC, Valentine MT, Weeks ER, Gisler T, Kaplan PD, Yodh AG, et al. Two-point microrheology of inhomogeneous soft materials. Phys Rev Lett 2000 Jul;85(4):888-891.

[17] Ashkin A. Forces of a single-beam gradient laser trap on a dielectric sphere in the ray optics regime. Biophys J 1992 Feb;61(2):569-582.

[18] Pullarkat PA, Fernandez PA, Ott A. Rheological properties of the eukaryotic cell cytoskeleton. Physics Repoarts 2007 Mar;449:29-22.

[19] Vadillo-Rodriguez V, Schooling SR, Dutcher JR. In situ characterization of differences in the viscoelastic response of individual gram-negative and gram-positive bacterial cells. J Bacteriol 2009 Sep;191(17):5518-5525.

[20] Zhao L, Schaefer D, Xu H, Modi SJ, LaCourse WR, Marten MR. Elastic properties of the cell wall of Aspergillus nidulans studied with atomic force microscopy. Biotechnol Prog 2005 Jan-Feb;21(1):292-299.

[21] Bidhendi AJ, Korhonen RK. A finite element study of micropipette aspiration of single cells: effect of compressibility. Comput Math Methods Med 2012 Oct;192618.

[22] Evans E, Yeung A. Apparent viscosity and cortical tension of blood granulocytes determined by micropipet aspiration. Biophys J 1989 Jul;56(1):151-160.

[23] Guck J, Ananthakrishnan R, Mahmood H, Moon TJ, Cunningham CC, Kas J. The optical stretcher: a novel laser tool to micromanipulate cells. Biophys J 2001 Aug;81[2]:767-784.

[24] Ebert S, Travis K, Lincoln B, Guck J. Fluorescence ratio thermometry in a microfluidic dual-beam laser trap. Optics Express 2007 Nov;15(23):15493-6.

[25] Wottawah F, Schinkinger S, Lincoln B, Ananthakrishnan R, Romeyke M, Guck J, et al. Optical Rheology of Biological Cells Physical Review Letters 2005 Mar;94:098103-3.

[26] Ou-Yang HD. Design and applications of oscillating optical tweezers for direct measurements of colloidal forces. In: Farinato RS, Dubin PL, editors. Colloid-Polymer Interactions: From Fundamentals to Practice. New York: Wiley; 1999. p. 385-20.

[27] Heilbronn A. Eine.neue Methode zur Bestimmung der Viskositat lebender Protoplasten. Jharb. f. wissensch. Botan., 1922;61,:284-38.

[28] Vonna L, Wiedemann A, Aepfelbacher M, Sackmann E. Local force induced conical protrusions of phagocytic cells. J Cell Sci 2003 Mar;116(Pt 5):785-790.

[29] Puig-de-Morales-Marinkovic M, Turner KT, Butler JP, Fredberg JJ, Suresh S. Viscoelasticity of the human red blood cell. Am J Physiol Cell Physiol 2007 Aug;293(2):C597-605.

[30] Amblard F, Maggs AC, Yurke B, Pargellis A, Leibler S. Subdiffusion and Anomalous Local Viscoelasticity in Actin Networks. Phys Rev Lett 1996 Nov;77(21):4470-4473.

[31] Bausch AR, Ziemann F, Boulbitch AA, Jacobson K, Sackmann E. Local measurements of viscoelastic parameters of adherent cell surfaces by magnetic bead microrheometry. Biophys J 1998 Oct;75(4):2038-2049.

[32] Valberg PA, Albertini DF. Cytoplasmic motions, rheology, and structure probed by a novel magnetic particle method. J Cell Biol 1985 Jul;101(1):130-140.

[33] Wang N, Ingber DE. Probing transmembrane mechanical coupling and cytomechanics using magnetic twisting cytometry. Biochem Cell Biol 1995 Jul-Aug;73(7-8):327-335.

[34] Zaner KS, Valberg PA. Viscoelasticity of F-actin measured with magnetic microparticles. J Cell Biol 1989 Nov;109(5):2233-2243.

[35] Tseng Y, Lee JS, Kole TP, Jiang I, Wirtz D. Micro-organization and visco-elasticity of the interphase nucleus revealed by particle nanotracking. J Cell Sci 2004 Apr;117(Pt 10):2159-2167.

[36] Yamada S, Wirtz D, Kuo SC. Mechanics of living cells measured by laser tracking microrheology. Biophys J 2000 Apr;78(4):1736-1747.

[37] Lee JS, Panorchan P, Hale CM, Khatau SB, Kole TP, Tseng Y, et al. Ballistic intracellular nanorheology reveals ROCK-hard cytoplasmic stiffening response to fluid flow. J Cell Sci 2006 May;119(Pt 9):1760-1768.

[38] Kapper I, Rupp CJ, Cargo R, Purvedorj B, Stoodley P. Viscoelastic fluid description of bacterial biofilm material properties. Biotechnol bioeng 2002 Nov;8: 190-7.

[39] Paul BC, El-Ganiny AM, Abbas M, Kaminskyj SG, Dahms TE. Quantifying the importance of galactofuranose in Aspergillus nidulans hyphal wall surface organization by atomic force microscopy. Eukaryot Cell 2011 May;10(5):646-653.

[40] Sneddon NI. The relation between load and penetration in the axisymmetric boussinesq problem for a punch of arbitrary profile. Int Eng Sci 1965;3(1):47-19.

[41] Wright JC, Armstrong I. The application of atomic force microscopy force measurements to the characterisation of microbial surfaces. Surf Interface Anal 2006 Oct;38:1419-9.

[42] Volle CB, Ferguson MA, Aidala KE, Spain EM, Nunez ME. Spring constants and adhesive properties of native bacterial biofilm cells measured by atomic force microscopy. Colloids Surf B Biointerfaces 2008 Nov;67(1):32-40.

[43] Zhao L, Schaefer D, Marten MR. Assessment of elasticity and topography of Aspergillus nidulans spores via atomic force microscopy. Appl Environ Microbiol 2005 Feb;71(2):955-960.

[44] Field JS, Swain MV. A simple predictive model for spherical indentation. J. Mater. Res., 1993 Feb ;8(2):297-9.

[45] Vadillo-Rodríguez V, Dutcher JR. Viscoelasticity of the bacterial cell envelope. Soft Matter 2011 Sep;7:4101-7.

[46] Zhu C, Bao G, Wang N. Cell mechanics: mechanical response, cell adhesion, and molecular deformation. Annu Rev Biomed Eng 2000 Aug;2:189-226.

[47] Shiu C, Zhang Z, Thomas CR. A comparison of the mechanical properties of different bacterial species Applied Microbiology 2002;2(4):155-6.
[48] Marquis RE. Salt-induced contraction of bacterial cell walls. J Bacteriol 1968 Mar;95(3):775-781.
[49] Mendelson NH, Thwaites JJ. Cell wall mechanical properties as measured with bacterial thread made from Bacillus subtilis. J Bacteriol 1989 Feb;171(2):1055-1062.
[50] Ou LT, Marquis RE. Electromechanical interactions in cell walls of gram-positive cocci. J Bacteriol 1970 Jan;101(1):92-101.
[51] Lederberg J. Bacterial Protoplasts Induced by Penicillin. Proc Natl Acad Sci U S A 1956 Sep;42[9]:574-577.
[52] WEIBULL C. The isolation of protoplasts from Bacillus megaterium by controlled treatment with lysozyme. J Bacteriol 1953 Dec;66(6):688-695.
[53] X. Yao X, Jericho M, Pink D, Beveridge T. Thickness and Elasticity of Gram-Negative Murein Sacculi Measured by Atomic Force Microscopy. Bacteriol 1999 Nov;181(22):6865-9.
[54] Lopez-Montero I, Mateos-Gil P, Sferrazza M, Navajas PL, Rivas G, Velez M, et al. Active membrane viscoelasticity by the bacterial FtsZ-division protein. Langmuir 2012 Mar;28(10):4744-4753.
[55] Borderd J, Trincia PJ. Fine structure of the germination of Aspergillus nidulans conidia. Transactions of the British Mycological Society 1970;54:143-3.
[56] Money NP. Osmotic adjustment and the role of turgor in mycelial fungi. In: Wessels JHG, Meinhardt F, editors. The Mycota: a Comprehensive Treatise on Fungi as Experimental Systems for Basic and Applied Research, vol. 1, Growth, Differentiation and Sexuality, Berlin: Springer; 1994. p. 67-21.
[57] Touhami A, Hoffmann B, Vasella A, Denis FA, Dufrene YF. Aggregation of yeast cells: direct measurement of discrete lectin-carbohydrate interactions. Microbiology 2003 Oct;149(Pt 10):2873-2878.
[58] Ma H, Snook LA, Kaminskyj SG, Dahms TE. Surface ultrastructure and elasticity in growing tips and mature regions of Aspergillus hyphae describe wall maturation. Microbiology 2005 Nov;151(Pt 11):3679-3688.
[59] Paul BC, El-Ganiny AM, Abbas M, Kaminskyj SG, Dahms TE. Quantifying the Importance of Galactofuranose in Aspergillus nidulans Hyphal Wall Surface Organization by Atomic Force Microscopy. Eukaryot Cell 2011 May;10(5):646-653.
[60] Kaminskyj SG, Dahms TE. High spatial resolution surface imaging and analysis of fungal cells using SEM and AFM. Micron 2008 Jun;39(4):349-361.
[61] Das SK, Das AR, Guha AK. Structural and nanomechanical properties of Termitomyces clypeatus cell wall and its interaction with chromium(VI). J Phys Chem B 2009 Feb ;113(5):1485-1492.
[62] Dufrene YF. Direct characterization of the physicochemical properties of fungal spores using functionalized AFM probes. Biophys J 2000 Jun;78(6):3286-3291.

[63] Dute RR, Peterson CM, Rushing AE. Ultrastructural Changes of the Egg Apparatus Associated with Fertilization and Proembryo Development of Soybean, *Glycine max*. Ann Bot 1989 Jan;64:123-12.

[64] van der Aa Kuhle A, Jesperen L, Glover RL, Diawara B, Jakobsen M. Identification and characterization of Saccharomyces cerevisiae strains isolated from West African sorghum beer. Yeast 2001 Aug;18(11):1069-1079.

[65] Alcaraz J, Buscemi L, Grabulosa M, Trepat X, Fabry B, Farre R, et al. Microrheology of human lung epithelial cells measured by atomic force microscopy. Biophys J 2003 Mar;84(3):2071-2079.

[66] Pelled G, Tai K, Sheyn D, Zilberman Y, Kumbar S, Nair LS, et al. Structural and nanoindentation studies of stem cell-based tissue-engineered bone. J Biomech 2007 Mar;40(2):399-411.

[67] Puig-de-Morales-Marinkovic M, Turner KT, Butler JP, Fredberg JJ, Suresh S. Viscoelasticity of the human red blood cell. Am J Physiol Cell Physiol 2007 Aug;293(2):C597-605.

[68] Cooper RA. Influence of increased membrane cholesterol on membrane fluidity and cell function in human red blood cells. J Supramol Struct 1978 Feb;8(4):413-430.

[69] Suresh S. Biomechanics and biophysics of cancer cells. Acta Biomater 2007 Jul;3(4):413-438.

[70] Li QS, Lee GY, Ong CN, Lim CT. AFM indentation study of breast cancer cells. Biochem Biophys Res Commun 2008 Oct;374(4):609-613.

[71] Valiron O, Caudron N, Job D. Microtubule dynamics. Cell. Mol. Life Sci. 2001 Dec;58:2069-24.

[72] Böhm JK, Stracke R, Mühlig P, Unger E. Motor protein-driven unidirectional transport of micrometer-sized cargoes across isopolar microtubule arrays. Nanotechnology 2001 Aug;12:238-6.

[73] Iyadurai SJ, Li MG, Gilbert SP, Hays TS. Evidence for cooperative interactions between the two motor domains of cytoplasmic dynein. Curr Biol 1999 Jul;9(14):771-774.

[74] Olmsted JB, Borisy GG. Ionic and nucleotide requirements for microtubule polymerization in vitro. Biochemistry 1975 Jul;14(13):2996-3005.

[75] Chen J, Kanai Y, Cowan NJ, Hirokawa N. Projection domains of MAP2 and tau determine spacings between microtubules in dendrites and axons. Nature 1992 Dec;360(6405):674-677.

[76] Weisshaar B, Matus A. Microtubule-associated protein 2 and the organization of cellular microtubules. Journal of Neurocytology 1993 Sep;22:727-7.

[77] Qiang L, Yu W, Andreadis A, Luo M, Baas PW. Tau protects microtubules in the axon from severing by katanin. J Neurosci 2006 Mar;26(12):3120-3129.

[78] Yang Y, Lin J, Kaytanli B, Salehbc OA, Valentine MT. Direct correlation between creep compliance and deformation in entangled and sparsely crosslinked microtubule networks. Soft Matter 2011 Dec;8:1776-8.

[79] Janson ME, Dogterom M. A bending mode analysis for growing microtubules: evidence for a velocity-dependent rigidity. Biophys J 2004 Oct;87(4):2723-2736.

[80] Pampaloni F, Lattanzi G, Jonas A, Surrey T, Frey E, Florin EL. Thermal fluctuations of grafted microtubules provide evidence of a length-dependent persistence length. Proc Natl Acad Sci U S A 2006 Jul;103(27):10248-10253.
[81] Lin Y, Koenderink GH, MacKintosh FC, Weitz DA. Viscoelastic Properties of Microtubule Networks. Macromolecules 2007 Sep;40:1714-6.
[82] Hawkins T, Mirigian M, Selcuk Yasar M, Ross JL. Mechanics of microtubules. J Biomech 2010 Jan;43(1):23-30.
[83] Renner W, Franke WW, Schmid E, Geisler N, Weber K, Mandelkow E. Reconstitution of intermediate-sized filaments from denatured monomeric vimentin. J Mol Biol 1981 Jun ;149(2):285-306.
[84] Zackroff RV, Goldman RD. In vitro assembly of intermediate filaments from baby hamster kidney (BHK-21) cells. Proc Natl Acad Sci U S A 1979 Dec;76(12):6226-6230.
[85] Hesse M, Magin TM, Weber K. Genes for intermediate filament proteins and the draft sequence of the human genome: novel keratin genes and a surprisingly high number of pseudogenes related to keratin genes 8 and 18. J Cell Sci 2001 Jul;114(Pt 14):2569-2575.
[86] Schweizer J, Bowden PE, Coulombe PA, Langbein L, Lane EB, Magin TM, et al. New consensus nomenclature for mammalian keratins. J Cell Biol 2006 Jul;174(2):169-174.
[87] Janmey PA. The cytoskeleton and cell signaling: component localization and mechanical coupling. Physiol Rev 1998 Jul;78(3):763-781.
[88] Fudge DS, Gardner KH, Forsyth VT, Riekel C, Gosline JM. The mechanical properties of hydrated intermediate filaments: insights from hagfish slime threads. Biophys J 2003 Sep;85(3):2015-2027.
[89] Coulombe PA, Omary MB. 'Hard' and 'soft' principles defining the structure, function and regulation of keratin intermediate filaments. Curr Opin Cell Biol 2002 Feb;14(1):110-122.
[90] Magin TM, Vijayaraj P, Leube RE. Structural and regulatory functions of keratins. Exp Cell Res 2007 Jun;313(10):2021-2032.
[91] Beil M, Braxmeier H, Fleischer F, Schmidt V, Walther P. Quantitative analysis of keratin filament networks in scanning electron microscopy images of cancer cells. J Microsc 2005 Nov;220[Pt 2]:84-95.
[92] Beil M, Eckel S, Fleischer F, Schmidt H, Schmidt V, Walther P. Fitting of random tessellation models to keratin filament networks. J Theor Biol 2006 Jul;241(1):62-72.
[93] Beil M, Micoulet A, von Wichert G, Paschke S, Walther P, Omary MB, et al. Sphingosylphosphorylcholine regulates keratin network architecture and visco-elastic properties of human cancer cells. Nat Cell Biol 2003 Sep;5(9):803-811.
[94] Strnad P, Windoffer R, Leube RE. Induction of rapid and reversible cytokeratin filament network remodeling by inhibition of tyrosine phosphatases. J Cell Sci 2002 Nov;115(Pt 21):4133-4148.

[95] Mackintosh C, Douglas P, Lillo C. Identification of a Protein That Inhibits the Phosphorylated Form of Nitrate Reductase from Spinach [Spinacia oleracea] Leaves. Plant Physiol 1995 Feb;107(2):451-457.

[96] Morse DC. Viscoelasticity of concentrated isotopic solutions of semiflexible polymers: 1. Model and stress tensor. Macromolecules 1998 Sep;31:7030-13.

[97] Fleischer F, Ananthakrishnan R, Eckel S, Schmidt H, Käs J, Svitkina T, et al. Actin network architecture and elasticity in lamellipodia of melanoma cells. New J. Phys. 2007 Nov;9:420-17.

[98] Heussinger C, Frey E. Stiff polymers, foams, and fiber networks. Phys Rev Lett 2006 Jan;96(1):017802.

[99] Bar H, Schopferer M, Sharma S, Hochstein B, Mucke N, Herrmann H, et al. Mutations in desmin's carboxy-terminal "tail" domain severely modify filament and network mechanics. J Mol Biol 2010 Apr;397(5):1188-1198.

[100] Lin YC, Yao NY, Broedersz CP, Herrmann H, Mackintosh FC, Weitz DA. Origins of elasticity in intermediate filament networks. Phys Rev Lett 2010 Feb;104(5): 058101.

[101] Otterbein LR, Cosio C, Graceffa P, Dominguez R. Crystal structures of the vitamin D-binding protein and its complex with actin: structural basis of the actin-scavenger system. Proc Natl Acad Sci U S A 2002 Jun;99(12):8003-8008.

[102] Gardel ML, Shin JH, MacKintosh FC, Mahadevan L, Matsudaira P, Weitz DA. Elastic behavior of cross-linked and bundled actin networks. Science 2004 May 28;304(5675):1301-1305.

[103] Lin CY, Koenderink GH, MacKintoshc FC, Weitz DA. Control of non-linear elasticity in F-actin networks with microtubules. Soft Matter 2011 Sep;7:902-4.

[104] Schmidt CF, Barmann M, Isenberg G, Sackmann E. Chain dynamics, mesh size, and diffusive transport in networks of polymerized actin: a quasielastic light scattering and microfluorescence study. Macromolecules 1989 Sep;22(9):3638-11.

[105] Kasza KE, Broedersz CP, Koenderink GH, Lin YC, Messner W, Millman EA, et al. Actin filament length tunes elasticity of flexibly cross-linked actin networks. Biophys J 2010 Aug 9;99(4):1091-1100.

[106] Janmey PA, Weitz DA. Dealing with mechanics: mechanisms of force transduction in cells. Trends Biochem Sci 2004 Jul;29(7):364-370.

[107] Gardel ML, Nakamura F, Hartwig J, Crocker JC, Stossel TP, Weitz DA. Stress-dependent elasticity of composite actin networks as a model for cell behavior. Phys Rev Lett 2006 Mar 3;96(8):088102.

[108] Tharmann R, Claessens MM, Bausch AR. Viscoelasticity of isotropically cross-linked actin networks. Phys Rev Lett 2007 Feb;98(8):088103.

[109] Stoodley P, Sauer K, Davies DG, Costerton JW. Biofilms as complex differentiated communities. Annu Rev Microbiol 2002 Jan;56:187-209.

[110] Liu Y, Tay JH. The essential role of hydrodynamic shear force in the formation of biofilm and granular sludge. Water Res 2002 Apr;36(7):1653-1665.

[111] Simoes T, Teixeira MC, Fernandes AR, Sa-Correia I. Adaptation of Saccharomyces cerevisiae to the herbicide 2,4-dichlorophenoxyacetic acid, mediated by Msn2p- and Msn4p-regulated genes: important role of SPI1. Appl Environ Microbiol 2003 Jul;69(7):4019-4028.

[112] Stoodley P, Wilson S, Hall-Stoodley L, Boyle JD, Lappin-Scott HM, Costerton JW. Growth and detachment of cell clusters from mature mixed-species biofilms. Appl Environ Microbiol 2001 Dec;67(12):5608-5613.

[113] Ohashi A, Harada H. Adhesion strength of biofilm developed in an attached-growth reactor. Water Science and Technology 1994;29(10):281-7.

[114] Stoodley P, Lewandowski Z, Boyle JD, Lappin-Scott HM. Structural deformation of bacterial biofilms caused by short-term fluctuations in fluid shear: an in situ investigation of biofilm rheology. Biotechnol Bioeng 1999 Oct;65(1): 83-92.

[115] Stoodley P, Lewandowski Z, Boyle JD, Lappin-Scott HM. Structural deformation of bacterial biofilms caused by short-term fluctuations in fluid shear: an in situ investigation of biofilm rheology. Biotechnol Bioeng 1999 Oct;65(1): 83-92.

[116] Klapper I, Rupp CJ, Cargo R, Purvedorj B, Stoodley P. Viscoelastic fluid description of bacterial biofilm material properties. Biotechnol Bioeng 2002 Nov;80(3):289-296.

[117] Lehtola MJ, Miettinen IT, Hirvonen A, Vartiainen T, Martikainen PJ. Estimates of microbial quality and concentration of copper in distributed drinking water are highly dependent on sampling strategy. Int J Hyg Environ Health 2007 Dec;210(6):725-732.

[118] Korstgens V, Flemming HC, Wingender J, Borchard W. Uniaxial compression measurement device for investigation of the mechanical stability of biofilms. J Microbiol Methods 2001 Jul;46(1):9-17.

[119] Towler BW, Rupp CJ, Cunningham AB, Stoodley P. Viscoelastic properties of a mixed culture biofilm from rheometer creep analysis. Biofouling 2003 Oct;19(5):279-285.

[120] Lau PC, Dutcher JR, Beveridge TJ, Lam JS. Absolute quantitation of bacterial biofilm adhesion and viscoelasticity by microbead force spectroscopy. Biophys J 2009 Apr;96(7):2935-2948.

[121] Lieleg O, Kayser J, Brambilla G, Cipelletti L, Bausch AR. Slow dynamics and internal stress relaxation in bundled cytoskeletal networks. Nat Mater 2011 Mar;10(3):236-242.

[122] Canse AW, Peters EAG, Gottenbose B, Baaijens EPT, Nuijs AM, Van Donjen MEH. Mechanical properties and failure of Streptococcus mutans biofilm studied using a microidentation device. J Microbio Methods 2006 Jul;(67):463-8.

[123] Lau AW, Hoffman BD, Davies A, Crocker JC, Lubensky TC. Microrheology, stress fluctuations, and active behavior of living cells. Phys Rev Lett 2003 Nov;91(19):198101.

[124] Aravas N, Laspidou CS. On the calculation of the elastic modulus of a biofilm streamer. Biotechnology and bioengineering 2008;101(1):196-3.

[125] Ohashi A, Koyama T, Syutsubo K, Harada H. A novel method for evaluation of biofilm tensile strength resisting erosion. Water sci technol 1999;39(7):261-7.

[126] Guélon T, Mathias JD, Stoodley P. Advances in biofilm mechanics. In: Hans-Curt, Wingender, Jost and Szewzyk, editor. *Biofilm Highlights.* Heidelberg, DE,: Springer; 2011. p. 111-38.

[127] Rheology of Biofilms. American Society of Civil Engineers 16th Engineering Mechanics Conference,; 2003.

[128] Picologlou BF, Zelver N, Characklis WG. Biofilm growth and hydraulic performance. J. Hydraul. Div., Am. Soc. Civ. Eng., 1980 May;106:733-12.

[129] Rogers SS, van der Walle C, Waigh TA. Microrheology of bacterial biofilms in vitro: Staphylococcus aureus and Pseudomonas aeruginosa. Langmuir 2008 Dec;24(23):13549-13555.

[130] Aggarwal S, Poppele EH, Hozalski RM. Development and testing of a novel microcantilever technique for measuring the cohesive strength of intact biofilms. Biotechnol Bioeng 2010 Apr;105(5):924-934.

[131] Poppele EH, Hozalski RM. Micro-cantilever method for measuring the tensile strength of biofilms and microbial flocs. J Microbiol Methods 2003 Dec;55(3):607-615.

[132] Stoodley P, Lewandowski Z, Boyle JD, Lappin-Scott HM. Structural deformation of bacterial biofilms caused by short-term fluctuations in fluid shear: an in situ investigation of biofilm rheology. Biotechnol Bioeng 1999 Oct;65(1): 83-92.

[133] Coufort C, Derlon N, Ochoa-Chaves J, Line A, Paul E. Cohesion and detachment in biofilm systems for different electron acceptor and donors. Water Sci Technol 2007;55(8-9):421-428.

[134] Rochex A, Godon JJ, Bernet N, Escudie R. Role of shear stress on composition, diversity and dynamics of biofilm bacterial communities. Water Res 2008 Dec;42(20):4915-4922.

[135] Schofield AL, Rudd TR, Martin DS, Fernig DG, Edwards C. Real-time monitoring of the development and stability of biofilms of Streptococcus mutans using the quartz crystal microbalance with dissipation monitoring. Biosens Bioelectron 2007 Oct;23(3):407-413.

[136] Thwaites JJ, Surana UC. Mechanical properties of Bacillus subtilis cell walls: effects of removing residual culture medium. J Bacteriol 1991 Jan;173(1):197-203.

[137] Cerf A, Cau JC, Vieu C, Dague E. Nanomechanical properties of dead or alive single-patterned bacteria. Langmuir 2009 May;25(10):5731-5736.

[138] Boylana RJ, Mendelson NH. Initial Characterization of a Temperature-Sensitive Rod$^-$ Mutant of *Bacillus subtilis.* J Bacteriol 1969 Dec;100(3):1316-5.

[139] Abu-Lail NI, Camesano TA. The effect of solvent polarity on the molecular surface properties and adhesion of Escherichia coli. Colloids Surf B Biointerfaces 2006 Aug;51(1):62-70.

[140] Lim CT, Zhou EH, Quek ST. Mechanical models for living cells--a review.. J Biomech 2006 Feb;39(2):195-21.

Chapter 7

Viscoelastic Properties of the Human Dermis and Other Connective Tissues and Its Relevance to Tissue Aging and Aging–Related Disease

Tetsuya Nemoto, Ryo Kubota, Yusuke Murasawa and Zenzo Isogai

Additional information is available at the end of the chapter

http://dx.doi.org/10.5772/50146

1. Introduction

Skin is the largest organ in humans and protects the body from environmental factors. The dermis is a layer that acts to protect the body from external physical force. Viscoelasticity is essential to facilitate the physical function of the skin. However, the clinical-biological-physical relevance of dermal connective tissue has not been fully investigated. In this chapter, we review past studies in this vast field and attempt to elucidate the combined clinical-biological-physical relevance of dermal connective tissue. In addition, we discuss Tarumi disease, which is characterized by impaired viscoelasticity and stiffness in connective tissues.

2. Extracellular matrices contribute to the viscoelastic properties of connective tissues including the dermis

The viscoelastic properties of human tissues are principally governed by the nature of the extracellular matrix (ECM). The ECM comprises secreted proteins that are deposited into the extracellular space. Although cell-ECM interaction and growth factor-ECM interactions have recently been investigated, the ECM represents the fundamental architecture of tissue. Some ECM proteins supply the connective tissues with hydration and resiliency through their characteristic molecular properties and assembly. ECM proteins such as collagen, proteoglycans, and glycoproteins are classified by their biochemical properties rather than their physical properties. Numerous ECM proteins are currently known and have been well characterized biologically. However, the contribution of each ECM protein to the viscoelastic properties of tissues has not been fully investigated due to the lack of a proper experimental system. Therefore, the contribution of each ECM protein to viscoelasticity in human tissue should be determined based on biological, physical, and clinical studies.

INTECH
open science | open minds

© 2012 The Author(s). Licensee InTech. This chapter is distributed under the terms of the Creative Commons Attribution License http://creativecommons.org/licenses/by/3.0), which permits unrestricted use, distribution, and reproduction in any medium, provided the original work is properly cited.

3. Connective tissue elements in the dermis: Collagen fiber, elastic fiber, and ground substance

As the dermis is the layer that protects the body from physical stress, understanding the nature of dermal connective tissues is vital. In this section, 3 components of connective tissues and the ECM are briefly introduced. ECM molecules are produced primarily by fibroblast cells.

Collagen fibers are major elements of the dermis and collagens are the most abundant protein in the human body; the dermis alone is composed of approximately 75% collagen proteins in dry weight. Twenty-eight collagen species have presently been identified. It has been reported that skin contains collagen types I, III, IV-VII, XIII, and XIV, with the major collagen in the dermis being type I collagen. Collagens that associate with the type I collagen fiber are classified as FACIT collagens and can provide additional mechanical properties to tissues. Collagens are characterized by repeated glycine-X-Y sequences and form triple-helical structures that are extensively modified after their secretion into the extracellular space. In immature tissues, such as those found in wound healing and fibrosis, type III collagen is expressed; however, it is not yet strong enough to support mature connective tissues. As the wound matures, type I collagen becomes dominant. Heterotypic type I and type III collagen fibrils are present in the dermis. Type VI collagen individually forms a unique filament called a microfilament (1,2).

Elastic fiber comprises elastin and microfibrils. As the dermis has to be stretched to adapt to the movement of body parts, elasticity is a critical property of the dermis. Elastin—a unique molecule that stretches and shrinks—is secreted as tropoelastin (the soluble precursor of mature elastin) and is subsequently processed and cross-linked within the extracellular space. Cross-linking by lysyl oxidase and desmosine formation is a crucial step for the stabilization of elastin within tissues. Another element in elastic fibers is fibrillin-microfibril. Microfibrils are fibrous elements that are 10 nm in width and are comprised mainly of fibrillins. Fibrillin is a large glycoprotein that is rich in cysteine residues and homotypically assembles into a microfibril in a well-regulated manner (3, 4). Fibrillins align in a parallel manner, from head to tail, in a staggered fashion within extracellular microfibrils (5). Other ECM molecules, including microfibril-associated glycoproteins (MAGPs), latent TGF-beta binding proteins (LTBPs), type XVI collagen, emilin, and versican, can associate with microfibrils through their binding affinity with fibrillins. Fibulins are yet another elastic fiber component, which can bridge elastin and microfibrils by their binding properties. Interestingly, fibulin-5 knock-out mice exhibit skin looseness (6, 7), indicating that this molecule may be essential for the development of elastic tissue.

Thick elastic fiber distributes horizontally in the reticular dermis, whereas thinner elastic fiber, including elaunin and oxytalan fibers, are seen to distribute in the papillary dermis. Oxytalan fibers are formed by bundled microfibrils without amorphous elastin. The staining of fibrillin in the skin shows horizontal distribution in the reticular dermis and vertical orientation in the papillary dermis (Figure 1). This complex elastic fiber meshwork confers the dermis with the ideal viscoelasticity to effectively protect the human body.

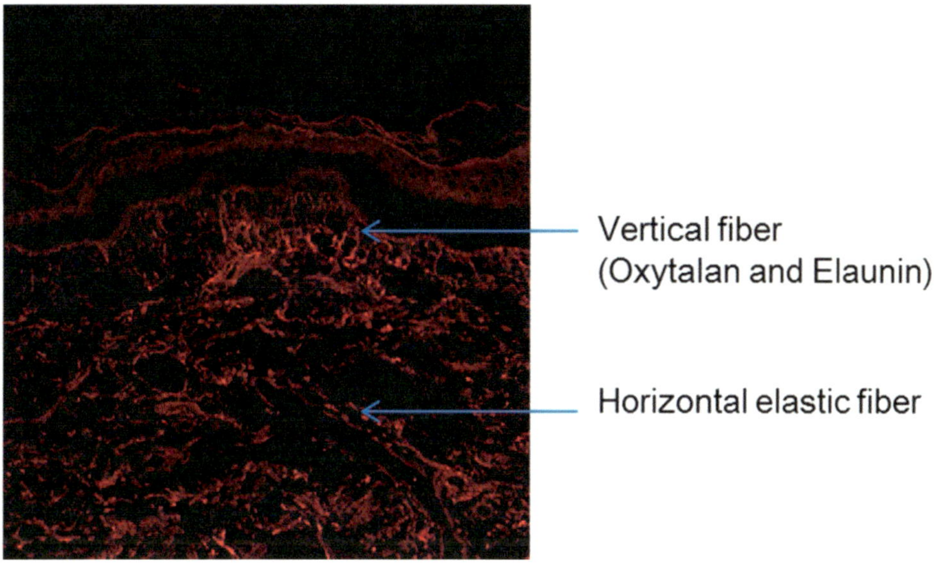

Figure 1. Immunohistochemical staining of fibrillin-1 in the human dermis. The distinct alignment of elastic fiber elements differs between the layers of the dermis

Proteoglycans are glycosaminoglycans (GAGs) that are covalently linked to a core protein. GAGs can be chondrotin/dermatan sulfate, heparan sulfate, heparin, and hyaluronan. GAGs hold a large amount of water within connective tissue, whereas free water in connective tissue is observed as edema. In particular, hyaluronan (HA) has a high affinity to water through its charge and, in general, is a high molecular weight linear GAG that distributes ubiquitously in connective tissue. Proteoglycans are major components of ground substance and are occasionally associated with fiber components in the dermis. Therefore, GAG is essential for maintaining the tissue viscosity of dermal connective tissue. In dermal connective tissues, decorin and versican are the major proteoglycans (8). Decorin is a small dermatan sulfate proteoglycan that binds to type I collagen. In the dermis, decorin is abundant in the papillary and the reticular dermis. Scott et al. have proposed that, through their charge, GAG chains of decorin can play a role in the viscoelastic property of connective tissue (9, 10). These models highlight the importance of GAG chains to the viscoelastic properties of connective tissues.

The supramolecular organization of the ECM in the dermis has been investigated using biochemical, biophysical, and ultrastructural methods. Connective tissues are not composed of a simple mixture of ECM molecules; therefore, the manner in which the ECM molecules assemble into fibrous components should be further investigated (11). In the dermis, each ECM molecule assembles into either elastic fiberous elements or ground substance. This can be observed in the electron micrograph of dermal connective tissue shown in Figure 2. Collagen and elastic fiber distribute distinctly, whereas the "empty space" is believed to be filled by ground substance. Thus, collagen and elastic fiber is embedded within ground substance, which itself is comprised of proteoglycans and hyaluronan.

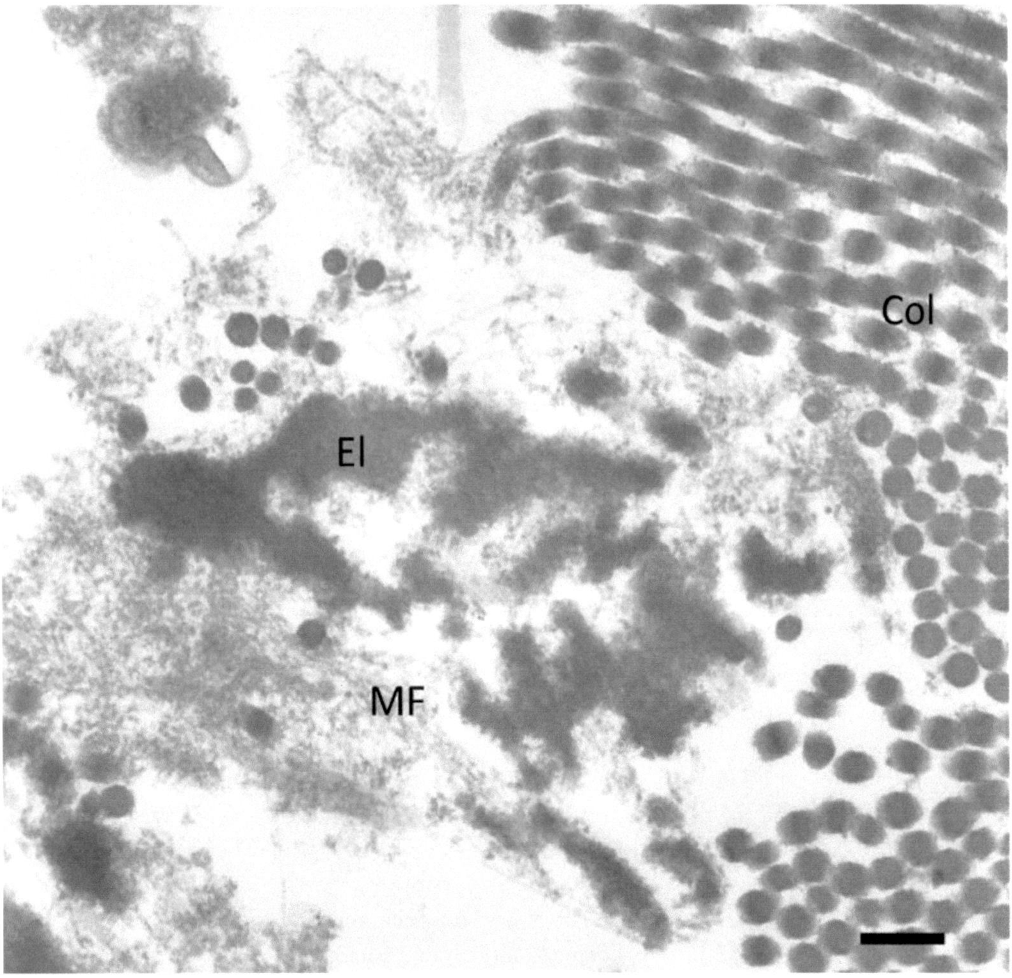

Figure 2. Electron microscopic image of dermal connective tissues. Col: collagen fiber, El: elastin core, MF: microfibrils (Bar = 200 nm)

4. Versican is a chondroitin sulfate proteoglycan and is critical for dermal viscoelasticity

Among the ECM molecules in the dermis, versican appears to be the most important molecule for tissue viscoelasticity. Versican (also called PG-M) is a large chondroitin sulfate proteoglycan that was originally characterized in a mesenchymal condensation in chick limb bud (12). Versican binds to hyaluronan via its amino-terminal G1 domain and to fibrillin-1, fibulin-1 and fibulin-2 via its carboxyl terminal G3 domain. The distribution of versican is similar to that of elastic fiber in the dermis and in other tissues (13, 14) and is immunolocalized to microfibrils through its binding affinity to fibrillin-1 (15) as observed in Figure 3.

Thus, versican plays critical roles in the viscoelastic properties of skin. Versican 1) connects with elastic microfibrils by binding to fibrillin via its G3 domain; 2) has chondroitin sulfate chains that hold a large amount of water within the ECM space; and 3) binds to HA, which holds a large quantity of water. Figure 4 shows the proposed structural model of the elastic-hydrated matrix in the dermis. Furthermore, the fibrillin-versican-hyaluronan network is also observed in the ciliary body (16).

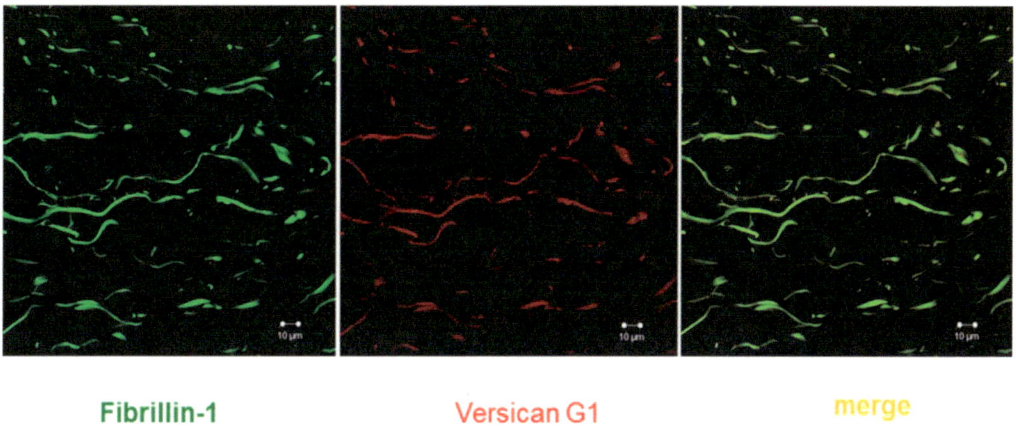

Figure 3. Versican co-localizes with fibrillin-1 in the dermis. Immunofluorescent staining using specific antibodies against versican and fibrillin-1 show co-localization.

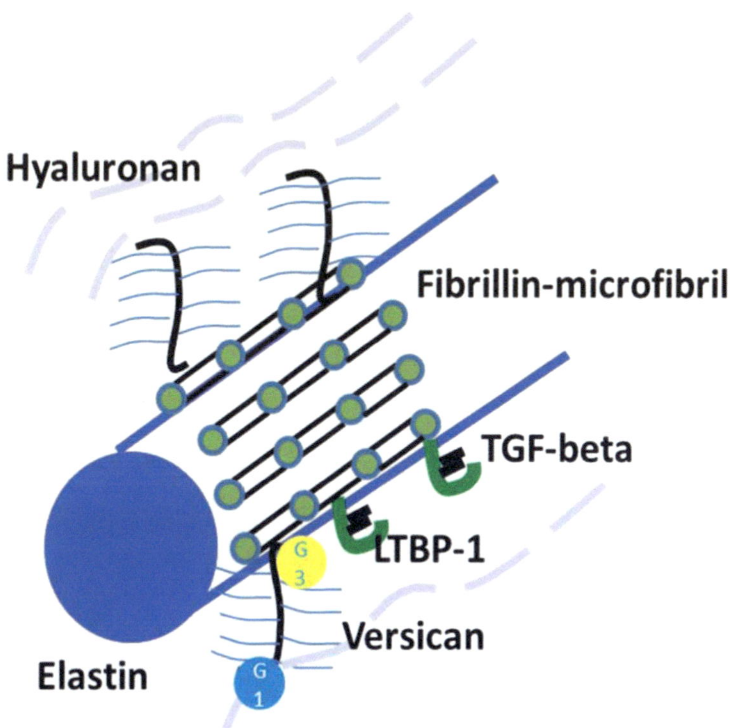

Figure 4. Schematic presentation of the dermal viscoelastic network linked by versican. Versican links elastic fibers to ground substance.

5. Biophysical examination of the skin and subcutaneous tissues

The fundamental role of human skin is to protect the body from invasion by external factors. Biological and chemical invasions of the body could be prevented by the skin, which include circulating cells of the innate immune system. Furthermore, protection from physical invasions—such as mechanical force and thermo injuries—are also important to maintain the homeostasis of the human body.

Skin consists of 3 layers, which include the epidermis, dermis, and subcutaneous tissue. Skin covers most of the body's surface, except for some "holes such as oral cavity". Thus, the physical barrier that skin provides is crucial to protect the human musculoskeletal system and internal organs. The physical properties of skin have been measured using several devices (17). In this study, the authors measured the mechanical properties of the skin by dynamic indentation. This study noted that the measurement of these mechanical properties by indentation is not well correlated with that by suction. (17). Furthermore, they also reported the aging-associated alteration of mechanical properties of the skin (18). The Cutemeter™ has been used to measure the viscoelasticity of skin. Additionally, we have recently established a novel method to measure the viscoelasticity of skin using a rheometer (AR instrument, AR 550) (Figure 5).

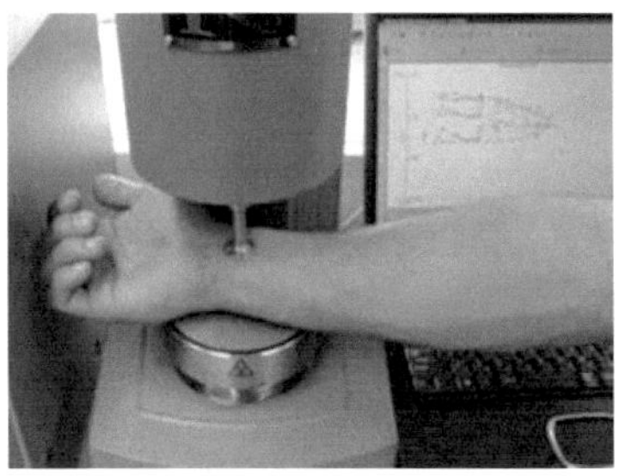

Figure 5. A rheometer can be used to determine the viscoelasticity of skin.

Using this method, skin is treated as a complex of different materials. The skin surface at the bottom of an appendage is immobilized so that deformity is only obtained by the external force generated from the upper probe. From the results shown in Figure 6, viscoelasticity of skin (and subcutaneous tissues) was estimated to be approximately 30 kPa. This data was not influenced by muscle contraction, thus indicating that the origin of the physical properties of skin could be the fascia (19).

Next, we developed a physical model for pressure ulcers and mechanical force around the ulcer was measured using our new device, real time skin strain monitor (RTSSM). A pressure ulcer is characterized as a skin and soft tissue injury caused by an external force on a bony prominence. However, it is not clear how a pressure ulcer is strained by external force. Previous studies have reported the similarity in strain properties of human soft tissue and industrial buffer materials. Therefore, we utilized a cell sponge as a testing material for its physically similar attributes to soft tissues. As shown in Figure 7, a physical model for pressure ulcers was developed. Strain gauge probes were stitched around the pressure ulcer model as indicated.

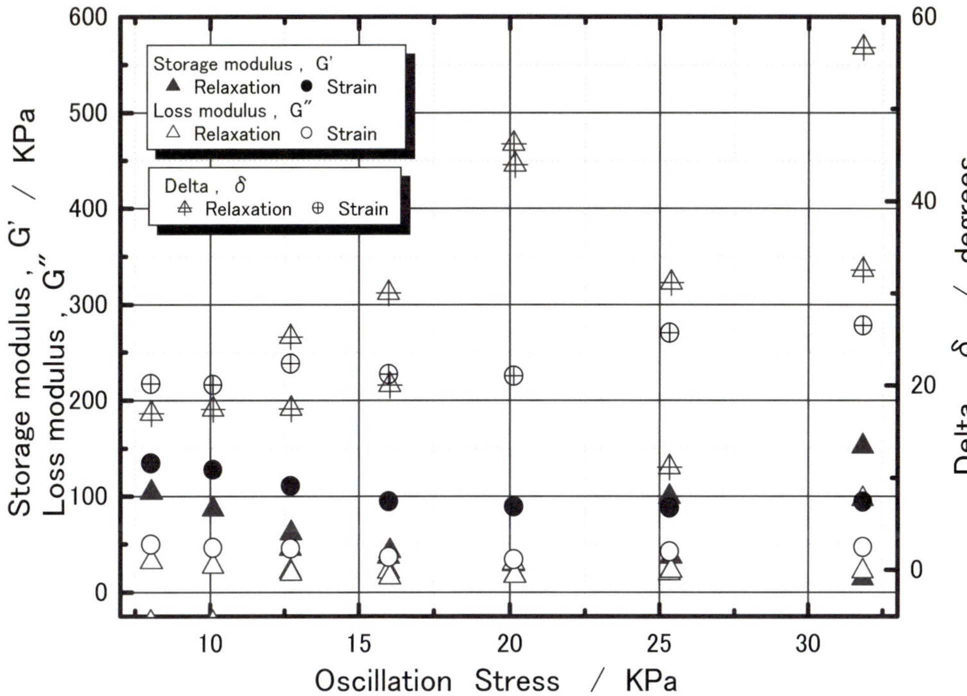

Figure 6. Viscoelasticity measurement by muscle contraction.

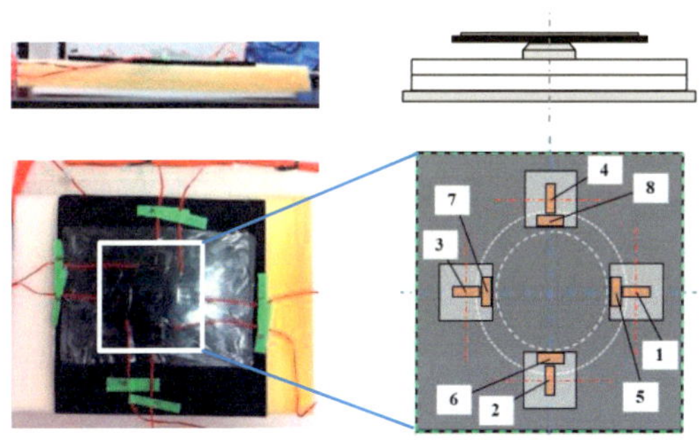

Figure 7. The composition of a pressure ulcer model and the positions of strain gauge probes. The physical model for the pressure ulcer is made of sponges; the probes are placed around the hole mimicking pressure ulcer.

Figure 8 shows the data observed from RTSSM when a tensile load is applied toward the channel 2–4 direction. From the results, it can be observed that this method is able to measure the strain force during the loaded state (0–0.3 seconds) and the relaxed state (after 0.3 seconds).

We further examined our model by testing the strain force around a pressure ulcer in a patient. This study was approved by the ethics committee of our institution and performed following written informed consent was obtained from the patients. As shown in Figure 9, the probe was adhered onto the dressing and the strained force was measured in the bedridden patients.

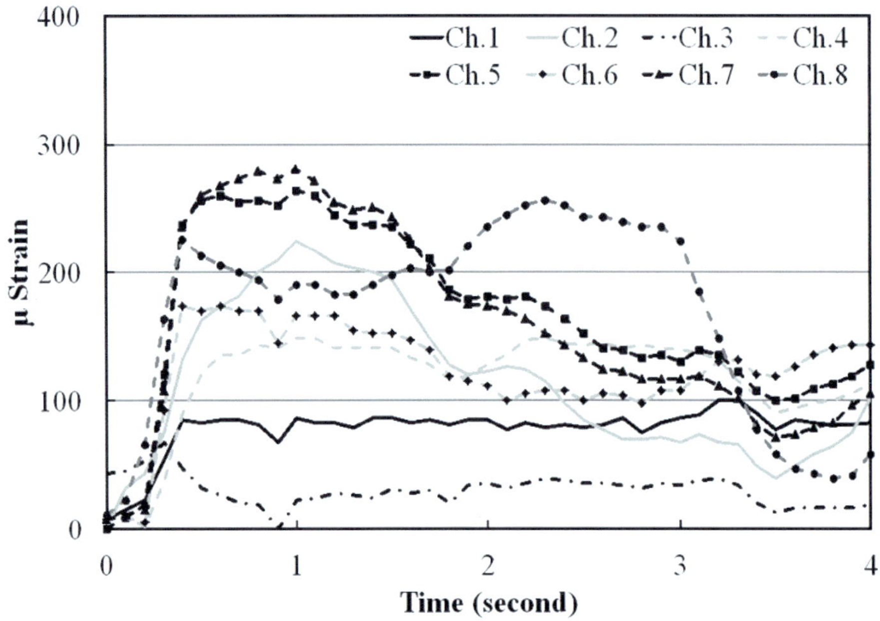

Figure 8. RTSSM measured by the strain distribution of a pressure ulcer model. The loading force is increased by 100 µ strain/s from the initial load at 50 µ strain/s and a maximum force of 250 µ strain/s is maintained.

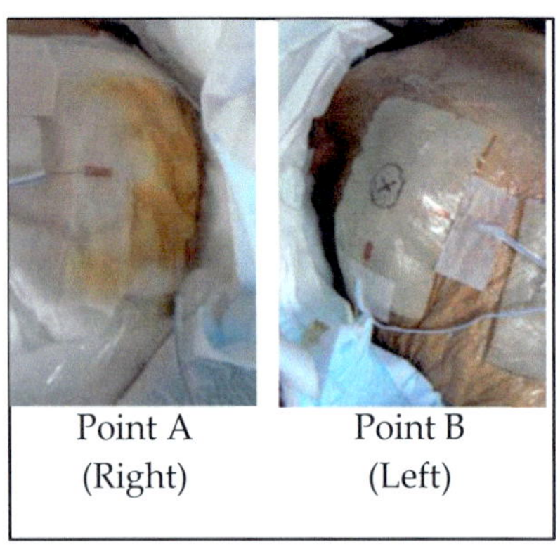

Figure 9. Measurement of strain force around pressure ulcers.

It has been noted that the head lifting position of the patient can occasionally worsen a pressure ulcer. Therefore, the manner in which positioning changes influence a pressure ulcer is an important issue for the care of a patient. To address this issue, we measured strain forces around the pressure ulcer during positioning changes. Measurement using RTSSM indicates that a positioning change can generate a strain force around the wound (Figure 10).

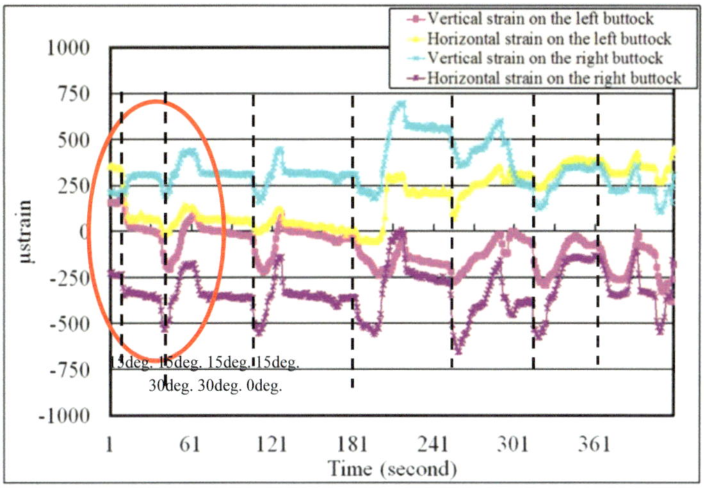

Figure 10. Changes in strain force at the buttocks are dependent on a positional change in head lifting.

Using the RTSSM, we next determined the direction of force by coordinating data from several probes. To this end, multiple probes can be adhered around the wound (Figure 11) and the measured force can be generated. In this case, it was reasoned that the different vectors, representing the strain force between the right and left sides, were generated due to the contracture of the right leg. Thus, the data obtained can be used to determine the positioning change that is ideal in the care of the patient with a pressure ulcer (20).

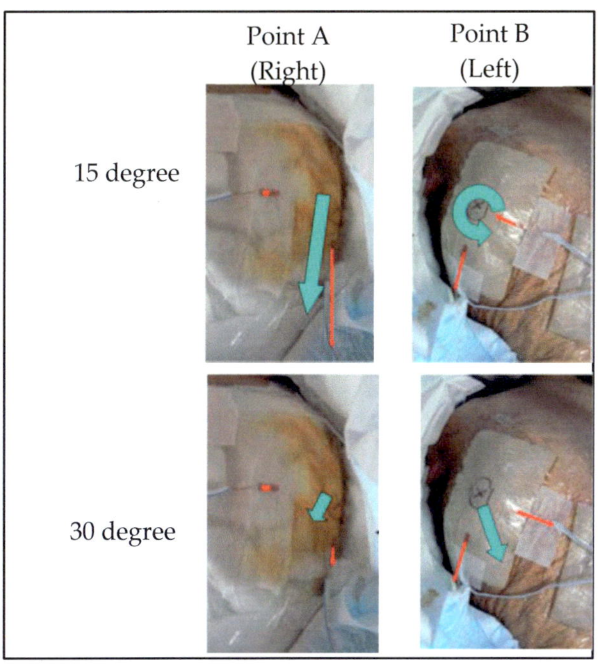

Figure 11. Positioning changes generate strain force on a pressure ulcer toward a specific direction. When the position of head is lifted at the indicated degree (15 or 30) a strain force is promptly changed.

6. Diseases caused by impaired viscoelastic properties of connective tissues

We discuss aging-associated diseases that result in impaired physical properties of connective tissues based on a review of genetic diseases that cause impairment in the physical properties of connective tissues.

Marfan syndrome (MFS) is a relatively common genetic connective tissue disease. MFS is an autosomal dominant connective tissue disease that affects the aorta, lungs, ciliary zonule, muscles, and other organs. However, most phenotypes appear only in the later stages of life. The primary cause of MFS appears to be due to mutations in the microfibrillar molecule fibrillin-1, although some phenotypes observed in various organs are believed to develop from the dysregulation of TGF-beta. One explanation for the genotype-phenotype correlation is due to the aberrant activation of TGF-beta stored within microfibrils through the binding between fibrillin-1 and LTBPs (21) (Figure 4). Recent studies have highlighted the importance of proper modulation of non-canonical TGF-beta signaling (22). The role of versican in MFS is currently unknown. Interestingly, the tissue phenotype resulting from MFS shows similarities to that of aging. For instance, aneurysm, emphysema, hernia, and muscle atrophy are all common features of MFS patients and also of elderly patients. However, the correlation between MFS and aging connective tissue phenotypes is currently unknown. MFS appears to be a model for impaired viscoelasticity of human tissues, which is discussed in the following section.

7. Aging-dependent changes of versican in the dermal connective tissues

The dermis changes prominently with age; for example, the thickness of the dermis becomes thin and wrinkles appear. Biochemical collagen content and histological density of collagen fiber is reduced (23). We have shown that versican is a key molecule for viscoelasticity of the dermis. The amount of versican extracted from the dermis decreases with age and its GAG composition is also altered (24, 25). Therefore, as described above, we hypothesize that loss or reduction of versican, or in the HA binding ability of versican, may lead to impaired viscoelasticity of the dermis. Versican is heavily accumulated within solar elastosis, which is a hallmark of photo-aged skin and where elastic fiber components, including elastin and fibrillin-1, have accumulated (26). Clinically, photo-aged skin is not viscoelastic and shows deep wrinkles, as observed in Figure 12.

Using recombinant versican G1 proteins and specific antibodies, we have indicated that a loss in the HA binding affinity of versican is characteristically observed in the region of solar elastosis (27). Versican specifically loses its HA binding domain (6084) in solar elastosis, whereas the carboxyl terminal domain (2B1) remains present. Therefore, the HA binding ability of elastic fibers is lost and microfibrils in solar elastosis are unable to bind to HA (Figure 13). Therefore, loss of the HA binding region of versican disrupts the fibrillin-versican-hyaluronan (Fi-Ver-Hy) network in the dermis.

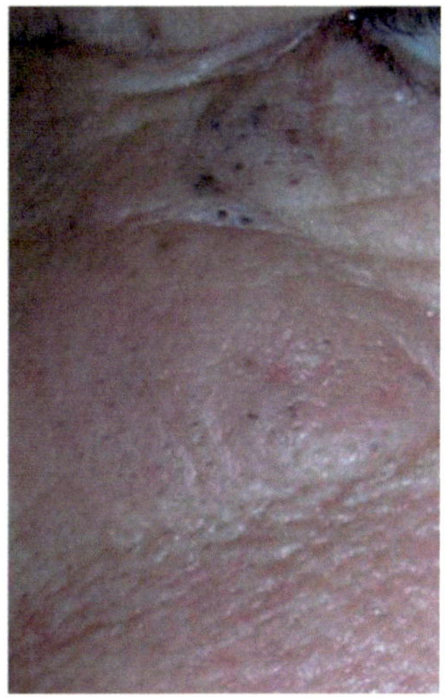

Figure 12. Clinical appearance of photo-aged skin. Deep wrinkles and comedo (black dots) are observed.

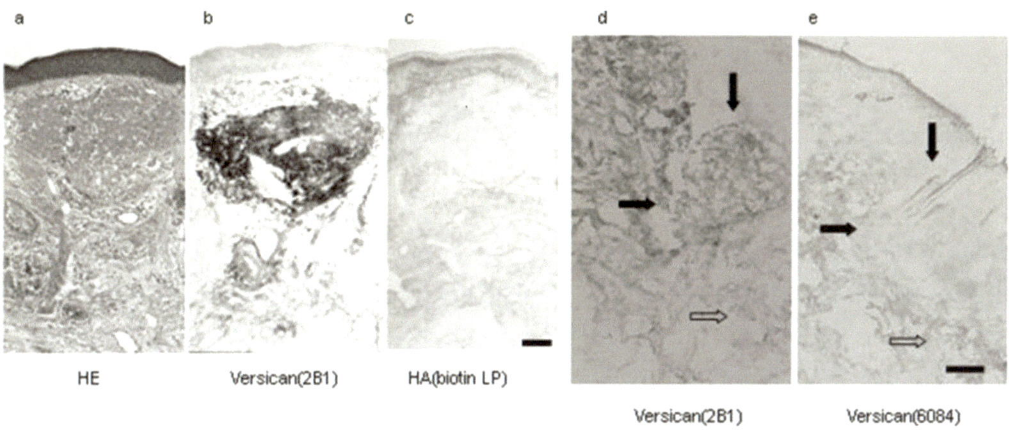

Figure 13. Histochemical results of solar elastosis. Versican is detected by its anti-fibrillin binding region (2B1) or its anti-HA binding region (6084). HA is detected using a biotin-conjugated link protein (biotin LP).

8. Tarumi disease

Based on the physical properties of skin and other connective tissues, we propose "Tarumi disease" as an aging-associated, connective tissue loosening disease. Tarumi diseases are preferentially found in the elderly population, with some exceptions. Tarumi is a Japanese word that represents tissue loosening. Aging-associated loosening of connective tissue is a major pathogenesis for emphysema, aneurysm, skin wrinkles, pelvic organ, and hernias. The Tarumi diseases that we are proposing are listed in the table below (Table 1). In 2001, an interesting association between pseudoexfoliation syndrome and abdominal aortic aneurysm was reported (28). However, it should be noted that this report has not been supported by the subsequent studies on the prevalence of these conditions.

Common pathogenesis among each Tarumi disease is currently unclear. Smoking is considered to be a precipitating factor in the common pathogenesis of aortic aneurysm and chronic obstructive pulmonary disease (COPD) (29, 30). However, other factors should be investigated for the Tarumi diseases. Tarumi disease may provide a novel perspective of tissue aging in geriatrics.

Therefore, studying Tarumi disease may be a useful step toward understanding common pathogenesis among these diseases. Future directions in Tarumi disease research require pathological, biochemical, and physical studies. Methods presented in this chapter may evaluate the looseness of tissues. Furthermore, surgical intervention using a tissue filler may be a useful method to improve these diseases. Finally, the phenotypical relationship between MFS and Tarumi disease may lead to understanding their common pathogenesis.

Organ	Proposed diseases
Aorta	Aortic aneurysm
Lung	Chronic obstructive pulmonary disease
Pelvis	Pelvis organ prolapse
Vein	Varicose vein
Skin	Wrinkles, Pressure ulcer
Abdomen	Hernia
Esophagus	Esophageal hiatal hernia

Table 1. Proposed Tarumi diseases.

9. Conclusion

Viscoelastic properties of the skin can be measured by various methods and are dependent on the connective tissue architecture formed by ECM molecules. In order to measure the actual viscoelasticity of the skin, we have developed a novel device that can monitor the external forces on the skin in real time. The device can be used for treatment and prevention of pressure ulcers that are affected by viscoelasticity and external force. Among the ECM molecules, versican—a chondroitin sulfate proteoglycan—is an important ECM molecule for viscoelasticity because it constitutes the fibrillin-versican-hyaluronan network. In human pathogenic conditions such as solar elastosis, loss of viscoelastic properties of the dermis is found to occur because of the loss of hyaluronan-binding versican. Marfan syndrome—a genetic connective tissue disease—is also characterized by loss of viscoelasticity in elastic tissues, such as those in the aorta. Finally, aging-associated loss of viscoelasticity and stiffness of connective tissue are proposed to be the common pathogenesis of Tarumi disease.

Author details

Tetsuya Nemoto and Ryo Kubota
Department of Gerontechnology, National Center for Geriatrics and Gerontology, Japan

Yusuke Murasawa and Zenzo Isogai
Department of Advanced Medicine, National Center for Geriatrics and Gerontology, Japan

Acknowledgement

Funding for this study was provided by The Research Funding for Longevity Sciences (21-18) from the National Center for Geriatrics and Gerontology (NCGG), Japan. We thank Hiroyuki Matsuura and Katsunori Furuta for their cooperation in this study. The authors declare that they have no conflicts of interest.

10. References

[1] Kielty, CM. and Shuttleworth, C. A. (1997). Microfibrillar Elements of the Dermal Matrix. Microsc. Res. Tech. 38, 407-427.
[2] Krieg T, and Aumailley M. (2011) The Extracellular Matrix of the Dermis: Flexible Structures with Dynamic Functions. Exp. Dermatol. 20, 689-695
[3] Sakai LY, Keene DR, Engvall E (1986) Fibrillin, A New 350-kD Glycoprotein, Is A Component of Extracellular Microfibrils. J Cell Biol. 103: 2499-2509.
[4] Hubmacher D, El-Hallous EI, Nelea V, Kaartinen, MT, Lee, ER., Reinhardt DP. (2008) Biogenesis of Extracellular Microfibrils: Multimerization of the Fibrillin-1 C Terminus into Bead-Like Structures Enables Self-Assembly. Proc. Natl. Acad. Sci. U. S. A. 105, 6548–6553.
[5] Kuo CL, Isogai Z, Keene DR, Hazeki N, Ono RN, Sengle G, Bächinger HP, Sakai LY. (2007) Effects of Fibrillin-1 Degradation on Microfibril Ultrastructure. J Biol Chem. 282: 4007-4020.
[6] Nakamura T, Lozano PR, Ikeda Y, Iwanaga Y, Hinek A, Minamisawa S, Cheng CF, Kobuke K, Dalton N, Takada Y, Tashiro K, Ross Jr J, Honjo T, Chien KR. (2002) Fibulin-5/DANCE is Essential for Elastogenesis in Vivo. Nature. 415: 171-175.
[7] Yanagisawa H, Davis EC, Starcher BC, Ouchi T, Yanagisawa M, Richardson JA, Olson EN. (2002) Fibulin-5 is an Elastin-Binding Protein Essential for Elastic Fibre Development in Vivo. Nature. 415: 168-171.
[8] Carrino DA, Sorrell JM, Caplan AI (2000) Age-Related Changes in the Proteoglycans of Human Skin. Arch Biochem Biophys. 373: 91-101.
[9] Scott JE (1995) Extracellular Matrix, Supramolecular Organisation and Shape. J Anat. 187: 259-269.
[10] Scott JE (2003) Elasticity in Extracellular Matrix 'Shape Modules' of Tendon, Cartilage, etc. A Sliding Proteoglycan-Filament Model. J Physiol. 553: 335-343.
[11] Keene DR, Marinkovich MP, Sakai LY (1997) Immunodissection of the Connective Tissue Matrix in Human Skin. Microsc Res Tech. 38: 394-406.
[12] Kimata K, Oike Y, Tani K, Shinomura T, Yamagata M, Uritani M, Suzuki S (1986) A Large Chondroitin Sulfate Proteoglycan (PG-M) Synthesized before Chondrogenesis in the Limb Bud of Chick Embryo. J Biol Chem. 261: 13517-13525.
[13] Zimmermann DR, Dours-Zimmermann MT, Schubert M, Bruckner-Tuderman L (1994) Versican is Expressed in the Proliferating Zone in the Epidermis and in Association with the Elastic Network of the Dermis. J Cell Biol. 124: 817-825.
[14] Bode-Lesniewska B, Dours-Zimmermann MT, Odermatt BF, Briner J, Heitz PU, Zimmermann DR (1996) Distribution of the Large Aggregating Proteoglycan Versican in Adult Human Tissues. J Histochem Cytochem. 44: 303-312.
[15] Isogai Z, Aspberg A, Keene DR, Ono RN, Reinhardt DP, Sakai LY (2002) Versican Interacts with Fibrillin-1 and Links Extracellular Microfibrils to Other Connective Tissue Networks. J Biol Chem. 277: 4565-4572.

[16] Ohno-Jinno A, Isogai Z, Yoneda M, Kasai K, Miyaishi O, Inoue Y, Kataoka T, Zhao JS, Li H, Takeyama M, Keene DR, Sakai LY, Kimata K, Iwaki M, Zako M (2008) Versican and Fibrillin-1 Form a Major Hyaluronan-Binding Complex in the Ciliary Body. Invest Ophthalmol Vis Sci. 49: 2870-2877.

[17] Boyer G, Laquièze L, Le Bot A, Laquièze S, Zahouani H (2009) Dynamic Indentation on Human Skin in Vivo: Ageing Effects. Skin Res Technol. 15: 55-67.

[18] Boyer G, Pailler Mattei C, Molimard J, Pericoi M, Laquieze S, Zahouani H (2012) Non Contact Method for in Vivo Assessment of Skin Mechanical Properties for Assessing Effect of Ageing. Med Eng Phys. 34: 172-178.

[19] Nemoto T, Isogai Z, Koide K, Itoh Y, Nogata F, Shimamoto A, Matsuura H (2007) Viscoelasticity Measurement of Skin in Vivo by Rheometer. J. Achievements in Mat. Manufactur. Eng. 21(2): 33-36

[20] Kubota R, Nemoto T, Ito Y, Isogai Z, Furuta K, Shimamoto, Matsuura H (2011) Examination of the Skin Characteristic Evaluation Using Skin Strain Measuring Method. Jpn. Soc. Exp. Mechanics. 11: 30-34 (in Japanese)

[21] Isogai Z, Ono RN, Ushiro S, Keene DR, Chen Y, Mazzieri R, Charbonneau NL, Reinhardt DP, Rifkin DB, Sakai LY. (2003) Latent Transforming Growth Factor Beta-Binding Protein 1 Interacts with Fibrillin and Is a Microfibril-Associated Protein. J Biol Chem. 24;278 :2750-2757.

[22] Holm TM, Habashi JP, Doyle JJ, Bedja D, Chen Y, van Erp C, Lindsay ME, Kim D, Schoenhoff F, Cohn RD, Loeys BL, Thomas CJ, Patnaik S, Marugan JJ, Judge DP, Dietz HC. (2011) Noncanonical TGFβ Signaling Contributes to Aortic Aneurysm Progression in Marfan Syndrome Mice. Science. 15: 358-361.

[23] Waller JM, Maibach HI (2006) Age and Skin Structure and Function, A Quantitative Approach (II): Protein, Glycosaminoglycan, Water, and lipid content and structure. Skin Res Technol. 12: 145-54.

[24] Carrino DA, Onnerfjord P, Sandy JD, Cs-Szabo G, Scott PG, Sorrell JM, Heinegård D, Caplan AI. (2003) Age-Related Changes in the Proteoglycans of Human Skin. Specific Cleavage of Decorin to Yield A Major Catabolic Fragment in Adult Skin. J Biol Chem. 278: 17566-17572.

[25] Carrino DA, Calabro A, Darr AB, Dours-Zimmermann MT, Sandy JD, Zimmermann DR, Sorrell JM, Hascall VC, Caplan AI (2011) Age-Related Differences in Human Skin Proteoglycans. Glycobiology. 21: 257-268.

[26] Bernstein EF, Fisher LW, Li K, LeBaron RG, Tan EM, Uitto J (1995) Differential Expression of the Versican and Decorin Genes in Photoaged and Sun-Protected Skin. Comparison by Immunohistochemical and Northern Analyses. Lab Invest. 72: 662-669.

[27] Hasegawa K, Yoneda M, Kuwabara H, Miyaishi O, Itano N, Ohno A, Zako M, Isogai Z (2007) Versican, A Major Hyaluronan-Binding Component in the Dermis, Loses its Hyaluronan-Binding Ability in Solar Elastosis. J Invest Dermatol. 127: 1657-1663.

[28] Schumacher S, Schlötzer-Schrehardt U, Martus P, Lang W, Naumann GO. (2001) Pseudoexfoliation Syndrome and Aneurysms of the Abdominal Aorta. Lancet. 357: 359-360.

[29] Aggarwal S, Qamar A, Sharma V, Sharma A (2011) Abdominal Aortic Aneurysm: A Comprehensive Review. Exp Clin Cardiol. ;16:11-5.

[30] Brusselle GG, Joos GF, Bracke KR (2011) New Insights into the Immunology of Chronic Obstructive Pulmonary Disease. Lancet. 378: 1015-1026.

Chapter 8

Dynamic Mechanical Response of Epithelial Cells to Epidermal Growth Factor

Jun Xi, Lynn S. Penn, Ning Xi, Jennifer Y. Chen and Ruiguo Yang

Additional information is available at the end of the chapter

http://dx.doi.org/10.5772/49977

1. Introduction

1.1. Mechanical properties of cells and their biological significances.

As a viscoelastic body, the cell exhibits both elastic and viscous characteristics (Kasza, 07). Although these mechanical properties have not been attributed wholly to a single element, such as the cytoskeletal network, the cytoplasm, the cell membrane, or the extracellular network (Janmey et al., 2007), it is agreed that they are determined predominantly by the cytoskeleton, a network of biopolymers in the form of actin filaments, microtubules, and intermediate filaments. The dynamic assembly and disassembly of these biopolymers give the cell the ability to move and to modulate its shape, elasticity, and mechanical strength in responses to mechanical and chemical stimuli from the external environment (Fletcher & Mullins, 2010). Among these cytoskeletal polymers, actin filaments are known to be primarily responsible for the rigidity of the cell. An increase in the concentration of actin filaments typically results in an increase in the rigidity of the cell, which can be characterized by Young's modulus (Satcher Jr & Dewey Jr, 1996).

The cytoskeleton is also essential in regulation of cell signaling and trafficking (Janmey, 1998; Papakonstanti & Stournaras, 2008). In particular, the structure of the cytoskeleton plays an essential role in EGFR signaling and trafficking that is initiated by the binding of epidermal growth factor (EGF) to the EGF receptor (EGFR) (Ridley, 1994; Song et al., 2008). EGF is a protein molecule known to play a crucial role in the regulation of cell growth, proliferation, differentiation and motility. EGFR is a transmembrane receptor that consists of an extracellular ligand-binding domain, a transmembrane domain, an intracellular tyrosine kinase domain, and a C-terminal regulatory domain (Scaltriti & Baselga, 2006). Binding of EGF to the extracellular domain of EGFR leads to the dimerization of EGFR, which in turn stimulates tyrosine kinase activity of the receptors and triggers autophosphorylation of

INTECH
open science | open minds

© 2012 The Author(s). Licensee InTech. This chapter is distributed under the terms of the Creative Commons Attribution License http://creativecommons.org/licenses/by/3.0), which permits unrestricted use, distribution, and reproduction in any medium, provided the original work is properly cited.

specific tyrosine residues within the cytoplasmic regulatory domain. The activation of tyrosine kinases initiates multiple downstream signaling pathways such as Ras/Raf-1/MAPK (Scaltriti & Baselga, 2006), PI3Kinase/Akt/mTOR (Ono & Kuwano, 2006), Src/NFKb (Lee C.-W. et al., 2007; Silva, 2004), catenin/cytoskeleton (Yasmeen et al., 2006) and PAK-1 /Rac pathways (McManus et al., 2000).

It is known that EGFR signaling induces drastic morphological changes, such as rounding of cells, induction of membrane ruffling and extension of filopodia (Bretscher, 1989; Chinkers et al., 1981). These changes can be attributed to the remodeling of cytoskeletal structures (Rijken et al., 1991), which may also alter mechanical properties of the cells (Kasza et al., 2007; Stamenovic, 2005). Currently, the connection between cell signalling and alterations of the mechanical properties of cells is still not fully understood in general. Information concerning the effects of EGF stimulation on the mechanical properties of cells will certainly provide insights into this connection. In addition, since EGFR is highly expressed in a variety of human tumors (Dei Tos & Ellis, 2005) and mutations in EGFR can produce aberrant cell signaling that often leads to uncontrolled cell growth and a malignant phenotype, such information will also shed light on the link between cell mechanical properties and human diseases (Bao & Suresh, 2003).

Many highly sensitive techniques have been developed over the years to assess mechanical properties of cells (Addae-Mensah & Wikswo, 2008). These include atomic force microscopy (Smith et al., 2005), magnetic twisting cytometry (Wang et al., 1993), micropipette aspiration (Alexopoulos et al., 2003), optical tweezers (Svoboda & Block, 1994), Shear-flow methods (Usami et al., 1993), particle-tracking microrheology (Wirtz, 2009), cantilever beams (Galbraith & Sheetz, 1997), and others. Each technique probes a cell or cells in a different manner and does not necessarily measure the same aspects of a cell as another technique. Thus, the use of more than one technique to study the same object (i.e., cell) may prove useful. This chapter describes the application of two sensitive techniques, the atomic force microscopy and the quartz crystal microbalance with dissipation monitoring, to the study of the mechanical properties of cells in response to exposure to EGF.

1.2. Probing mechanical response of cells with atomic force microscopy

Atomic force microscopy (AFM) is one of the most popular choices for probing the mechanical properties of cells, because individual cells can be probed in high sensitivity and resolution with a minimum of force (Radmacher Manfred, 2007). To measure the mechanical properties of the cell with AFM, the top surface of a live cell is indented with the sharp tip located at the end of a cantilever (a probe). The cantilever is mounted on a piezoelectric tube that moves the cantilever down and up in the vertical direction toward and away from the surface of the cell. The deflection in the cantilever is typically measured by a laser that tracks a spot on the tip of the cantilever. From the position of the cantilever and its deflection, force-displacement curves during the indentation of the cell by the probe are generated as shown in Figure 1 (Radmacher M., 1997).

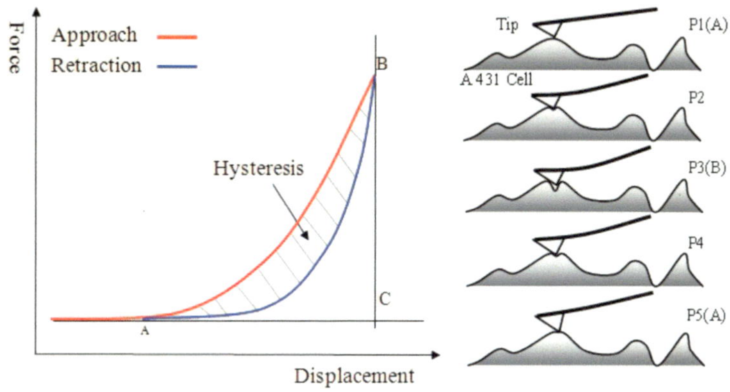

Figure 1. Typical force-displacement curves generated for the approach and retraction of the AFM probe. Approach and retraction correspond to mechanical loading and unloading of the AFM probe on the top surface of the cell.

To indent epithelial cells that are being treated with biologically active molecules, the force applied on the cells by the AFM probe is often kept approximately 50 nN or slightly lower to minimize the adverse effects on the cells caused by the probe. With the force at this level, the probe can have the probing depth that is sufficiently deep (100 to 500 nm) to register the cytoskeleton remodeling (Schillers et al., 2010) but is still shallow enough to avoid influence from the nucleus and the solid substrate on which the cells rest (Melzak et al., 2011). In addition to a low magnitude of loading, the velocity of loading should be kept low enough so that the transient friction interactions between the probe tip and the cell surface are avoided (Alcaraz et al., 2003).

Force-displacement curves acquired with all of these precautions in place then can be used to estimate values for Young's modulus and energy dissipation of the cell. The Young's modulus of a cell can be extracted from a curve of unloading force displacement with the aid of the Herzian elastic contact model for a conically shaped tip indenting an elastic body (Touhami et al., 2003):

$$F = kd = \frac{2}{\pi} \bullet \frac{E}{1 - \nu^2} \delta^2 \tan\alpha \tag{1}$$

where F is the applied force, d is the deflection of the cantilever, and k is the spring constant of the cantilever. Also, α is the half angle of the cone-shaped tip, υ is the Poisson ratio (taken to be 0.5, for an incompressible material), δ is the indentation depth, and E is Young's modulus. It should be noted that the force-displacement curves are dependent on the frequency of the probe (Hoffman & Crocker, 2009); this means that the estimated values of modulus are not unique.

In the estimation of the Young's modulus, the cell is assumed to be an elastic body, i.e., to return all of the energy deposited during the loading portion of the indentation process. However, in reality, the cell is not perfectly elastic but exhibits some dissipative behaviour.

This dissipative behavior is manifested as a loss (as heat to the surroundings) of some of the energy stored during loading, and can be seen in the indentation process as hysteresis in a cycle of force displacement (Figure 1). In a cell, energy dissipation is believed to be accomplished by internal friction and/or viscous damping mechanisms (Alcaraz et al., 2003; Smith et al., 2005). In AFM, the mechanical energy dissipated per cycle of indentation is given quantitatively by the area of the hysteresis loop enclosed by the approach and retraction curves (Alcaraz et al., 2003), as shown in Figure 1.

1.3. Probing mechanical response of cells with quartz crystal microbalance with dissipation monitoring

In contrast to AFM, the quartz crystal microbalance with dissipation monitoring (QCM-D) has not been widely used in characterization of cell mechanics. The QCM-D is an ultrasensitive piezoelectric device (Hook F. et al., 1998; Rodahl Michael et al., 1996; Rodahl M. & Kasemo, 1996) that is able to detect mass coupled (adsorbed or adhered) to the surface of the sensing element. The sensing element is a single piezoelectric quartz crystal in the form of a thin disc with a metal electrode deposited on its underside. This sensor crystal is set into free vibration in shear mode by means of a pulse of current. In air, the sensor crystal has a characteristic resonant frequency; this frequency is changed when any material, liquid or solid, is coupled or attached to it by adsorption or adhesion.

The mass of an ultrathin and elastic adsorbed layer, which exhibits negligible dissipation, is given by the Sauerbrey equation:

$$\Delta m = \frac{C}{n \times \Delta f} \tag{2}$$

where Δm is mass, C is the instrument sensitivity constant in ng/cm^2 of the crystal surface area, n is the frequency overtone number (n = 1, 3, 5, ...), and Δf is the simplified representation of Δf_n, the change in resonant frequency at the overtone number n, caused by the attached mass (Sauerbrey, 1959). The acoustic shear wave from the vibrating sensor penetrates the attached layer without disrupting it and the instrument monitors Δf of the layer at multiple overtones (Hook Fredrik et al., 2001) as a function of time. The instrument simultaneously monitors the change in dissipation factor, ΔD, which is defined as the ratio of the dissipated energy to the stored (elastic) energy per vibrational cycle. An ultrathin, elastic layer exhibits a negligible value of ΔD, but a layer that is viscoelastic exhibits a non-negligible and measureable value of ΔD. Both Δf and ΔD provide information about the mechanical response of a layer attached to the surface of the sensor crystal, and any type of change in the layer produces changes in these quantities.

The QCM has been used to assess the changes in mass and mechanical properties of a layer of biomolecules immobilized on the surface of the quartz crystal (Dixon, 2008). Specific examples include the use of the QCM to show protein adsorption (Hook F. et al., 1998), to indicate changes in an immobilized layer caused by ligand–receptor interactions (Janshoff et al., 1997; Lee H. et al., 2010), to detect nucleic acid hybridization (Furtado & Thompson, 1998), and to study immunoresponse (Aizawa et al., 2001).

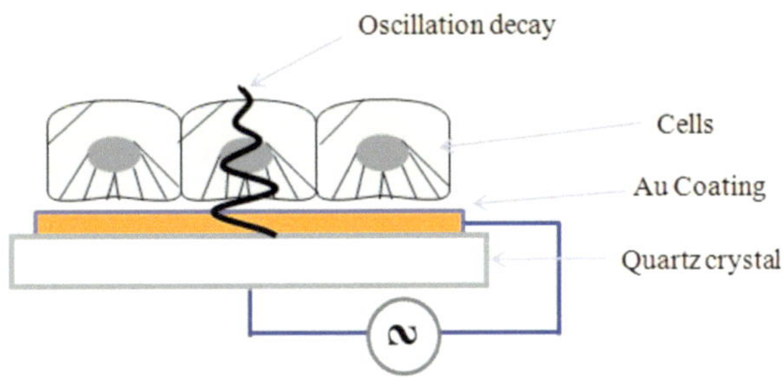

Figure 2. Diagram of a layer of cells attached to the surface of a sensor crystal in the QCM-D technique. The vibrational wave originating from the piezoelectric sensor crystal penetrates the cells from the bottom and diminishes with distance above the sensor surface. This figure is adapted with permission from (Chen et al., 2011). Copyright 2011, American Chemical Society.

In the field of cell biology, the QCM technique has become particularly attractive for its capability to study cells in a label-free manner (Heitmann et al., 2007; Janshoff et al., 1996; Matsuda et al., 1992; Redepenning et al., 1993; Wegener et al., 1998). Importantly, the technique is non-invasive to mammalian cells when the amplitude of shear oscillation is kept under 1 nm (Heitmann & Wegener, 2007). It has been used for determining the kinetics of cell attachment and spreading (Fredriksson et al., 1998; Nimeri et al., 1998) and for monitoring the long term growth of cells (Otto et al., 1999; Reipa et al., 2006). More recently, the QCM has been applied to characterization of cell viscoelasticity (Alessandrini et al., 2006; Galli Marxer et al., 2003; Li et al., 2008; Marx et al., 2005; Pax et al., 2005; Voinova et al., 2004). When the instrument used has the capability for monitoring the change in dissipation factor as well as frequency, the technique is termed QCM-D (QCM with dissipation monitoring). Because the acoustic signal diminishes exponentially with distance above the surface of the quartz crystal oscillator on which the cells are deposited, the QCM probes primarily the basal area of the cell monolayer (Heitmann et al., 2007; Le Guillou-Buffello et al., 2011). The test configuration is shown in Figure 2. Thus, Δf and ΔD, for adherent cells measured by the QCM-D technique can be expected to be related to mechanical changes in basal area of the cells and may be related to strength and quality of adhesion between the cell and the surface it contacts (Fredriksson et al., 1998; Rodahl M. et al., 1997).

2. Investigation of the dynamic mechanical response of an A431 cell monolayer in response to EGF

2.1. Mechanical behavior of an A431 cell monolayer probed with AFM

An example of the use of the AFM to assess change in the mechanical response of cells upon exposure to a biologically active molecule is illustrated in this section. Figure 3 shows the Young's moduli, obtained at a probe speed of 5.8 μm/s and indentation depth of ~500 nm, for two hundred randomly selected A431 cells before and after the treatment with a 40-nM

EGF solution (Yang et al., 2012). The focus is on the comparison, rather than on the modulus values themselves. The two samples show a statistically significant increase ($p < 0.05$) in average modulus as a result of the treatment: 11.2 ± 2.8 kPa for untreated cells and 18.7 ± 2.0 kPa for treated cells.

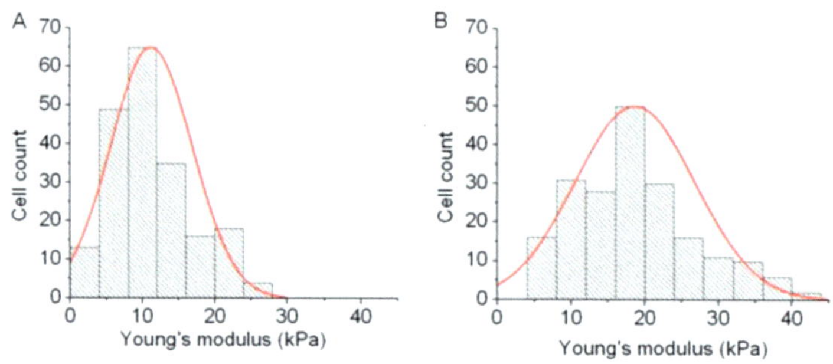

Figure 3. Histograms of the distributions of the Young's modulus of two hundred randomly selected A431 cells before (A) and after (B) the treatment with a 40-nM EGF solution in buffer (Yang et al., 2012).

Figure 4 shows changes in amount of the dissipated energy of cells upon EGF stimulation. It is evident that energy dissipation of a cell, as indicated by the area of the hysteresis loop, increases after the treatment with 40 nM EGF (Figures 4A and 4B). Figure 4C summarize the distribution of such differences exhibited by one hundred randomly selected cells. A statistically significant difference ($p < 0.05$) in energy dissipation per cycle is shown: 3.09 ± 0.79 fJ before the treatment and 5.10 ± 0.71 fJ after the treatment.

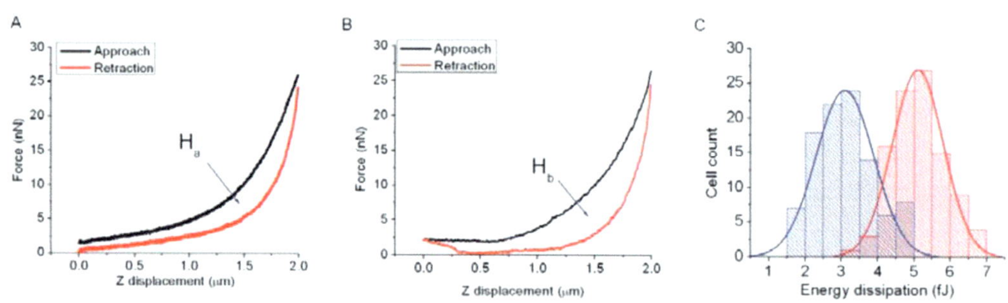

Figure 4. Force-displacement curves for a single cell before (A) and after (B) the treatment with 40 nM EGF. Histograms (C) of energy dissipation for one hundred randomly selected cells before (blue) and after (red) the treatment with EGF.

The underlying meaning of a simultaneous increase in both stiffness (Young's modulus) and energy dissipation of the cells upon the EGF treatment can be better understood with the soft glassy rheology (SGR) model (Fabry et al., 2001). In this model, the cell is considered as a soft glassy material that is structurally disordered and metastable (Sollich, 1998). The hysteresivity of the cell, η, defined as the fraction of the elastic (input) energy that is dissipated as heat, is frequency insensitive and remains constant in the low frequency regime (Fredberg & Stamenovic, 1989).

Although both stiffness and dissipation of the cells increase simultaneously, the hysteresivity of the cells, which is determined based on the ratio of area within the hysteresis loop to area under the approach curve (Figure 1) (Collinsworth et al., 2002; Fung, 1984; Smith et al., 2005), is not constant but increases as a function of time, shown in Figure 5. According to the SGR model, upon a non-thermal stimulation (e.g., ATP depletion, or cell relaxing agent), the cell can undergo a change in mechanical ordering state either toward the glass transition as hysteresivity decreases or away from the glass transition as hysteresivity increases. So the cell can modulate its mechanical state between a more solid-like state and a more liquid-like state (Smith et al., 2005). The increase in hysteresivity in Figure 5 implies that the mechanical state of the EGF-treated cells moves away from the glass transition and possibly takes on a more fluidic behavior. This interpretation is consistent with morphological changes observed in A431 cells responding to EGF stimulation, where the cells undergo cell rounding, membrane ruffling, and filopodia extension, all of which might be facilitated by a more fluid-like state of the cells (Chinkers et al., 1981).

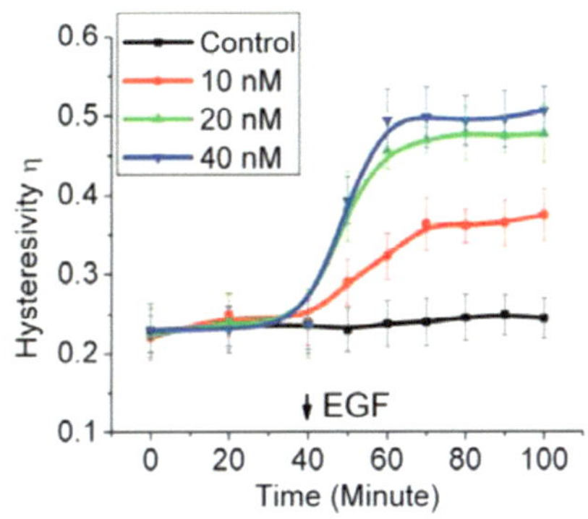

Figure 5. Hysteresivity versus time for exposure of A431 cells to EGF at 0, 10 nM, 20 nM, and 40 nM (Yang et al., 2012). A 30-min baseline was established for each measurement prior to the addition of EGF. Trend lines were used to connect each data point to illustrate the trend for each response curve.

2.2. Mechanical behavior of an A431 cell monolayer probed with QCM-D

An example of the use of QCM-D to assess change in the mechanical response of cells upon exposure to a biologically active molecule is shown in Figure 6. This figure shows Δf and ΔD obtained simultaneously in response to addition of a solution of EGF. The spike at 5 min is an artifact of the manual addition of the solution with a pipette. After the initial spike, ΔD goes down and then slowly recovers, while Δf steadily increases with time (Chen et al., 2011). Because ΔD was substantially more sensitive than Δf to dosage of EGF, we focus the descriptions primarily on the changes in ΔD.

Figure 7 shows the QCM-D measurement of ΔD for a monolayer of A431 cells exposed to buffer alone and to EGF. Figure 7A reveals that, after the spike caused by manual addition of the solution, the control cells (0 nM EGF) showed a small decrease in Δ*D*, while the experimental cells (10 nM EGF) exhibited a large decrease. This difference represents the change in mechanical energy dissipation of the basal areas of the cells induced by EGF.

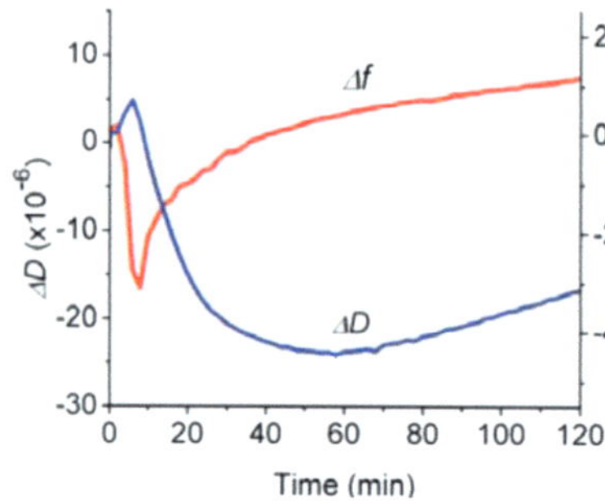

Figure 6. Typical response of A431 cells to EGF. Both Δ*f* and Δ*D* are shown as a function of time.

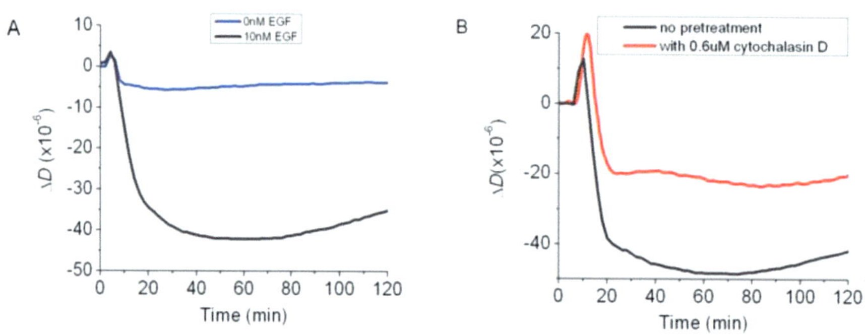

Figure 7. Real-time QCM-D measurements of the Δ*D* response of a monolayer of A431 cells to exposure to EGF (Chen et al., 2011). (A) Δ*D* response of the cells in the presence (bottom) and absence (top) of 10 nM EGF. (B) Δ*D* response of the cells pretreated with an inhibitor of actin polymerization, cytochalasin D, showing suppression of the EGF-induced response. Both figures are adapted with permission from (Chen et al., 2011). Copyright 2011, American Chemical Society.

A likely cause of the reduction in Δ*D* upon exposure to EGF was the remodeling of the cytoskeleton, a process integral to the mechanical response of the cell. This remodeling can be interfered with if the polymerization of actin is inhibited. Figure 7B shows that pretreatment of the monolayer of cells with a 0.6 μM solution of cytochalasin D, a potent, cell-permeable inhibitor of actin polymerization (Schliwa, 1982), diminished the magnitude of the response of the cells to EGF. This result supports the remodeling of the cytoskeleton as the cause of the Δ*D* response, which echoes the previous finding of Heitmann and coworkers' (Heitmann et al., 2007).

It is desirable to connect the EGF-induced changes in Δ*D* to EGFR-mediated cell signaling. As already mentioned, one of the domains of EGFR, the receptor for EGF, is the intracellular tyrosine kinase domain. This domain is integral to the signaling pathway within the cell that transmits the effects of EGF to other parts of the cell, and if this domain is inhibited, the signaling pathway is disrupted. PD158780 has been shown to be a potent inhibitor of the tyrosine kinase domain for EGFR (Rewcastle et al., 1996). With a pretreatment with this inhibitor, the cells showed a greatly suppressed ΔD response upon exposure to EGF, while the cells without pretreatment showed the expected decrease in ΔD (Figure 8). These results are evidence that the EGF-induced the ΔD response is due to EGFR-mediated cell signaling.

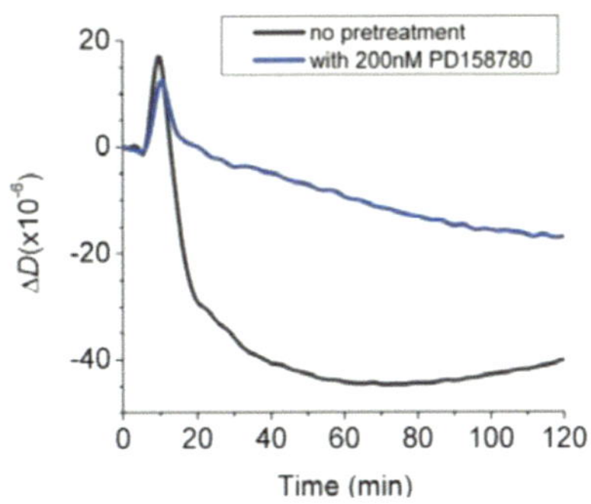

Figure 8. The ΔD response of the cells pretreated with EGFR tyrosine kinase inhibitor, PD158780, showing suppression of 10 nM EGF-induced response. This figure is adapted with permission from (Chen et al., 2011). Copyright 2011, American Chemical Society.

2.3. Remodeling of cytoskeleton probed with fluorescence imaging

The change in mechanical properties of cells has been attributed to remodeling of the cytoskeleton (Kuznetsova et al., 2007), which can be induced by EGF treatment (Rijken et al., 1995; Rijken et al., 1998). Direct evidence of remodeling of the cytoskeleton can be obtained with fluorescence imaging. For this, the cells were first treated with a 10-nM EGF solution for 60 min under the same conditions used for both AFM and QCM-D measurements. The actin filaments of the cytoskeleton were then stained with fluorescently labeled phalloidin and imaged with an inverted fluorescence microscope.

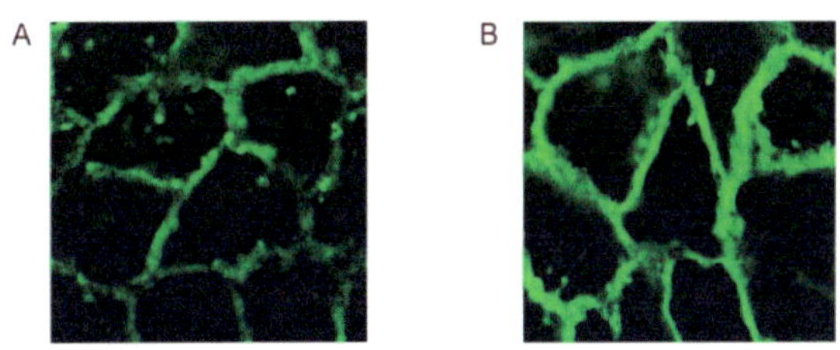

Figure 9. Remodeling of actin filaments in a monolayer of A431 cells induced by the treatment of 10 nM of EGF at 37°C. (A) and (B) show fluorescence-stained actin cortex at the top-half of the cell layer before and after the 60-min treatment with EGF solution, respectively (Yang et al., 2012).

As shown in Figures 9A and B, the top portion of the membrane skeleton exhibited an increase in brightness after the 60-min EGF treatment, indicating an increase of the number and size of the cortical actin filaments. Considering that the cortical actin provides cells with a structural framework, the increase in cortical actin can be assumed responsible for the increased rigidity of the cells, which is manifested as the increase in Young's modulus measured by AFM. In addition, because both energy dissipation and hysteresivity were derived from the same set of force-displacement curves used for determining Young's modulus of the top region of the cells, it is reasonable to assume that the increases in dissipation and hysteresivity were also related to the increase in cortical actin filaments.

The basal area of the cell monolayer, i.e., the area probed by the QCM-D, was also examined by means of fluorescence imaging. In this case, the stress fibers, which are actin filaments that reside in the bottom portion of the cells and are involved in the formation of focal adhesions attaching the cell to the substrate, were studied. As shown in Figures 10A and 10B, the cells displayed a decrease in amount and size of stress fibers after 60 min of exposure to EGF. Thus the decrease in dissipation observed by means of the QCM-D technique can be reasonably assumed to be related to a reduction in number and size of the actin stress fibers in the basal region of the cells. In addition, this reduction in stress fibers suggests a loss of adhesion and/or contact between the cells and the solid substrate, which is consistent with the EGF-induced cell rounding and retracting that has been reported previously (Chinkers et al., 1979; 1981).

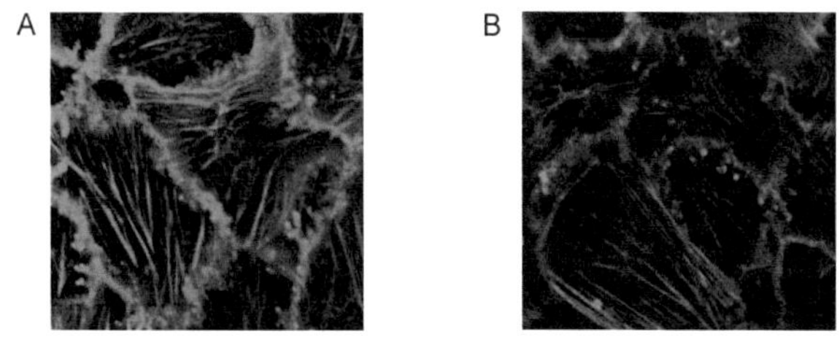

Figure 10. Remodeling of actin filaments in a monolayer of A431 cells induced by the treatment of 10 nM of EGF at 37°C. (A) and (B) show fluorescence-stained stress fibers at the basal area of the cell layer before and after the 60-min EGF treatment, respectively.

2.4. Comparison of AFM with QCM-D results

The two sensitive techniques, AFM and QCM-D, described in this chapter probe two different regions of cells in a monolayer on a solid surface. The nature of the AFM limits it to probing the top surface and immediately underlying volume of individual cells. The nature of the QCM-D limits it to probing the basal areas of the cells in a monolayer. The AFM technique revealed an increase in Young's modulus, energy dissipation, and hysteresivity in response to EGF, while the QCM-D technique revealed a decrease in energy dissipation factor. The fluorescence studies of the changes in actin in the top and bottom surfaces upon exposure to EGF corroborate the differences observed in mechanical properties.

Because both hysteresivity (η) and the change in energy dissipation factor (ΔD) represent the ratio between the dissipated energy and the elastic energy, these two quantities are analogous to each other (Collinsworth et al., 2002). The fact that the opposite trends in them were obtained upon exposure of the cells to EGF is perhaps not unexpected (Figure 5 and 7A). For a monolayer of cells on a solid surface, the bottoms of the cells are restricted in shape by contact with a solid surface, while the tops of the cells have no such restrictions. The bottom surfaces of the cells may therefore exhibit unique responses to the presence of the solid surface and the tops of the cells may not. Thus, contrasts in mechanical properties and in the number, size, and form of actin filaments signify that the mechanical responses of cells to EGFR signaling are regionally specific.

Interestingly, both hysteresivity (η) and the change in energy dissipation factor (ΔD) required about approximately the same amount of time (30 to 40 min) to reach each respective maxima or minimum after the addition of 10 nM of EGF. Such a similarity suggests that both of these regional mechanical responses (ΔD and η) were mediated by closely coupled cell signaling pathway(s) of EGFR.

3. Conclusions

In this chapter, the examples were provided to illustrate how to employ a combination of AFM and QCM-D to characterize the mechanical behavior of cells in response to exposure to EGF. This unique combination allowed the comparative assessment of the upper volume of the cell bodies as well as the basal areas of the cells. Results from both parts of the A431 cells reveal a regionally specific mechanical behavior of the cells, which can be attributed to the distinct cytoskeleton structures utilized by the cells to alter the local structure in response to EGF stimulation. The signaling pathways that mediate the remodeling of the cytoskeleton in the upper volume of the cell bodies are likely closely coupled to those in the basal areas of the cells. There is a clear correlation between the time-dependent mechanical response and the dynamic process of EGFR signaling.

Overall, the combination of AFM and QCM-D is able to provide a more complete and refined mechanical profile of the cells during the dynamic cell signaling process than either technique alone. The use of combined techniques to track real-time cell signaling based on the measurement of cellular mechanical response in a label-free manner is a powerful approach for investigating the role of EGFR in causing abnormal cell behavior. This combined, real-time approach may have the potential to be applied to the study of other types of receptor-mediated cell signaling and trafficking. This approach should contribute to the fundamental understanding of the correlation between cell function and cell mechanical properties.

Author details

Jun Xi, Lynn S. Penn and Jennifer Y. Chen
Drexel University, Department of Chemistry, Philadelphia, PA 19104, USA

Ning Xi and Ruiguo Yang
Michigan State University, Department of Electrical and Computer Engineering, East Lansing, MI 48824, USA

Acknowledgement

This manuscript recapitulates some information from original works that have previously been published in *Analytical Chemistry* (Chen et al., 2011) and *Experimental Cell Research* (Yang et al., 2012). This work was supported in part by NSF Grants IIS-0713346 and DMI-0500372, ONR Grants N00014-04-1-0799 and N00014-07-1-0935, and NIH Grant R43 GM084520. We also appreciate the financial support from Drexel University.

4. References

Addae-Mensah, K. A. & Wikswo, J. P. (2008). Measurement techniques for cellular biomechanics in vitro. *Exp. Biol. Med.*, Vol.233, No.7, pp. 792-809,

Aizawa, H., et al. (2001). Conventional diagnosis of treponema pallidum in serum using latex piezoelectric immunoassay. *Materials Science and Engineering: C*, Vol.17, No.1-2, pp. 127-132, ISSN 0928-4931

Alcaraz, J., Buscemi, L., Grabulosa, M., Trepat, X., Fabry, B., Farr, R. & Navajas, D. (2003). Microrheology of human lung epithelial cells measured by atomic force microscopy. *Biophysical Journal*, Vol.84, No.3, pp. 2071-2079, ISSN 0006-3495

Alessandrini, A., Croce, M. A., Tiozzo, R. & Facci, P. (2006). Monitoring cell-cycle-related viscoelasticity by a quartz crystal microbalance. *Appl. Phys. Lett.*, Vol.88, No.8, pp. 083905/083901-083905/083903, ISSN 0003-6951

Alexopoulos, L. G., Haider, M. A., Vail, T. P. & Guilak, F. (2003). Alterations in the mechanical properties of the human chondrocyte pericellular matrix with osteoarthritis. *Journal of Biomechanical Engineering*, Vol.125, No.3, pp. 323-333,

Bao, G. & Suresh, S. (2003). Cell and molecular mechanics of biological materials. *Nat Mater*, Vol.2, No.11, pp. 715-725, ISSN 1476-1122

Bretscher, A. (1989). Rapid phosphorylation and reorganization of ezrin and spectrin accompany morphological changes induced in a-431 cells by epidermal growth factor. *The Journal of Cell Biology*, Vol.108, No.3, pp. 921-930,

Chen, J. Y., Li, M., Penn, L. S. & Xi, J. (2011). Real-time and label-free detection of cellular response to signaling mediated by distinct subclasses of epidermal growth factor receptors. *Analytical Chemistry*, Vol.83, No.8, pp. 3141-3146, ISSN 0003-2700

Chinkers, M., McKanna, J. A. & Cohen, S. (1979). Rapid induction of morphological changes in human carcinoma cells a-431 by epidermal growth factors. *J. Cell Biol.*, Vol.83, No.1, pp. 260-265,

Chinkers, M., McKanna, J. A. & Cohen, S. (1981). Rapid rounding of human epidermoid carcinoma cells a-431 induced by epidermal growth factor. *J. Cell Biol.*, Vol.88, No.2, pp. 422-429,

Collinsworth, A. M., Zhang, S., Kraus, W. E. & Truskey, G. A. (2002). Apparent elastic modulus and hysteresis of skeletal muscle cells throughout differentiation. *American Journal of Physiology - Cell Physiology*, Vol.283, No.4, pp. C1219-C1227,

Dei Tos, A. P. & Ellis, I. (2005). Assessing epidermal growth factor receptor expression in tumours: What is the value of current test methods. *Eur. J. Cancer*, Vol.41, No.10, pp. 1383-1392, ISSN 0959-8049

Dixon, M. C. (2008). Quartz crystal microbalance with dissipation monitoring: Enabling real-time characterization of biological materials and their interactions. *J. Biomol. Tech.*, Vol.19, No.3, pp. 151-158, ISSN 1943-4731 (Electronic)

Fabry, B., Maksym, G. N., Butler, J. P., Glogauer, M., Navajas, D. & Fredberg, J. J. (2001). Scaling the microrheology of living cells. *Physical Review Letters*, Vol.87, No.14, pp. 148102,

Fletcher, D. A. & Mullins, R. D. (2010). Cell mechanics and the cytoskeleton. *Nature*, Vol.463, No.7280, pp. 485-492, ISSN 0028-0836

Fredberg, J. J. & Stamenovic, D. (1989). On the imperfect elasticity of lung tissue. *Journal of Applied Physiology*, Vol.67, No.6, pp. 2408-2419,

Fredriksson, C., Kihlman, S., Rodahl, M. & Kasemo, B. (1998). The piezoelectric quartz crystal mass and dissipation sensor: A means of studying cell adhesion. *Langmuir,* Vol.14, No.2, pp. 248-251, ISSN 0743-7463

Fung, Y. C. (1984). Structure and stress-strain relationship of soft tissues. *American Zoologist,* Vol.24, No.1, pp. 13-22,

Furtado, L. M. & Thompson, M. (1998). Hybridization of complementary strand and single-base mutated oligonucleotides detected with an on-line acoustic wave sensor. *Analyst,* Vol.123 No.10, pp. 1937 - 1945,

Galbraith, C. G. & Sheetz, M. P. (1997). A micromachined device provides a new bend on fibroblast traction forces. *Proceedings of the National Academy of Sciences of the United States of America,* Vol.94, No.17, pp. 9114-9118,

Galli Marxer, C., Collaud Coen, M., Greber, T., Greber, U. & Schlapbach, L. (2003). Cell spreading on quartz crystal microbalance elicits positive frequency shifts indicative of viscosity changes. *Anal. Bioanal. Chem.,* Vol.377, No.3, pp. 578-586,

Heitmann, V., Reiss, B. & Wegener, J. (2007). The quartz crystal microbalance in cell biology: Basics and applications. *Springer Ser. Chem. Sens. Biosens.,* Vol.5, No.Piezoelectric Sensors, pp. 303-338,

Heitmann, V. & Wegener, J. (2007). Monitoring cell adhesion by piezoresonators: Impact of increasing oscillation amplitudes. *Analytical Chemistry,* Vol.79, No.9, pp. 3392-3400, ISSN 0003-2700

Hoffman, B. D. & Crocker, J. C. (2009). Cell mechanics: Dissecting the physical responses of cells to force. *Annual Review of Biomedical Engineering,* Vol.11, No.1, pp. 259-288,

Hook, F., Rodahl, M., Brzezinski, P. & Kasemo, B. (1998). Energy dissipation kinetics for protein and antibody-antigen adsorption under shear oscillation on a quartz crystal microbalance. *Langmuir,* Vol.14, No.4, pp. 729-734, ISSN 0743-7463

Hook, F., Kasemo, B., Nylander, T., Fant, C., Sott, K. & Elwing, H. (2001). Variations in coupled water, viscoelastic properties, and film thickness of a mefp-1 protein film during adsorption and cross-linking: A quartz crystal microbalance with dissipation monitoring, ellipsometry, and surface plasmon resonance study. *Analytical Chemistry,* Vol.73, No.24, pp. 5796-5804, ISSN 0003-2700

Janmey, P. A. (1998). The cytoskeleton and cell signaling: Component localization and mechanical coupling. *Physiol. Rev.,* Vol.78, No.3, pp. 763-781,

Janmey, P. A., Georges, P. C., Hvidt, S., YuLi, W. & Dennis, E. D. (2007). Basic rheology for biologists, In: *Methods in cell biology,* pp. 1, 3-27, Academic Press, ISBN 0091-679X,

Janshoff, A., Wegener, J., Sieber, M. & Galla, H. J. (1996). Double-mode impedance analysis of epithelial cell monolayers cultured on shear wave resonators. *European Biophysics Journal,* Vol.25, No.2, pp. 93-103,

Janshoff, A., Steinem, C., Sieber, M., el Bayâ, A., Schmidt, M. A. & Galla, H. J. (1997). Quartz crystal microbalance investigation of the interaction of bacterial toxins with ganglioside containing solid supported membranes. *European Biophysics Journal,* Vol.26, No.3, pp. 261-270,

Kasza, K. E., Rowat, A. C., Liu, J., Angelini, T. E., Brangwynne, C. P., Koenderink, G. H. & Weitz, D. A. (2007). The cell as a material. *Curr. Opin. Cell Biol.,* Vol.19, No.1, pp. 101-107, ISSN 0955-0674

Kuznetsova, T. G., Starodubtseva, M. N., Yegorenkov, N. I., Chizhik, S. A. & Zhdanov, R. I. (2007). Atomic force microscopy probing of cell elasticity. *Micron,* Vol.38, No.8, pp. 824-833, ISSN 0968-4328

Le Guillou-Buffello, D., Gindre, M., Johnson, P., Laugier, P. & Migonney, V. (2011). An alternative quantitative acoustical and electrical method for detection of cell adhesion process in real-time. *Biotechnology and Bioengineering,* Vol.108, No.4, pp. 947-962, ISSN 1097-0290

Lee, C.-W., Lin, C.-C., Lin, W.-N., Liang, K.-C., Luo, S.-F., Wu, C.-B., Wang, S.-W. & Yang, C.-M. (2007). Tnf-{alpha} induces mmp-9 expression via activation of src/egfr, pdgfr/pi3k/akt cascade and promotion of nf-{kappa}b/p300 binding in human tracheal smooth muscle cells. *Am J Physiol Lung Cell Mol Physiol,* Vol.292, No.3, pp. L799-812,

Lee, H., Contarino, M., Umashankara, M., Schön, A., Freire, E., Smith, A., Chaiken, I. & Penn, L. (2010). Use of the quartz crystal microbalance to monitor ligand-induced conformational rearrangements in hiv-1 envelope protein gp120. *Anal. Bioanal. Chem.,* Vol.396, No.3, pp. 1143-1152, ISSN 1618-2642

Li, F., Wang, J. H. C. & Wang, Q.-M. (2008). Thickness shear mode acoustic wave sensors for characterizing the viscoelastic properties of cell monolayer. *Sensors and Actuators B: Chemical,* Vol.128, No.2, pp. 399-406, ISSN 0925-4005

Marx, K. A., Zhou, T., Montrone, A., McIntosh, D. & Braunhut, S. J. (2005). Quartz crystal microbalance biosensor study of endothelial cells and their extracellular matrix following cell removal: Evidence for transient cellular stress and viscoelastic changes during detachment and the elastic behavior of the pure matrix. *Anal. Biochem.,* Vol.343, No.1, pp. 23-34, ISSN 0003-2697

Matsuda, T., Kishida, A., Ebato, H. & Okahata, Y. (1992). Novel instrumentation monitoring in situ platelet adhesivity with a quartz crystal microbalance. *ASAIO Journal,* Vol.38, No.3, pp. M171-M173, ISSN 1058-2916

McManus, M. J., Boerner, J. L., Danielsen, A. J., Wang, Z., Matsumura, F. & Maihle, N. J. (2000). An oncogenic epidermal growth factor receptor signals via a p21-activated kinase-caldesmon-myosin phosphotyrosine complex. *Journal of Biological Chemistry,* Vol.275, No.45, pp. 35328-35334,

Melzak, K. A., Moreno-Flores, S., Lopez, A. E. & Toca-Herrera, J. L. (2011). Why size and speed matter: Frequency dependence and the mechanical properties of biomolecules. *Soft Matter,* Vol.7, No.2, pp. 332-342, ISSN 1744-683X

Nimeri, G., Fredriksson, C., Elwing, H., Liu, L., Rodahl, M. & Kasemo, B. (1998). Neutrophil interaction with protein-coated surfaces studied by an extended quartz crystal microbalance technique. *Colloids and Surfaces B: Biointerfaces,* Vol.11, No.5, pp. 255-264, ISSN 0927-7765

Ono, M. & Kuwano, M. (2006). Molecular mechanisms of epidermal growth factor receptor (egfr) activation and response to gefitinib and other egfr-targeting drugs. *Clinical Cancer Research,* Vol.12, No.24, pp. 7242-7251,

Otto, K., Elwing, H. & Hermansson, M. (1999). Effect of ionic strength on initial interactions of escherichia coli with surfaces, studied on-line by a novel quartz crystal microbalance technique. *J. Bacteriol.,* Vol.181, No.17, pp. 5210-5218,

Papakonstanti, E. A. & Stournaras, C. (2008). Cell responses regulated by early reorganization of actin cytoskeleton. *FEBS Lett.,* Vol.582, No.14, pp. 2120-2127, ISSN 0014-5793

Pax, M., Rieger, J., Eibl, R. H., Thielemann, C. & Johannsmann, D. (2005). Measurements of fast fluctuations of viscoelastic properties with the quartz crystal microbalance. *The Analyst*, Vol.130, No.11, pp. 1474-1477,

Radmacher, M. (1997). Measuring the elastic properties of biological samples with the afm. *IEEE Engineering in Medicine and Biology Magazine*, Vol.16, No.2, pp. 47,

Radmacher, M. (2007). Studying the mechanics of cellular processes by atomic force microscopy, In: *Methods in cell biology*, YuLi W. & Dennis E. D., pp. 347-372, Academic Press, ISBN 0091-679X,

Redepenning, J., Schlesinger, T. K., Mechalke, E. J., Puleo, D. A. & Bizios, R. (1993). Osteoblast attachment monitored with a quartz crystal microbalance. *Analytical Chemistry*, Vol.65, No.23, pp. 3378-3381, ISSN 0003-2700

Reipa, V., Almeida, J. & Cole, K. D. (2006). Long-term monitoring of biofilm growth and disinfection using a quartz crystal microbalance and reflectance measurements. *Journal of Microbiological Methods*, Vol.66, No.3, pp. 449-459, ISSN 0167-7012

Rewcastle, G. W., Palmer, B. D., Thompson, A. M., Bridges, A. J., Cody, D. R., Zhou, H., Fry, D. W., McMichael, A. & Denny, W. A. (1996). Tyrosine kinase inhibitors. 10. Isomeric 4-[(3-bromophenyl)amino]pyrido[d]- pyrimidines are potent atp binding site inhibitors of the tyrosine kinase function of the epidermal growth factor receptor. *J. Med. Chem.*, Vol.39, No.9, pp. 1823-1835, ISSN 0022-2623

Ridley, A. J. (1994). Membrane ruffling and signal transduction. *BioEssays*, Vol.16, No.5, pp. 321-327, ISSN 1521-1878

Rijken, P. J., Hage, W. J., van Bergen en Henegouwen, P. M., Verkleij, A. J. & Boonstra, J. (1991). Epidermal growth factor induces rapid reorganization of the actin microfilament system in human a431 cells. *J. Cell Sci.*, Vol.100, No.3, pp. 491-499,

Rijken, P. J., Post, S. M., Hage, W. J., van Bergen en Henegouwen, P. M. P., Verkleij, A. J. & Boonstra, J. (1995). Actin polymerization localizes to the activated epidermal growth factor receptor in the plasma membrane, independent of the cytosolic free calcium transient. *Experimental Cell Research*, Vol.218, No.1, pp. 223-232, ISSN 0014-4827

Rijken, P. J., van Hal, G. J., van der Heyden, M. A. G., Verkleij, A. J. & Boonstra, J. (1998). Actin polymerization is required for negative feedback regulation of epidermal growth factor-induced signal transduction. *Experimental Cell Research*, Vol.243, No.2, pp. 254-262, ISSN 0014-4827

Rodahl, M., Hook, F. & Kasemo, B. (1996). Qcm operation in liquids: An explanation of measured variations in frequency and q factor with liquid conductivity. *Analytical Chemistry*, Vol.68, No.13, pp. 2219-2227, ISSN 0003-2700

Rodahl, M. & Kasemo, B. (1996). A simple setup to simultaneously measure the resonant frequency and the absolute dissipation factor of a quartz crystal microbalance. *Rev. Sci. Instrum.*, Vol.67, No.9, pp. 3238-3241, ISSN 0034-6748

Rodahl, M., Hook, F., Fredriksson, C., Keller, C. A., Krozer, A., Brzezinski, P., Voinova, M. & Kasemo, B. (1997). Simultaneous frequency and dissipation factor qcm measurements of biomolecular adsorption and cell adhesion. *Faraday Discuss*, No.107, pp. 229-246, ISSN 1359-6640 (Print)

Satcher Jr, R. L. & Dewey Jr, C. F. (1996). Theoretical estimates of mechanical properties of the endothelial cell cytoskeleton. *Biophysical Journal*, Vol.71, No.1, pp. 109-118, ISSN 0006-3495

Sauerbrey, G. (1959). Verwendung von schwingquarzen zur wägung dünner schichten und zur mikrowägung. *Zeitschrift für Physik A Hadrons and Nuclei,* Vol.155, No.2, pp. 206-222,

Scaltriti, M. & Baselga, J. (2006). The epidermal growth factor receptor pathway: A model for targeted therapy. *Clinical Cancer Research,* Vol.12, No.18, pp. 5268-5272,

Schillers, H., Wälte, M., Urbanova, K. & Oberleithner, H. (2010). Real-time monitoring of cell elasticity reveals oscillating myosin activity. *Biophysical Journal,* Vol.99, No.11, pp. 3639-3646, ISSN 0006-3495

Schliwa, M. (1982). Action of cytochalasin d on cytoskeletal networks. *J. Cell Biol.,* Vol.92, No.1, pp. 79-91,

Silva, C. M. (2004). Role of stats as downstream signal transducers in src family kinase-mediated tumorigenesis. *Oncogene,* Vol.23, No.48, pp. 8017-8023, ISSN 0950-9232

Smith, B. A., Tolloczko, B., Martin, J. G. & Grutter, P. (2005). Probing the viscoelastic behavior of cultured airway smooth muscle cells with atomic force microscopy: Stiffening induced by contractile agonist. *Biophysical Journal,* Vol.88, No.4, pp. 2994-3007, ISSN 0006-3495

Sollich, P. (1998). Rheological constitutive equation for a model of soft glassy materials. *Physical Review E,* Vol.58, No.1, pp. 738,

Song, W., Wu, J., Ge, G. & Lin, Q. (2008). Two domains of the epidermal growth factor receptor are involved in cytoskeletal interactions. *Biochemical and Biophysical Research Communications,* Vol.370, No.4, pp. 589-593, ISSN 0006-291X

Stamenovic, D. (2005). Effects of cytoskeletal prestress on cell rheological behavior. *Acta Biomaterialia,* Vol.1, No.3, pp. 255-262, ISSN 1742-7061

Svoboda, K. & Block, S. M. (1994). Biological applications of optical forces. *Annual Review of Biophysics and Biomolecular Structure,* Vol.23, No.1, pp. 247-285,

Touhami, A., Nysten, B. & Dufrene, Y. F. (2003). Nanoscale mapping of the elasticity of microbial cells by atomic force microscopy. *Langmuir,* Vol.19, No.11, pp. 4539-4543, ISSN 0743-7463

Usami, S., Chen, H.-H., Zhao, Y., Chien, S. & Skalak, R. (1993). Design and construction of a linear shear stress flow chamber. *Annals of Biomedical Engineering,* Vol.21, No.1, pp. 77-83,

Voinova, M. V., Jonson, M. & Kasemo, B. (2004). On dissipation of quartz crystal microbalance as a mechanical spectroscopy tool. *Spectroscopy,* Vol.18, No.4, pp. 537-544,

Wang, N., Butler, J. P. & Ingber, D. E. (1993). Mechanotransduction across the cell surface and through the cytoskeleton. *Science,* Vol.260, No.5111, pp. 1124-1127,

Wegener, J., Janshoff, A. & Galla, H. J. (1998). Cell adhesion monitoring using a quartz crystal microbalance: Comparative analysis of different mammalian cell lines. *European Biophysics Journal,* Vol.28, No.1, pp. 26-37,

Wirtz, D. (2009). Particle-tracking microrheology of living cells: Principles and applications. *Annual Review of Biophysics,* Vol.38, No.1, pp. 301-326,

Yang, R., Chen, J. Y., Xi, N., Lai, K. W. C., Qu, C., Fung, C. K. M., Penn, L. S. & Xi, J. (2012). Characterization of mechanical behavior of an epithelial monolayer in response to epidermal growth factor stimulation. *Experimental Cell Research,* Vol.318, No.5, pp. 521-526, ISSN 0014-4827

Yasmeen, A., Bismar, T. A. & Al Moustafa, A.-E. (2006). Erbb receptors and epithelial-cadherin-catenin complex in human carcinomas. *Future Oncology,* Vol.2, No.6, pp. 765-781,

Chapter 9

Numerical Simulation Model with Viscoelasticity of Arterial Wall

Tomoki Kitawaki

Additional information is available at the end of the chapter

http://dx.doi.org/10.5772/49976

1. Introduction

1.1. Numerical analysis of intravascular flow

Numerical flow simulation is useful for understanding fluid phenomena such as blood flow or pulse wave propagation in the systemic arteries. For numerical analysis of intravascular flow, it is important to consider not only incompressible assumption and blood viscosity but also the viscoelasticity of the blood vessel wall; however, blood flow *in vivo* is complicated because of the unsteadiness of pulsatile flow and complex viscoelastic properties of the blood vessel wall. In order to conduct such numerical flow analysis in a viscoelastic blood vessel, it is effective to use the one-dimensional distributed parameter model, which can analyze flow along with the blood vessel axis. This distributed parameter model, pressure, flow volume and cross-section of the tube for every section element are defined and the time change is analyzed.

According to previous research, quantitative numerical simulation requires a model which take in both effects of unsteady viscous friction and viscoelasticity of the vessel wall in case flow unsteadiness is large (Reuderink et al., 1989). Conventional one-dimensional numerical simulation models can be classified into a linear distributed parameter model (Snyder et al., 1968; Avolio et al., 1980) and a nonlinear distributed parameter model (Anliker & Rockwall, 1971; Schaaf & Abbrecht, 1972; Porenta et al., 1986). The linear distributed parameter model has the feature that is easy to take in the influence of viscoelasticity and to conduct numerical analysis of the flow unsteadiness, since superposition of a periodic solution is possible; however, the influence of fluid nonlinearity cannot be disregarded. On the other hand, the conventional nonlinear distributed parameter model does not involve the effects of such flow unsteadiness and the viscoelastic behavior of the blood vessel wall concurrently with the difficulty of model construction. Hence, these models can be used only for qualitative discussion.

INTECH
open science | open minds

© 2012 The Author(s). Licensee InTech. This chapter is distributed under the terms of the Creative Commons Attribution License http://creativecommons.org/licenses/by/3.0), which permits unrestricted use, distribution, and reproduction in any medium, provided the original work is properly cited.

Consequently, in order to construct a numerical simulation model of intravascular flow for quantitative analysis with viscoelasticity of the blood vessel wall, it is necessary to use a nonlinear distributed parameter model to be able to include the viscoelasticity.

1.2. Clinical significance of blood vessel stiffness

Arterial stiffness indexes, such as PWV (pulse wave velocity: velocity of pressure wave propagation in circulatory systems), PP (pulse pressure: systolic blood pressure minus diastolic blood pressure), and AIx (augmentation index: index showing the effect of reflecting waves coming from peripheries) have received a lot of attention because they can indicate the risks for cardiovascular diseases (Oliver & Webb, 2003). These clinical indexes, which are obtained by analyzing blood pressure pulse waveforms *in vivo,* are used for predicting the prognosis of cardiovascular diseases and thus analyses of pulse waveform are clinically important. In order to understand the basis of these indexes, it is necessary to understand how changes in mechanical properties, such as blood vessel viscoelasticity, can affect pulse waveforms in the systemic arteries.

1.3. Research purpose and contents of this chapter

We thereby proposed a nonlinear one-dimensional numerical simulation model which can accurately calculate the viscous resistance of unsteady flow and the viscoelasticity of the tube wall. We have shown that the numerical model can describe well wave propagation in silicone tubes representing blood vessels (Kitawaki et al., 2003). The result showed that the viscoelasticity of the vessel wall plays an important role in the form of a pulsatile wave (Kitawaki & Shimizu, 2005); therefore, it is necessity to consider the viscoelastic effect accurately in the quantitative investigation of changes in pulsatile waves *in vivo*.

In **section 2**, we construct a one-dimensional numerical simulation model that takes into account unsteady viscosity and the generalized viscoelastic model from the theory of the fluid equation (Kitawaki et al., 2003; Kitawaki & Shimizu, 2006). In addition, the high-speed calculation method is described. When the wall of a blood vessel deforms finitely due to changes in the internal pressure, the wall's physical properties, such as deformation compliance and viscoelasticity, change nonlinearly (Hayashi et al., 1980; Sato & Ohshima, 1985). In **section 3**, it was checked whether our one-dimensional model would be able to simulate the pulse wave propagation of small pressure waves in silicone tubes even when their deformation compliance and viscoelasticity changed independently, using appropriate values of the viscoelastic parameter of the silicone rubber tube for numerical simulation (Kitawaki & Shimizu, 2006). In **section 4**, the influence of viscoelasticity change on periodic pulsatile wave propagation was studied (Kitawaki & Shimizu, 2009). The purpose of the section was to investigate the effect of viscoelasticity change on periodic pulsatile wave quantitatively. For this purpose, we studied the pulse wave propagation of periodic small pressure waves using a silicone tube connected with terminal resistance, and obtained the waveform changes of pulsatile waves due to the change of mechanical properties, including the viscoelasticity of the tube. We then compared the experimental results with numerically calculated results using a one-dimensional numerical simulation model with terminal resistance treatment. Finally, we studied the effect of vessel wall viscoelasticity on the propagation of a periodic pulsatile wave by comparing numerical simulation results between the difference of viscoelastic models and viscoelastic parameters. The final **section 5**, describes the conclusion of this chapter.

2. Theory

2.1. Basic equations

When constructing the numerical model, we neglect the effects of bends of vessels. We also assume that the tube does not leak and that the flow is axisymmetric and incompressible. Under these conditions, the equations of continuity and momentum conservation of the one-dimensional model are given by (Olufsen, 1999),

$$\frac{\partial A}{\partial t}+\frac{\partial Q}{\partial x}=0 \tag{1}$$

$$\frac{\partial Q}{\partial t}+\frac{\partial}{\partial x}\left(\frac{Q^2}{A}\right)+\frac{A}{\rho}\frac{\partial p}{\partial x}+F_t=0 \tag{2}$$

where A is the cross-section of the tube, Q is the mean sectional flow volume, p is the mean pressure, t is time, x is distance along the vessel axis, ρ is the fluid density, and F_t is the viscous resistance.

By assuming that we are dealing with a Newtonian fluid, and an oscillating flow velocity distribution in a cylindrical tube, using the Womersley model, the viscous resistance F_t is given by (Zielke, 1968),

$$F_t=4\pi\nu\left\{2V+\int_0^t\frac{\partial V}{\partial t}W(t-\tau)d\tau\right\} \tag{3}$$

where V is the mean sectional velocity, ν is the kinematic viscosity, and $W(t)$ is a weight function. In a rigid cylindrical tube,

$$W(t)=\lim_{\omega\to\infty}\int_{-\pi}^{\pi}\frac{1}{F_J(\alpha i^{3/2})-2}e^{i\omega t}d\omega \tag{4}$$

where $F_J(z)=zJ_0(z)/J_1(z)$, and J_0, J_1 are 0th and 1st order Bessel functions of the complex number z. $\omega(=2\pi f)$ is the angular frequency of the flow oscillating at frequency f, $\alpha=R\sqrt{\frac{\omega}{\nu}}$ is the Womersley number and R is the tube radius.

For a long wavelength, the flow velocity distribution in a distensible tube is similar to that in a rigid tube. Therefore, the weight function in a distensible tube can be approximated by Eq. (4), regardless of changes in the tube's cross-section induced by the internal pressure. Note that the radius expressed as the Womersley number becomes a function of position and time, and consequently the weight function itself also becomes a function of position and time.

2.2. Deformation models of a wall

The tube law that describes the relationship between the tube cross-section and the internal pressure can be described by an elastic model and two viscoelastic models as below. In actuality, for a silicone tube, the relationship varies with internal pressure change, and shows nonlinearity. However, by assuming small local transformations, the following models can be used to approximate a linear model.

2.2.1. Elastic model

When the tube is perfectly elastic, the tube law of the elastic model can be expressed by,

$$\Delta p = \frac{1}{Cs}\Delta A \tag{5}$$

where Δp and ΔA are the differences in the pressure and cross-section relative to reference values, and $Cs = \left.\frac{\partial A}{\partial p}\right|_{A_0}$ is tube deformation compliance at reference cross-section A_0.

2.2.2. Voigt model

By assuming that the viscoelasticity of the tube causes a phase lag between the applied pressure and resulting change in cross-section of the tube, from the Voigt model, the tube deformation law can be expressed by,

$$\Delta p = \frac{1}{Cs}\left\{\Delta A + \tau_V \frac{\partial A}{\partial t}\right\} \tag{6}$$

where τ_V is the relaxation time that accounts for the phase lag.

2.2.3. Generalized viscoelastic model

For complex viscoelasticity, as is the case for blood vessels, the generalized viscoelastic model can be applied (Westerhof and Noordergraaf, 1970). The following tube law of the generalized viscoelastic model can be derived, as shown in the next paragraph.

$$\Delta p = \frac{1}{Cs}\left\{\Delta A + \int_0^t \sum_{i=1}^{n} f_i e^{-(t-u)/\tau_i} \frac{dA(u)}{dt} du\right\} \tag{7}$$

The viscoelastic property of the tube wall is reflected in the second term of Eq. (7), which contains both the dynamic viscoelasticity parameter f_i and the relaxation time parameter τ_i.

2.3. Generalized viscoelastic model

In order to obtain the tube law in one-dimensional flow analysis, the tube law of the generalized viscoelastic model is derived from the complex viscoelastic coefficient as follows. The complex viscoelastic coefficient of the generalized viscoelastic model *E(s)* can be expressed by (Westerhof and Noordergraaf, 1970),

$$E(s) = E_0 \prod_{i=1}^{n}(1+s\tau'_i) \Big/ \prod_{i=1}^{n}(1+s\tau_i) \tag{8}$$

where E_0 is the static elastic coefficient, s is angular frequency ($=i\omega$), and τ_i and, τ'_i are relaxation time parameters which express the viscoelasticity of the tube wall. This equation can be transformed:

$$E(s) = E_0\left(1+\sum_{i=1}^{n} f_i \frac{s}{s+1/\tau_i}\right) \tag{9}$$

where $f_i = -\prod_{k=1}^{n}(\tau_i - \tau'_k) \Big/ \left\{\tau_i \prod_{\substack{k=1\\ i\neq k}}^{n}(\tau_i - \tau_k)\right\}$.

Using this formula, the tube law in the frequency domain becomes

$$\Delta p(s) = \frac{hE_0}{2RA_0}\cdot \Delta A(s)\cdot\left\{1+\sum_{i=1}^{n} f_i \frac{s}{s+1/\tau_i}\right\} \tag{10}$$

where h is the tube wall thickness, $\Delta p(s)$ and $\Delta A(s)$ are Laplace transformations of the pressure and cross-section from a reference value. The compliance of the tube deformation is $C_s = \dfrac{2RA_0}{hE_0}$.

By transforming Eq. (10) back to the time domain by inverse Laplace transformation, we obtain the tube law of the generalized viscoelastic model.

$$\Delta p = \frac{1}{Cs}\left\{\Delta A + \int_0^t \sum_{i=1}^{n} f_i e^{-(t-u)/\tau_i} \frac{dA(u)}{dt} du\right\} \tag{11}$$

2.4. High-speed calculation method

The following high-speed calculation method that was obtained for the viscous resistance term of rigid tube (Kagawa et al., 1983) was applied to the convolution integrals which appeared in the viscous resistance term in Eq. (3) and the viscoelastic terms in Eq. (7). By approximating the weight function in Eq. (4) by the sum of the exponential function, $W(t) = \sum_{i=0}^{k} m_i e^{-n_i \tau}$; the viscous resistance in Eq. (3) can be expressed by recursive formulations as follows,

$$F_t(t) = 4\pi\nu\left\{2V(t) + \sum_{i=0}^{k} y_i(t)\right\} \tag{12}$$

$$\begin{cases} y_i(t) = 0 & (t = 0) \\ y_i(t+\Delta t) = e^{-n_i \Delta\tau} y_i(t) & \\ \quad + m_i e^{-n_i(\Delta\tau/2)}\{V(t+\Delta t) - V(t)\} & (t > 0) \end{cases} \tag{13}$$

where $\tau = \nu t/R^2$ and $\Delta\tau = \nu\Delta t/R^2$ are normalized time and time step, respectively.

We need to determine the value of term number k in the Eq. (12) from the value of $\Delta\tau$ with consideration of approximation accuracy. In present experimental condition, the term number k was 10, because $\Delta\tau$ was calculated to be $1.2\text{x}10^{-5}$.

The weight function of the convolution integral in the tube law of Eq. (7) can be expressed as a summation of the exponential function. Therefore, the tube law of the recursive formulations is as follows,

$$p(t) - p_0 = \frac{1}{Cs}\left\{(A - A_0) + \sum_{i=1}^{n} z_i\right\} \tag{14}$$

$$\begin{cases} z_i(t) = 0 & (t = 0) \\ z_i(t+\Delta t) = e^{-\Delta t/\tau_i} z_i(t) & \\ \quad + f_i e^{-\Delta t/2\tau_i}\{A(t+\Delta t) - A(t)\} & (t > 0) \end{cases} \tag{15}$$

It is possible to determine the term number n in Eq. (14) from the viscoelastic characteristics.

3. Application to the fluid experiment of a computational model: In case of a difference in viscoelasticity

3.1. Method

3.1.1. Experimental model

Experimental apparatus

Figure 1 shows a schema of the experimental apparatus used in this study. The experiment tube consisted of two silicone tubes which was different form previous experiments (Kitawaki et al., 2003; Kitawaki & Shimizu 2005) with an inner diameter of 9 mm, a thickness of 0.5 mm, and a length of 1.45 m, connected by a rigid brass tube 5 cm in length and 9 mm in inner diameter. A piston pump was connected to one end of the tube via a 5-cm long brass tube, and a water tank with a valve was connected to the other end via another 5-cm long brass tube. The three brass tubes were fixed to a metal plate upon which the complete tube rested without longitudinal tension, allowing the silicone tubes to change shape freely. A pressure sensor (Nihon Koden, DX-100) was connected to the middle of each brass tube.

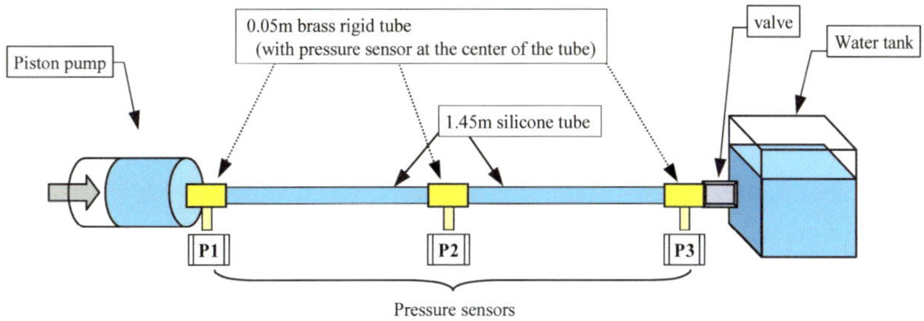

Figure 1. Experimental apparatus

Figure 2(a) is a schema of the piston pump. The piston pump was driven by a computer-controlled stepping motor, and was capable of generating various waveforms with various flow volumes. The piston cylinder receives the backpressure from the water inside the tube. Therefore, displacement of the inner cylinder was measured by a laser displacement sensor (Keyence, LK-030).

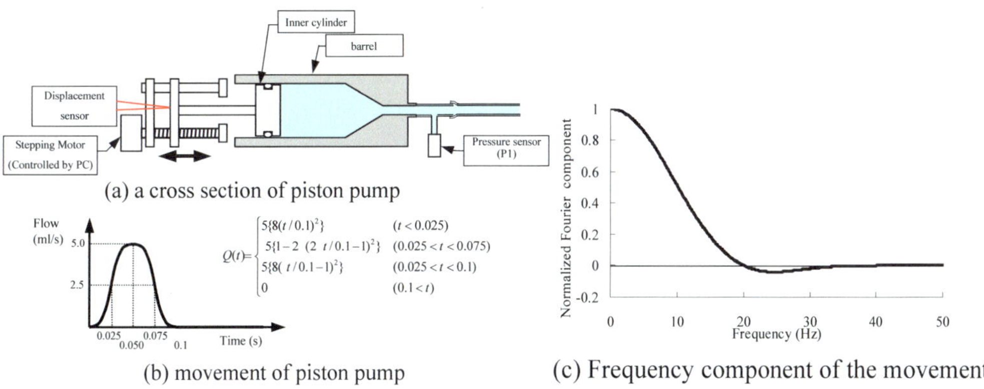

Figure 2. Piston pump

Experimental conditions

The tube, piston pump, and water tank were filled with water. Baseline internal pressures were set by adjusting the water head of the tank. The piston pump generated a single impulse. The flow volume at the inlet of the experiment tube was determined as shown in Fig. 2(b); when the impulse time t_i was set to 0.1 s, and the maximum flow volume Q_m was 5.0 ml/s. As seen in Fig. 2(c) which shows the Fourier transformation of the flow volume when a single impulse was generated, the highest frequency component of the flow volume was 30 Hz. Signals from the three pressure sensors and the displacement sensor were recorded for 10 seconds by a PC at a sampling rate of 1 kHz. The trials, that is, generation of a single impulse by the pump, were performed at increasingly higher baseline pressures of 2.5, 5.0, 7.5 and 10.0 kPa at 90-minute intervals. The trials were generated more than 70 minutes after changing the baseline pressures because 30 minutes were necessary for the viscoelasticity to return to a steady state. The baseline pressures were established by opening the valve, and the valve was closed just before the start of each trial.

3.1.2. *Tube wall properties*

Static tube law

Tube wall properties were determined for recent silicone tube. The static tube law and compliance of the silicone tube was obtained from the relationship between the volume of the piston pump and the internal pressure, as shown in Fig. 3, while performing one stroke of the piston pump over a period of about 30 minutes. The compliances, reference pressure, and reference cross-section of local tube deformation to the numerical calculation of each trial were determined from the collinear approximation of the pressure range of the experimental conditions. For example, a range of 2.5 - 4.3 kPa was used when the baseline pressure was 2.5 kPa, because the pressure pulse amplitude of a single impulse was about 1.8 kPa. These local compliances, assumed constant under the flow experimental conditions, are shown by squares in Fig. 3. These local compliances were used in all the wall viscoelastic models.

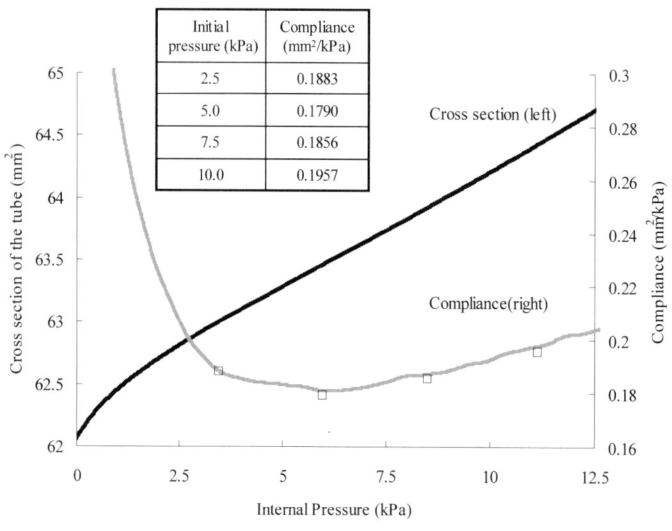

Figure 3. Static relation between pressure and cross section of the silicone tube

Dynamic viscoelastic property of the tube

Dynamic viscoelasticity of a strip of the silicone tube was measured by a device (TA Instruments DMA2980). The measurement conditions were determined as initial strain of 0.18% and amplitude of 0.145%. These values correspond to a baseline pressure of 0.28 kPa and a wave amplitude of 1.8 kPa, respectively. The measured viscoelastic properties are in Fig. 4. This figure shows the dynamic modules of viscoelasticity normalized against the static modules in Fig. 4(a) and the loss tangent in Fig. 4(b), respectively. These results mean that, in the measurement range 0.1~30 Hz, the dynamic viscoelasticity property of the silicon tube, both the real parts and loss tangent, tends to increase gradually with increase in the frequency.

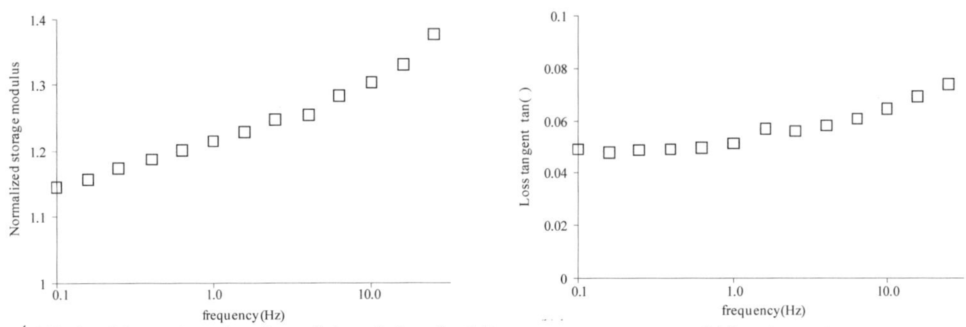

(a) Ratio of dynamic and static modulus of viscoelasticity (b) loss tangent

Figure 4. Dynamic viscoelastic modules of silicone tube.(a) Ratio of dynamic and static modulus of viscoelasticity

The procedure used to decide the value of the viscoelastic tube parameter for the experimental condition change (Kitawaki & Shimizu, 2006) is described below. (1) First, the relaxation time parameter τ_i was determined so that it could cover the frequency range 0.1-30 Hz and it could increase between 0.01 and 200 Hz at regular intervals on a logarithmic scale. The term number n in Eq. (7) was set to 7 to keep the term number to a minimum and to show that the viscoelastic properties tended to increase smoothly. (2) The values of the dynamic viscoelasticity parameter f_i were then determined to keep the difference between the experimental results and calculation results to a minimum from the low frequency term (i=1) to the high frequency term (i=7), using the different effect of each term on the numerical calculation results. (3) By changing the experimental conditions, the viscoelastic parameter was fitted by fixing the relaxation time parameter and changing only the dynamic viscoelasticity parameter.

Relaxation time τ_V of the Voigt model was determined as 2.5 ms from measurements of the delay time of the displacement of the outer diameter for the internal pressure change of the silicon tube.

3.1.3. *Numerical calculation*

The basic equations of the numerical calculation were digitized using a staggered grid system in space. For the calculation, Jameson-Baker's 4th order 4 step method as a time differential and 4th order central differential with numerical friction as space was used (Jameson & Baker, 1983). Flow volume and cross-section of the next time step were obtained from an equation of continuity and momentum conservation, and then the pressure was calculated from the tube law as a function of time (Kitawaki et al., 2003). Convolution integrals appear in the viscous resistance term in Eq. (3) and in the viscoelastic term in Eq. (7). Since calculation of the convolution integrals requires a lot of computer memory to hold past velocity and cross-section values, and takes a lot of computational time, a high-speed calculation method for the viscous resistance term of a rigid tube (Kagawa et al., 1983) was applied, as shown in *section 2.4*.

The following conditions were used in the calculation.

1. Baseline pressure was determined as an initial pressure by the water head of the tank.
2. No flow in the initial state.
3. Flow volume calculated by displacement sensor was used as an input boundary condition.
4. No flow boundary condition was applied to the distal end.

Time step Δt and grid interval Δx were set at 0.5 ms and 0.05 m, respectively. The Courant number was 0.18~0.21 because the propagation velocity of the pressure wave was about 18~21 m/s, and the CFL condition (numerical stability condition) was satisfied. Actual calculation was performed on a workstation computer.

3.2. Results and discussions

3.2.1. Difference between calculated and experimental results

Experimental results of pressure propagation

The internal pressure waves measured at the 3 positions in the tube, and the flow volume are shown in Fig. 5, for a baseline pressure of 2.5 kPa. The flow volume was calculated from the measured displacement data of the piston pump by using LPF (FIR 25 Hz) and derivative filter.

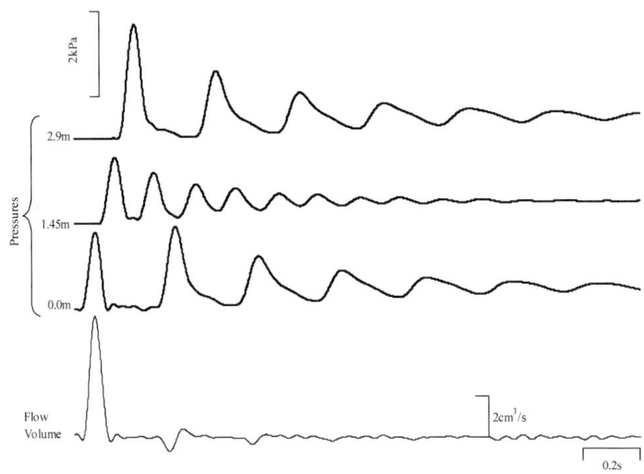

Figure 5. Time profiles of the pressure waves at three positions in the tube. The distance between input piston pump and measurement position are indicated on the left. The flow volume into the tube are given in the lower curve.

We can see that the pressure wave of the single impulse, generated by the movement of the piston pump, propagates towards the distal end, from where it is reflected. Upon returning to the proximal end, it is reflected again, back towards the distal end. The amplitude of the propagating pressure wave gradually attenuates, and the width of the pressure wave gradually increases because of the viscosity of the fluid and the viscoelasticity of the tube. The mean pressure rises for the fluid pushed by the piston pump. Both ends of the experimental tube are closed after the first pressure wave is generated by the piston pump. However, we can see an oscillatory wave when the pressure wave is returning back to the proximal end. This wave shows that the piston cylinder receives backpressure from the reflected pressure wave inside the tube.

Reynold's numbers and the wavelengths of the pressure waves were calculated from the measured value of the maximum velocity and pressure data from these experiments. Inspection of the Reynold's numbers (about 700) shows that laminar flow occurred under all experimental conditions. The wavelengths of the pressure waves (2.1 m) were sufficiently longer than the 4.5-mm radius of the tube, validating the long wavelength assumption and the assumptions of the Womersley model. Compared with the wavelength of the pressure wave, the length of the central rigid brass tube (5 cm) is sufficiently short, and because the rigid brass tube has a very small effect on the propagating pressure wave, it can be neglected in the calculation. The reproducibility of the pressure and flow volumes was good. Additionally, repeated measurements were virtually identical, so viscoelasticity changes such as memory effect did not happen.

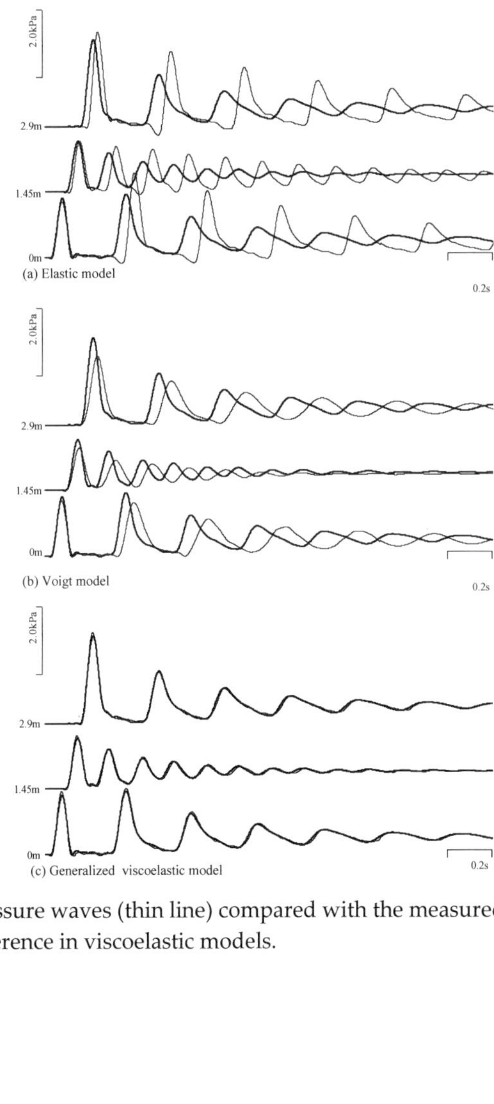

Figure 6. Calculated pressure waves (thin line) compared with the measured waves (bold line) at three location in the tube. Difference in viscoelastic models.

Comparison between calculated and experimental results

The numerical calculations were performed for a baseline pressure of 2.5 kPa and then compared with the experimental values of measurements of the pressure at the three positions along the experimental tube. The calculation results of combinations of the Womersley model with the elastic model, Voigt model and generalized viscoelastic model were compared, as shown in Fig. 6.

When the Womersley model was combined with all models, the initial pressure waves were close to the experimental value. However, when the Womersley model was combined with the elastic or Voigt models, the difference between the calculated and experimental results gradually increased because the calculated propagation velocities and attenuation level were underestimated. On the other hand, when combined with the generalized viscoelastic model together with the optimized viscoelastic parameter, the calculated results agreed well with the experimental results for all the experimental period. In the elastic model or Voigt model, when the tube deformation compliance or relaxation time τ_v was changed, the calculation result did not agree with the experimental result. These results show that the one-dimensional model using Womersley model combined with the generalized viscoelastic model are necessary in order for the numerically calculated result to agree with the experimental result.

3.2.2. *Difference of viscoelasticity by the change of baseline pressure*

Difference between pressure propagation experiments of baseline pressure difference

The flow volume of the inlet and the pressure waveform at the inlet position with changes in the baseline pressure are shown in Fig. 7. The flow volume in Fig. 7(a) are well controlled, and the movements of the piston are almost the same. As seen in Fig. 7(b), the pressure waveform changed differently when the baseline pressure was 2.5 kPa compared with other baseline pressures, because the propagation velocity was different. At baseline pressures of 5.0, 7.5 and10.0 kPa, the propagation is very similar during the first 0.8 seconds.

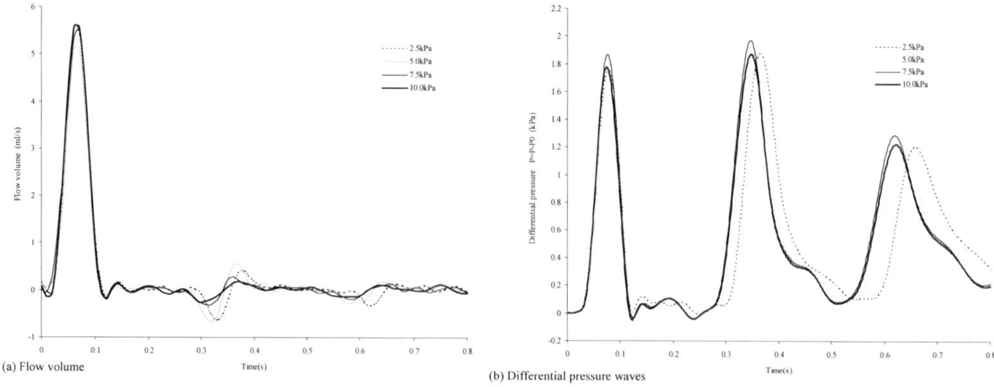

Figure 7. Flow volume and pressure waves at beginning of the tube. Difference in initial pressures

Determination of viscoelastic properties

Because the experimental results for the baseline pressures of 5.0, 7.5 and 10.0 kPa are very similar, the optimized value of the viscoelastic parameter for baseline pressure of 5.0 kPa was used for the three experimental conditions, and comparisons of the calculated results using this viscoelastic parameter and the experimental results are shown in Fig. 8. As can be seen, the experimental and calculated results agreed well throughout the experimental period when the baseline pressure was 5.0 kPa. In contrast, at baseline pressures of 7.5 and 10.0 kPa, the agreement is good for the first 0.2 seconds, and gradually depreciates thereafter, because of the slight difference in their propagation velocities. Therefore, the experimental result for baseline pressure of 7.5 and 10.0 kPa cannot be simulated using the viscoelastic parameter optimized for baseline pressure of 5.0 kPa.

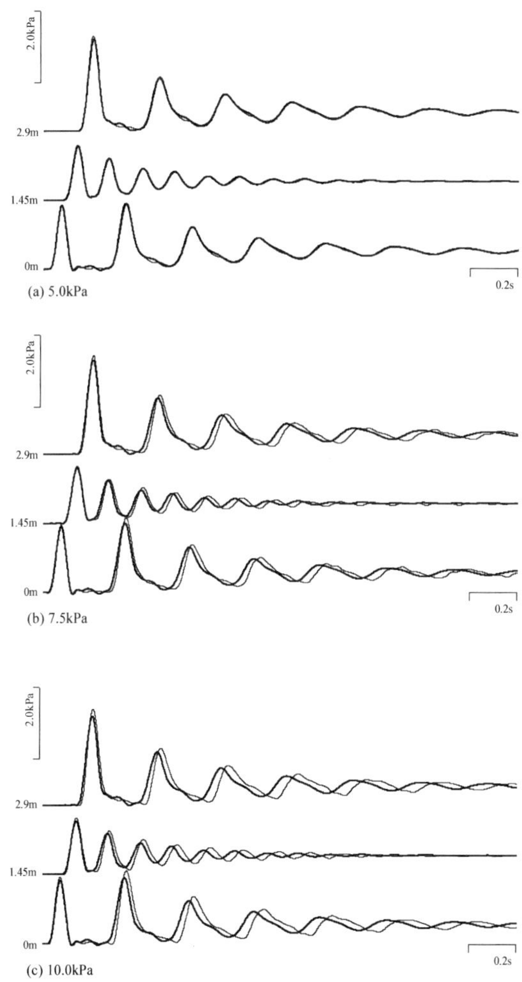

Figure 8. Calculated pressure waves (thin line) compared with the measured waves (bold line) using viscoelastic parameter optimized for a baseline pressure of 5.0 kPa.

According to this result, the optimized value of the viscoelastic parameter for baseline pressure of 7.5 and 10.0 kPa were obtained respectively and calculations were conducted again. As shown in Fig. 9, the calculated and experimental results agree well. In the first 0.4 seconds, a clear difference between the two results can be seen, especially at higher baseline pressures. The differences could not be decreased by changing the viscoelastic parameters. Even with these differences, the agreement between the two results is good.

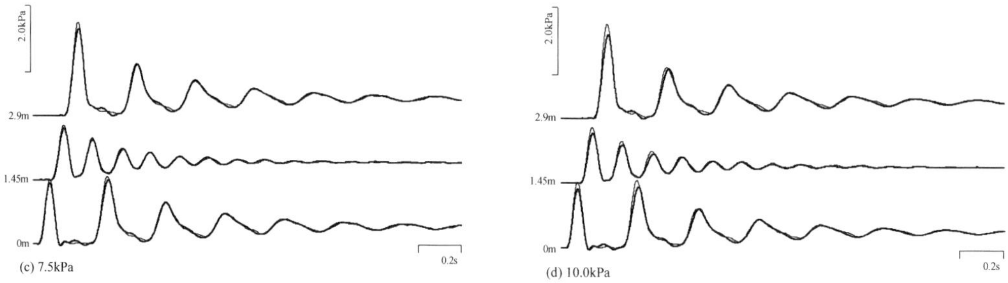

Figure 9. Calculated pressure waves (thin line) compared with the measured waves (bold line) using optimized viscoelastic parameter for each experimental condition.

These results show that the one-dimensional model using the Womersley model combined with the generalized viscoelastic model can accurately simulate the effect of changes in the silicon tube's viscoelasticity due to changes in the internal pressure when the viscoelastic parameter is appropriately determined. Additionally, the viscoelastic properties express the viscoelastic property change which depend the internal pressure of viscoelastic tube.

3.2.3. Relationship between viscoelasticity and static elasticity

The viscoelastic properties, both of the dynamic viscoelasticity parameters and the relaxation time parameters shown in Table 1 including the frequency *f* calculated from the relaxation time parameters, and the viscoelastic properties calculated from optimized values are plotted in Fig. 10. The squares in Fig. 10 indicate the measured values of the silicone tube viscoelasticity, and are close to the values obtained when the baseline pressure was 2.5 kPa, though a small error in the high frequency area of the loss tangent. This may be because the corresponding baseline pressure of the dynamic viscoelasticity measurement (0.28 kPa) is closer to the experimental baseline pressure of 2.5kPa than to the other experimental conditions.

From Fig. 10, it seems that the viscoelastic property change may be related to the baseline pressure. For example, when the baseline pressure increases from 2.5 to 5.0 kPa, the increase in the dynamic modules is steady at all frequencies. On the other hand, when the baseline pressure is 5.0 kPa or greater, the dynamic modules increase with increasing baseline pressure by increasing the frequency. Because the normalized dynamic modulus in the high frequency region changes with baseline pressure, the viscoelastic change can be said to be independent of the deformation compliance.

i	τ_i(s)	f(Hz)	f_i			
			2.5kPa	5.0kPa	7.5kPa	10.0kPa
1	17.0	0.009	0.089	0.165	0.130	0.112
2	3.75	0.042	0.031	0.033	0.058	0.087
3	0.7	0.227	0.057	0.061	0.086	0.110
4	0.15	1.061	0.068	0.073	0.098	0.119
5	0.03	5.305	0.090	0.096	0.121	0.138
6	0.005	31.83	0.170	0.170	0.183	0.190
7	0.0008	198.9	0.220	0.220	0.223	0.224

Table 1. Optimized viscoelastic parameters for each baseline pressure

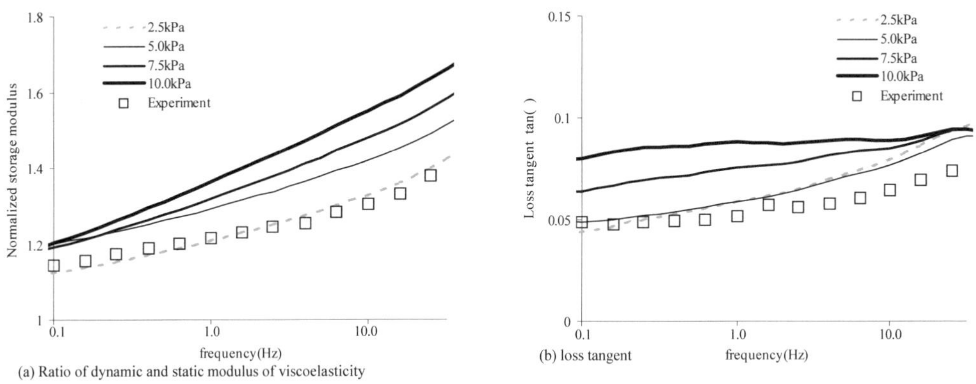

Figure 10. Change in dynamic viscoelastic module.

Accordingly, the one-dimensional numerical model, which takes into account unsteady viscosity and the generalized viscoelastic model, is good for simulating the propagation of small pressure waves in silicone tubes even when their deformation compliance and viscoelastic properties change independently.

3.3. Conclusion of this section

For the numerical analysis of the viscoelastic tubes, a nonlinear one-dimensional numerical model was investigated by including the unsteady viscous resistance and the effect of the tube wall viscoelasticity. By comparing the calculated results using these models with experimental results of a viscoelastic silicone tube, we can make the following conclusions.

1. The approximation error of the numerical simulation model is satisfactory small when the Womersley model is combined with the generalized viscoelastic model for the flow

analysis of the viscoelastic tube even their deformation compliance and viscoelasticity change independently.

2. The relationship between the deformation compliance and viscoelasticity, which depends on the internal pressure, can be analyzed using this numerical model by appropriately choosing the value of the viscoelastic parameter.

4. Application to the fluid experiment of a computational model: In case of periodic pulsatile flow

4.1. Method

4.1.1. Experimental model

Experimental apparatus

Figure 11 shows a schema of the experimental apparatus used in this study in order to generate the periodic pressure pulse wave. A silicone tube with an inner diameter of 9.0 mm, 0.5 mm thick, and a 2.04 m long (*L*) was connected by three pressure sensors (Nihon Koden, DX-200), one each 2 cm from both ends of the tube and one in the center. These sensors were connected at intervals of 1.0 m. The experiment tube and each pressure sensor were connected by a narrow connecting tube made of silicone with an inner diameter of 3.0 mm, 0.5 mm thick, and 3 cm long. The frequency characteristics of each pressure sensor system, including these connecting tubes were sufficiently higher than those of the frequency component of the periodic pulsatile flow rate described later; hence, the sensor system can accurately measure the internal pressure of the experiment tube. A piston pump was connected to the proximal end of the experiment tube through the flow sensor (Nihon Koden, MFV-1100) and a tank was connected to the distal end of the tube through the terminal resistance. The experiment tube was placed on a metal plate without longitudinal tension, allowing the silicone tube to change shape freely. Figure 11(b) is a scheme of the piston pump connected to the proximal end of the experiment tube. The piston pump was driven by a computer-controlled stepping motor and was capable of generating various waveforms with various flow rates. Actual flow rate was measured at the proximal end of the tube by the flow sensor, as shown in Fig. 11.

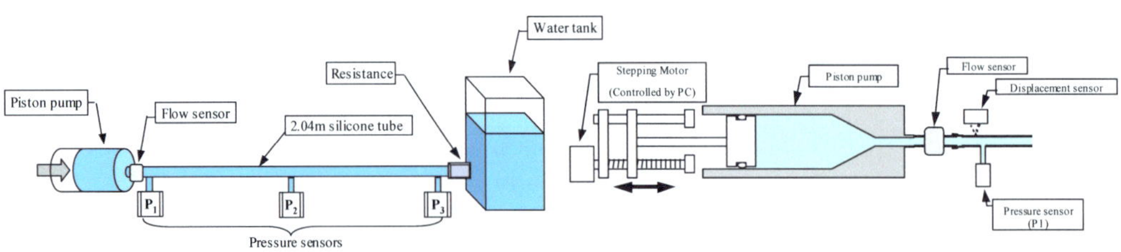

(a) Conceptual scheme of experiment (b) Cross-section of piston pump

Figure 11. Experimental apparatus

The terminal resistance installed at the distal end of the tube simulated the peripheral arterioles *in vivo*. This terminal resistance, shown in Fig. 12, was filled with a bundle of around 200 thin stainless capillary tubes (26 G: with an outer diameter of 0.51 mm, inner diameter of 0.3 mm, and 50 mm long) in a rigid brass tube with an inner diameter of 8.0 mm.

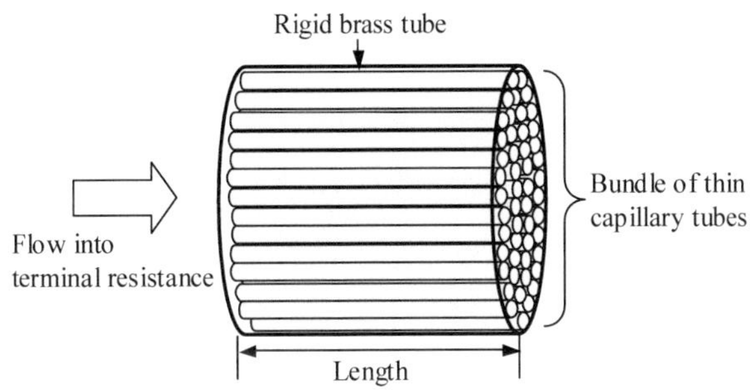

Figure 12. Terminal resistance

Experimental method

The experiment tube and piston pump were filled with water. Baseline internal pressure was set by adjusting the water head of the tank. Experimental trials were performed at increasingly higher baseline internal pressures of 5.6, 8.4 and 11.2 kPa (40, 60, 80 mmHg) respectively, because the deformation compliance and viscoelasticity of the experiment tube changed depending on the baseline internal pressure of the tube, as described later. The trials were generated at more than 60-minute intervals after changing the internal pressure, because 30-60 minutes were necessary for the viscoelasticity of the experiment tube to return to a steady relaxation state. Thus, the effect of a change in tube viscoelasticity on pulse wave propagation was examined.

Figure 13 shows the waveform of the flow rate ($Q(t)$) by the piston pump and the frequency component of the flow rate. In order to simulate periodic pulsation from the heart, we generated a pulsatile flow rate with 0.4-second ejection time (t_i) and about 2.8 ml/s maximum flow rate (Q_m) eight times in a 1.0-second period (t_p) at the proximal end of the experiment tube. In this case, the laminar flow condition was satisfied, as Reynold's number and Womersley's number were 400 and 11.3, respectively. By the movement of the piston pump, the filling fluid from the piston pump flowed into the tank through the terminal resistance. The tank has a hole at the height of the baseline internal pressure head; thereby, the water head pressure always equaled the baseline internal pressure during each experimental trial. In each trial, signals from the three pressure sensors and the flow sensor were recorded for 10 seconds by a PC at a sampling rate of 1 kHz. The experiment was performed several times under the same conditions, and it was confirmed that reproducibility was high.

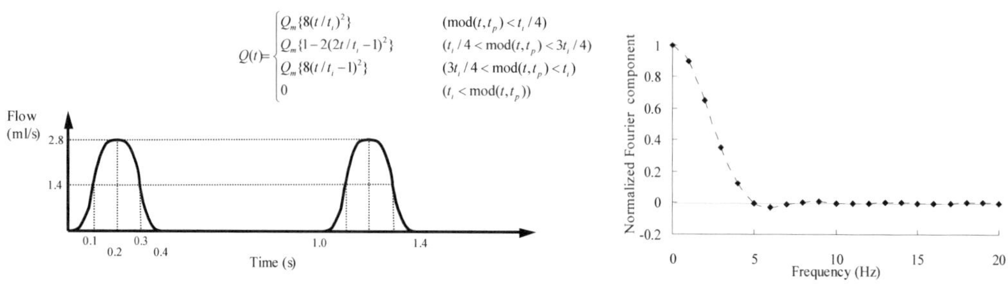

(a) Flow rate profile (b) Normalized Fourier component

Figure 13. Movement of piston pump

4.1.2. Numerical simulation

Mechanical properties of experiment tube

The static tube law and compliance of the experiment tube were obtained from the relationship between the increasing volume of the tube and the internal pressure while performing one slow stroke of the piston pump over a period of about 2.5 hours. This relationship between the static cross-section and the pressure, as well as the calculated tube deformation compliance are shown in Fig. 14. The static tube law was not exactly linear at any pressure. The measurement results in a range of pressures in each experimental trial were approximated to a linear model, because the nonlinear effects of tube law were not incorporated to the calculation in the present numerical simulation model. Tube deformation compliance at each baseline internal pressure is shown by squares in Fig. 14. At over 5 kPa, the tube deformation compliance increased with the increase in the internal pressure, showing countertrend to the tube law of blood vessels (Hayashi et al., 1980).

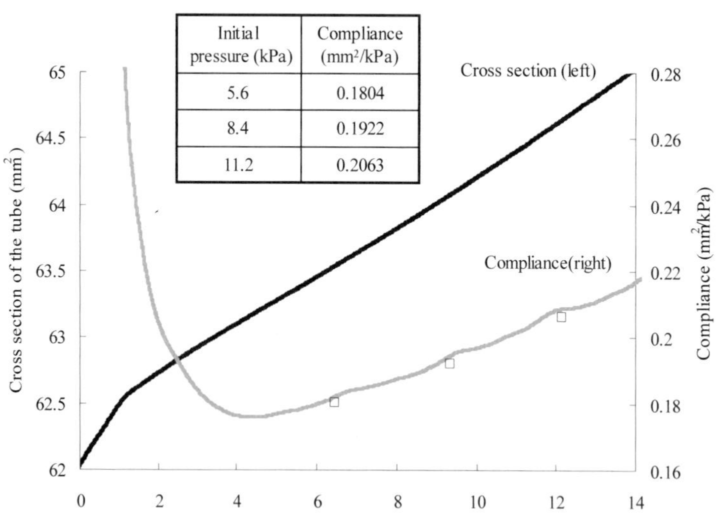

Figure 14. Static relation between pressure and cross-section of the silicone tube

Viscoelasticity of a strip of the silicone tube was measured by a dynamic viscoelasticity measuring device. The measured viscoelastic properties are same as shown in section 3.1.2. This result indicates that both the dynamic modulus viscoelasticity ratio and loss tangent, tends to increase gradually with the increase in frequency. This trend is similar to the viscoelasticity property of blood vessels (Learoyd et al., 1966), and the dynamic modulus viscoelasticity ratio of the blood vessels is similar to or slightly more than that of the experiment tube. The procedure used to decide the values of the viscoelastic tube parameter are same as shown in section 3.1.2.

Terminal resistance

The viscous resistance of flow in the capillary tube of terminal resistance can be approximated by Poiseuille flow, as the terminal resistance occurred from rigid pipe flow and its inside diameter was sufficiently small. The ratio of the flow rate in the capillary tube to the pressure difference between both ends of the capillary tube (terminal resistance) seemed to be constant regardless of the flow rate; therefore, terminal resistance R_T (1.80 kPa/ml/s) was determined from the pressure difference and the flow rate obtained by moving the piston pump at a constant speed so as to generate a constant flow (about 1.1 ml/s).

Numerical simulation method

As initial conditions, baseline internal pressure and a cross-section which corresponded to the baseline internal pressure were used. It was assumed that there was no flow in the initial state inside the tube. Hence, the initial conditions can be shown as follows,

$$\begin{cases} p(x,0) = p_0 \\ A(x,0) = A_0 \\ Q(x,0) = 0 \end{cases} \tag{16}$$

The flow rate measured by a flow sensor was used as the input boundary condition, and terminal resistance R_T was used as the output boundary condition. Hence, the boundary conditions can be shown as follows,

$$\begin{cases} Q(0,t) = Q(t) \\ Q(L,t) = \dfrac{p(L,t)}{R_T} \end{cases} \tag{17}$$

We used the same numerical simulation methods as in a previous paper (Kitawaki & Shimizu, 2006), including calculation schemes, computational algorithms, and fast calculation methods. The basic equations of the numerical simulation were digitized using a staggered grid system in space. For the calculation, Jameson-Baker's 4th order 4-step method as a time differential and 4th order central differential as space with numerical friction was used (Jameson & Baker, 1983). The flow rate and cross-section of the next time step were obtained from equations of continuity and momentum conservation, and then pressure was calculated from the tube law as a function of time. Convolution integrals appear in the viscous resistance term in Eq. (3) and in the viscoelastic term in Eq. (7). A high-speed calculation method was applied, since calculation of the convolution integrals requires significant computer memory to hold past velocity and cross-section values and requires much computational time.

Time step Δt and grid interval Δx were set at 0.5 ms and 0.04 m, respectively. The Courant number was 0.325 because the propagation velocity of the pressure wave was about 26 m/s at maximum. As a result, the CFL condition (numerical stability condition) was satisfied.

4.2. Results and discussions

4.2.1. Experimental results of pressure propagation

As an example of experimental results, the internal pressure measured at three points in the experiment tube, and the flow rate at the proximal end are shown in Fig. 15, at the baseline pressure of 5.6 kPa. The change in the whole time experiment is shown in Fig. 15(a), and an enlarged view of the last but one full-cycle pulsation is shown in Fig. 15(b). The pressure value shown here was set to be a differential pressure from the baseline internal pressure.

As shown in Fig. 15(a), there is a rise in the mean pressure or waveform change up to the first three pulses, but then it converges in the steady state. Also, there are oscillating pressure waves at the proximal and distal ends of the initial pulsation, due to wave reflections at both ends of the experiment tube; however, these oscillatory waves disappeared after the second pulsation, because this phenomenon is caused by the remaining reflected wave of the previous pulse.

As shown in Fig. 15(b) of the steady state of the pressure waveform, part of the waveform measured at the proximal end becomes almost flat in its upper portion for about 0.3 seconds, and then the pressure exponentially decreases. It is thought that this flat portion is generated by the progressive wave, which is produced by the movement of the piston pump, overlapping the reflected wave coming from the distal end of the tube. At the same time, the pressure waveform measured at the distal end has a pointed peak shape and a small reflected wave at the shoulder. The maximum pressure gradually increases from the proximal end of the tube towards the distal end, and the phenomenon is similar to the Peaking phenomenon *in vivo*.

4.2.2. Comparison between calculated results and experimental results

The calculated results in each tube law model were compared with experimental results in order to confirm the effect of the difference of the tube law model on the experimental results. Figure 16 shows examples of steady-state pulse waveforms at the baseline internal pressure of 5.6 kPa. Experimental results are shown by a bold black line, and calculated results are shown as a thin red line. In the Voigt model, calculated results in which the relaxation time parameter is multiplied by 4 (10.0ms) are also shown as a dotted line.

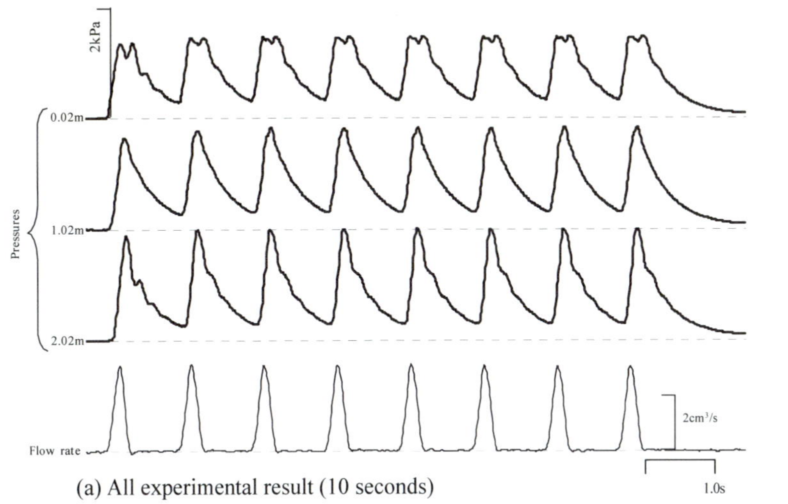

(a) All experimental result (10 seconds)

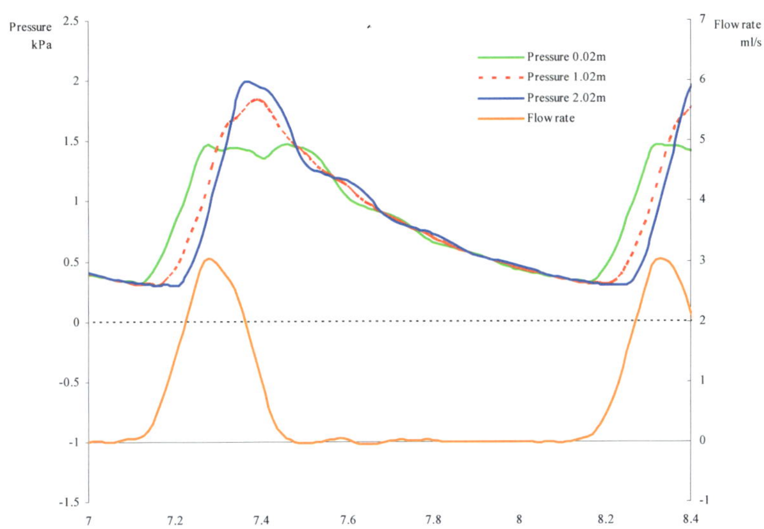

(b) Enlarged view in a period

Figure 15. Time profiles of pressure waves at three positions in the tube and flow volume into the tube.

In the calculated results using the elastic model and the Voigt model, the calculated maximum pressure is lower than that of the experimental results, and the calculated minimum pressure is higher than that of the experimental results; therefore, the difference between the calculated results and experimental results is large. Also, it is shown that the difference between the elastic model and the Voigt model is small, indicating that in the periodic change of waveforms, the Voigt model is not sufficiently effective. In addition, the calculated results did not agree with the experimental results in the Voigt model, because the effect is too small even when the value of the relaxation time parameter is varied as shown in Fig. 16(b).

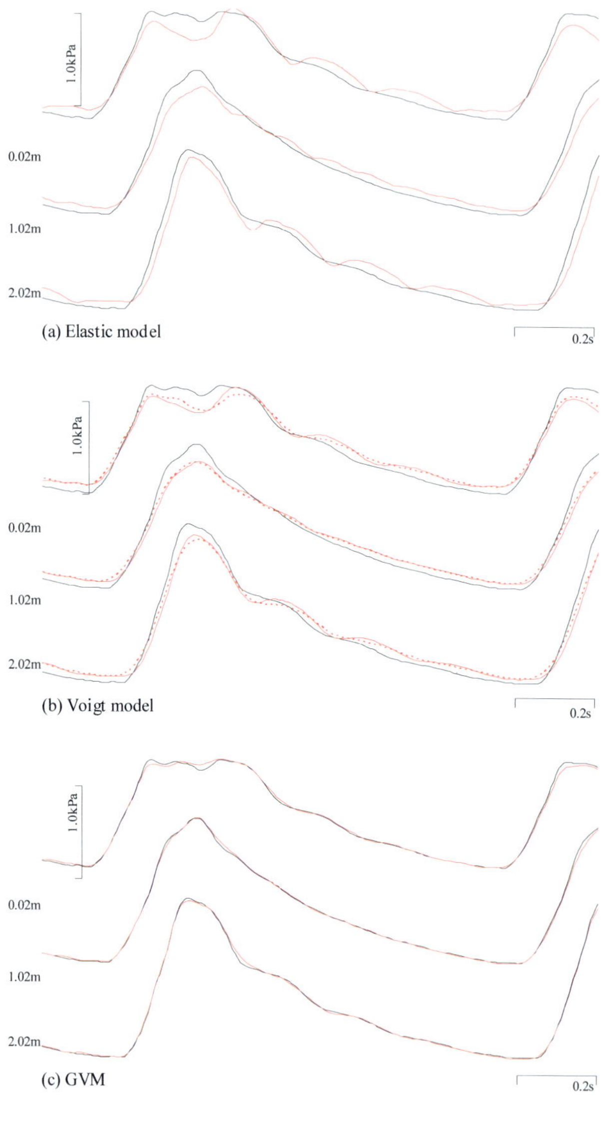

Figure 16. Calculated pressure waves (thin red line) compared with the measured waves (bold black line) at three locations in the tube

On the other hand, the calculated results using the generalized viscoelastic model almost completely agreed with the experimental results in the whole region. As mentioned above, it was proven that numerical simulation using the generalized viscoelastic model could express the experimental result accurately.

4.2.3. *Difference in the viscoelastic parameter*

In order to find out how many changes in the viscoelastic property of the tube were accompanied by baseline internal pressure change, and how this change influenced pulse wave propagation, the value of the viscoelastic parameter was determined under different conditions of baseline internal pressure. Initially, pressure waveforms at the baseline internal pressure of 8.4 kPa and 11.2 kPa were calculated using the value of the viscoelastic parameter at a baseline internal pressure of 5.6 kPa (hereinafter, this viscoelastic parameter is referred to as GVM(1)). The change in tube deformation compliance was incorporated into these calculations. As an example, a comparison of the calculated results and experimental results at the proximal and distal ends at a baseline internal pressure of 11.2 kPa is shown in Fig. 17. PP at the proximal end using GVM(1) was smaller than the experimental value, and it was proven that the viscoelastic effect of GVM(1) was not sufficient.

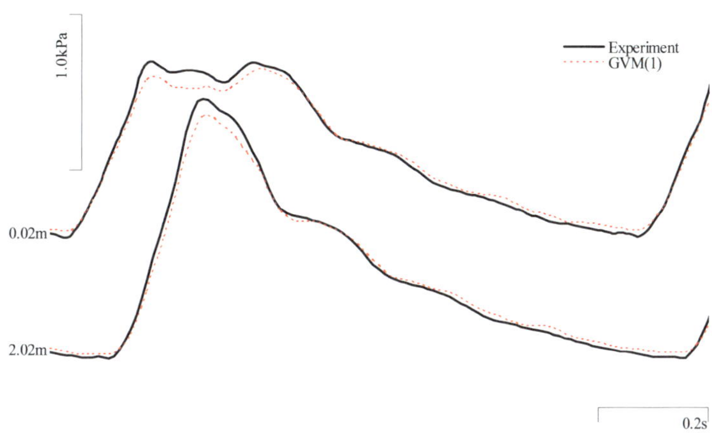

Figure 17. Calculated pressure waves (dotted red line) compared with the measured waves (bold black line) at two locations in the tube

Based on this result, the value of the viscoelastic parameter was adjusted. We changed the value of the dynamic viscoelastic parameter in the low frequency region (i=1) so that the dynamic modulus elasticity ratio increased with the increase in baseline internal pressure in the experiment tube, and the value of the dynamic viscoelastic parameter in other frequency regions (i=2-7) was adjusted to a fixed ratio. The special evaluation function was not used to decide parameters but we took it into consideration to agree with the pulse waveform and the PWV. The viscoelastic parameter value decided in this way (hereinafter referred to as GVM(2)) is described in Table 2 and Fig. 18. The squares in Fig. 18 are the measured viscoelastic values of Fig. 4.

Agreement of the calculated results and experimental results using GVM(2), although not shown, was the same level as that in Fig. 16 (c).

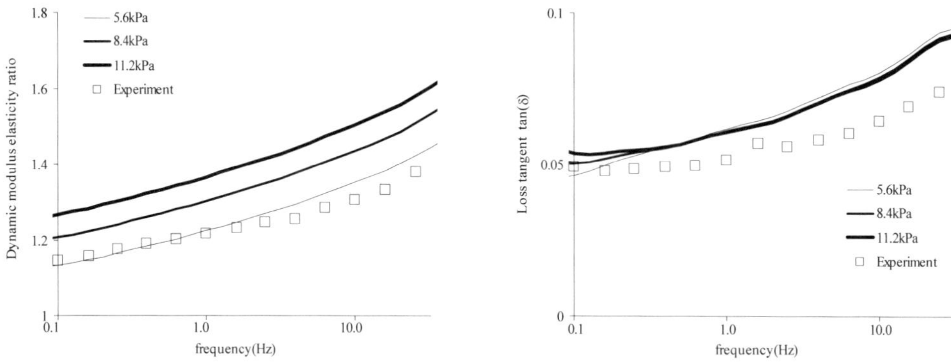

(a) Dynamic modulus viscoelasticity ratio (b) loss tangent

Figure 18. Change in viscoelastic properties of silicone tube

i	τ_i (s)	f(Hz)	f_i		
			5.6kPa	8.4kPa	11.2kPa
1	17.0	0.009	0.095	0.169	0.227
2	3.75	0.042	0.033	0.034	0.036
3	0.7	0.227	0.061	0.063	0.066
4	0.15	1.061	0.073	0.076	0.079
5	0.03	5.305	0.096	0.100	0.104
6	0.005	31.83	0.170	0.177	0.184
7	0.0008	198.9	0.220	0.229	0.238

Table 2. Optimized viscoelastic parameters for each baseline pressure

4.2.4. Effect of difference in viscoelasticity on pulse wave propagation

The effect of the difference in viscoelastic properties on pulse wave propagation was considered by analyzing the calculated results in which the viscoelastic model and the viscoelastic parameter were changed. PWV and PP were obtained from the calculated and experimental results. The relationship of these indexes with the baseline internal pressure is shown in Fig. 19. These indexes were obtained from mean values in four pulsations in the latter half in the steady state. Defining the maximum value of the second order differential waveform of the pulse waveform as an initial rise of the pulse wave, PWV was calculated by dividing the distance between calculation points (at the proximal and distal ends) by the time difference in the initial rise of these pulse waves. The error range of PWV is shown as an error in propagation velocity by a quantize error in calculating the propagation time. PP was obtained from the pulse wave at the proximal end.

Effect of dynamic elasticity modulus ratio

Figure 19(a) also shows the elastic tube theoretical velocity (Moens-Korteweg's theoretical velocity) calculated from the tube inner diameter and tube deformation compliance. The theoretical velocity decreased by 5.8% as a result of the influence of the increase in tube deformation compliance, when the baseline internal pressure increased from 5.6 kPa to 11.2 kPa. On the other hand, PWV calculated from the experimental results was bigger than the elastic tube theoretical velocity, and the increasing ratio against the elastic theoretical velocity gradually increased with the rise in baseline internal pressure.

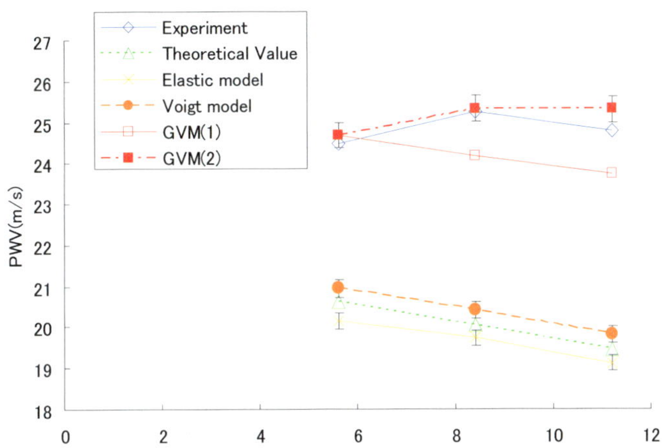

(a) PWV change

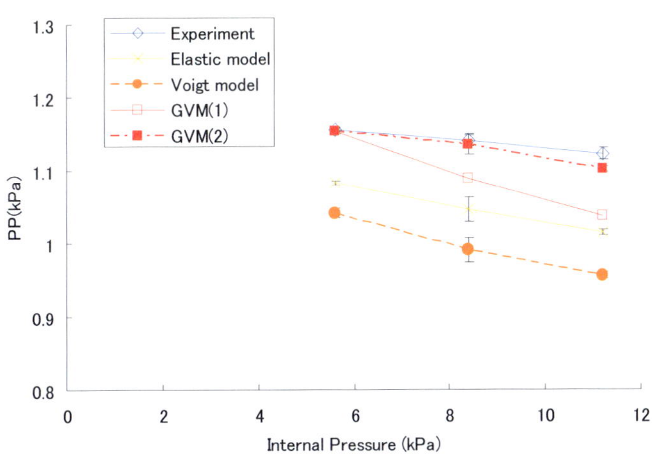

(b) PP change

Figure 19. PWV and PP change with internal pressure change

PWV using the elastic model approximately agreed with the theoretical elastic tube velocity, since the viscosity of the tube was not included. Using the Voigt model, the small effect by the viscoelasticity on the increase in PWV and PWV was slightly larger than the theoretical elastic tube velocity; however, the calculated and experimental results agreed well with the generalized viscoelastic model in GVM(2).

PP in the calculated results of the elastic model and the Voigt model was 90% of that in the experimental results. In the generalized viscoelastic model, the calculated results of GVM(2) agreed well with the experimental results.

This results show that, with the influence of viscoelastic change shown in Fig. 18, PWV increased by around 20-30% more than theoretical elastic tube velocity and PP increased by approximately 10% because increasing the elastic modulus by the effect of the low frequency component of viscoelasticity is more effective than the energy dissipation effect by the high frequency component of viscoelasticity. Furthermore, it can be said that the increase of viscoelasticity compensated for the influence caused by the change in tube deformation compliance.

Effect of the viscoelastic parameter

The effect of the viscoelastic parameter was examined from the difference in the calculated results between GVM(2) and GVM(1) for the baseline internal pressure of 11.2 kPa. The dynamic modulus elasticity ratio of GVM(1) was nearly 10.1% lower than that of GVM(2) within 1–10 Hz, as shown in Fig. 18. By this effect, PWV using GVM(1) was 6.6% lower than that using GVM(2), as shown in Fig. 19, and PP with GVM(1) was 6.4% lower than that with GVM(2). Considering the results with a baseline internal pressure of 8.4kPa, the difference of PWV and PP caused by the viscoelastic parameters was almost proportional to the change of baseline internal pressure; therefore, it might be necessary to determine the dynamic modulus elasticity ratio with accuracy below 4% in order to keep the accuracy of PWV and PP below 5%.

As described above, it was clarified that tube viscoelasticity played an important role in pulse wave propagation under periodic pulsatile conditions and had a significant effect on clinical arterial stiffness indexes, including PWV and PP.

4.3. Conclusion of this section

Using the one-dimensional numerical simulation model of a viscoelastic tube, the effect of tube viscoelasticity on periodic pulse wave propagation was examined by comparing the experimental results with the calculated results with a viscoelastic silicone tube. As a result, we can reach the following conclusions.

1. Numerical simulation using the generalized viscoelastic model can accurately express the experimental results with periodic pulsation of pressure waves.
2. When the dynamic modulus elasticity ratio increases, both PWV and PP increase, because increasing the elastic modulus is more effective than the energy dissipation effect by viscoelasticity change.

3. It is necessary to measure the dynamic modulus elasticity ratio with accuracy below 4% in order to estimate the values of PWV and PP with accuracy below 5%..

5. Conclusion

In this chapter, a nonlinear one-dimensional numerical model was constructed by including the effect of tube wall viscoelasticity and unsteady viscous resistance for numerical analysis of the viscoelastic tubes. This model can analyze the effect of viscoelasticity on intravascular flow that few previous papers have considered. Using a one-dimensional numerical simulation model of a viscoelastic tube, the effect of tube viscoelasticity on a pulsatile wave and periodic pulse wave propagation were examined by comparing the experimental results with the calculated results using a viscoelastic silicone tube.

Clinical blood vessel stiffness indexes vary with viscoelasticity change, because the elastic modulus is more effective than the energy dissipation effect by viscoelasticity change. This result showed that the viscoelasticity of the vessel wall plays an important role in the form of a pulsatile wave; therefore, it is important to consider the viscoelastic effect accurately in quantitative investigations using intravascular flow analysis.

Author details

Tomoki Kitawaki
Graduate School of Health Sciences, Okayama University, Japan

6. References

Anliker, M., & Rockwall, R. L. (1971) Nonlinear Analysis of Flow Pulses and Shock Waves in Arteries, *ZAMP*, Vol.22, pp. 217-246, ISSN 0044-2275

Avolio, A. P. (1980) Multi-Branched Model of the Human Arterial System, *Med. & Biol. Eng. & Comput.*, Vol.18, pp. 709-18, ISSN 0140-0118

Hayashi, K., Handa, H., Nagasawa, S. (1980) Okumura, A., & Moritake, K., Stiffness and Elastic Behavior of Human Intracranial and Extracranial Arteries, *J. Biomech.*, Vol.13, pp. 175-184, ISSN 0021-9290

Jameson, A., & Baker, T. J. (1983) Solution of the Euler Equations for Complex Configurations, *AIAA paper*, 83-1929, pp. 293-302, ISSN 0146-3705

Kagawa, T., Lee, I., Kitagawa, A., & Takenaka, T. (1983) High Speed and Accurate Computing Method of Frequency-Dependent Friction in Laminar Pipe Flow for Characteristics Method, *Trans. Jpn. Soc. Mech. Eng.*, Vol.B49, pp. 2638-44, ISSN 0387-5016

Kitawaki, T., Shimizu, M., Himeno, R., & Liu, H. (2003) One-Dimensional Numerical Simulation of Visco-Elastic Tube, *Trans. Jpn. Soc. Mech. Eng.*, Vol.A69-677, pp. 55-61, ISSN 0387-5008

Kitawaki, T., & Shimizu, M. (2005) Effect of the Blood Vessel Viscoelasticity on the Blood Pressure Wave Propagation, *Trans. Jpn. Soc. Mech. Eng.*, Vol.B71-707, pp. 1768-1774, ISSN 0387-5016

Kitawaki, T., & Shimizu, M. (2006) Flow Analysis of Viscoelastic Tube Using One-Dimensional Numerical Simulation Model, *Journal of Biomechanical Science and Engineering*, Vol. 1, No. 1, pp.183-194, ISSN 1880-9863

Kitawaki, T., & Shimizu, M. (2009) Effect of the Blood Vessel Viscoelasticity on Periodic Blood Pressure Wave Propagation, *Journal of Fluid Science and Technology*, Vol. 4, No. 1, pp. 234-245, ISSN 1880-5558

Learoyd, B. M., Taylor, M. G. (1966) Alterations with age in the viscoelastic properties of human arterial walls, *Circ Res.*, Vol.18 No. 3, pp. 278-292, ISSN 0009-7330

Oliver, J. J., & Webb, D. J. (2003) Noninvasive Assessment of Arterial Stiffness and Risk of Atherosclerotic Events, *Arterioscler. Thromb. Vasc. Biol.*, 23, pp. 554-66, ISSN 1079-5642

Olufsen, M. (1999) Structured Tree Outflow Condition for Blood Flow in Larger Systemic Arteries, *Am. J. Physiology*, Vol.276 No. 1, pp. H257-268, ISSN 0002-9513

Porenta, G., Young, D. F. & Rogg, T. R. (1986) A Finite-Element Model of Blood Flow in Arteries Including Taper, Branches, and Obstructions, *J. Biomech. Eng.*, Vol.108, pp. 161-7, ISSN 0148-0731

Reuderink P. J., Hoogstraten HW, Sipkema P, Hillen B, Westerhof N. (1989) Linear and nonlinear one-dimensional models of pulse wave transmission at high Womersley numbers, *J Biomech.*, Vol.22, pp.819-827, ISSN 0021-9290

Sato, M., & Ohshima (1985) Nonlinear Viscoelastic Behaviour of Canine Arterial Walls, *Med. & Biol. Eng. & Comput.*, Vol.23, pp. 565-571, ISSN 0140-0118

Schaaf, B. W., & Abbrecht, P. H. (1972) Digital Computer Simulation of Human Systemic Arterial Pulse Transmission: a Non-Linear Model, *J. Biomech.*, Vol.5, pp. 345-364, ISSN 0021-9290

Snyder, M. F., Rideout, V. C., & Hillestad, R. J. (1968) Computer Modeling of the Human Systemic Arterial Tree, *J. Biomech.*, Vol.1, pp. 341-53, ISSN 0021-9290

Westerhof, N., & Noordergraaf, A. (1970) Arterial Viscoelasticity: a Generalized Model. Effect on Input Impedance and Wave Travel in the Systematic Tree. *J. Biomech.*, Vol.3, pp. 357-79, ISSN 0021-9290

Zielke, W. (1968). Frequency-Dependent Friction in Transient Pipe Flow. *Trans. ASME, Ser.D*, Vol.90, pp. 109-115, ISSN 0021-9223

Section 3

Food Colloids

Chapter 10

Viscoelastic Properties of Starch and Non-Starch Thickeners in Simple Mixtures or Model Food

Ioanna G. Mandala

Additional information is available at the end of the chapter

http://dx.doi.org/10.5772/50221

1. Introduction

Design of new food products is one of the most important tasks in the food industry. Improving or controlling texture of foodstuffs leads to products with advanced functionalities e.g. creation of nursing-care foods, ''ready-to swallow foods'', soft or hard gels etc. (Funami, 2011).

Texture change can be achieved by adding hydrocolloids that in small quantities bind large amounts of water and can then control both structure and texture. Starches belong to the same category of hydrocolloids, although they are used in a wide range of products either as raw materials or as food additives. Starches can differ with respect to the amylose content depending on their origin, or can be structurally modified. Native starches could have negative aspects such as gel syneresis, retrogradation, breakdown, cohesive, rubbery pastes and undesirable gels formation (Whistler & BeMiller, 1997), but this is not the case with modified starches. Moreover, modified food starches are less expensive and are more widely available than gums or other food stabilizers. A way to overcome shortcomings of native starches is their blending with polysaccharide hydrocolloids. Native or modified starches, and non-starch hydrocolloids are increasingly important ingredients in the modern health-conscious food industry (Techawipharat et al., 2008), considering that specific starch types such as resistant starch can be considered insoluble fibers as well.

This chapter aims at highlighting recent research in the field of viscoelastic properties of starches and their mixtures with some selected hydrocolloids. Furthermore, these interactions will be linked to the final rheological characteristics of specific products aimed at successful product development.

The control of texture in real foods with several ingredients can be achieved through viscoelasticity measurements of carbohydrate mixtures at low concentrations. This research

INTECH
open science | open minds

© 2012 The Author(s). Licensee InTech. This chapter is distributed under the terms of the Creative Commons Attribution License http://creativecommons.org/licenses/by/3.0), which permits unrestricted use, distribution, and reproduction in any medium, provided the original work is properly cited.

can result in new products development that aims at specific texture characteristics. Examples are focused on oscillatory experiments and in some case correlations to viscous properties are presented. The role of rheology in current research is further shown and discussed through these examples.

2. Oscillatory rheological method and hydrocolloids behavior

Among different techniques used to distinguish between the solid and liquid-like characteristics of a food colloid, the best technique is to use an oscillatory rheological method (Dickinson, 1992). In an oscillatory rheological experiment, both stress and strain commonly present a sinusoidal variation. This is the most popular method to characterize viscoelasticity, since relative contributions of viscous and elastic response of materials can be measured. The cycle time, or frequency of oscillation, defines the timescale of these tests. By observing material response as a function of frequency, material can be tested at different timescales. The observation of material response at different frequencies is also referred to as mechanical spectroscopy (Stanley et al., 1996). Linear viscoelasticity is known as the region where stress and strain waves are set at such low values that stress is proportional to strain. This type of tests is also known as small amplitude oscillatory shear (SAOS). The relationship between stress and strain is then described and storage, loss modulus, complex shear modulus as well as dynamic viscosity can be measured. The storage dynamic modulus (G′) is a measure of the energy stored in the material and recovered from it per cycle while the loss modulus (G′′) is a measure of the energy dissipated or lost per cycle of sinusoidal deformation (Ferry, 1980, Stanley et al., 1996). The ratio of the energy lost to the energy stored for each cycle can be defined by tanδ.

The viscoelastic behavior of a simple or more complex structure can be determined in the above way. Furthermore, as structure is not disrupted, changes including sol-gel transition, gel curing, aggregation, flocs creation etc. can be monitored. Generally speaking, rheological properties could be of high interest in a) product quality characterization b) process design and flow conditions analysis (e.g. pump sizing, filtration, extrusion etc.) c) design of new foodstuffs d) basic research on ingredients' interactions.

According to Roos-Murphy (1984) solutions and gels belong to the categories of entanglement solutions, weak gels and strong gels. Hydrocolloids including starch can belong to all of the three categories revealing the wide spectrum of structures they can adopt according to their own natural state and the environmental conditions found.

Entanglement solutions (e.g. guar gum solutions) present a strong dependence of both storage and loss modulus on frequency. Weak gels behavior (e.g. xanthan gum) is characterized by gel-type mechanical spectrum, whereas strong gels (e.g. amylopectin, amylose gels) present high storage modulus values irrespective of frequency, as junction zones among macromolecules are stable on a relatively long time scale. The spectrum of hydrocolloids can vary at different concentrations.

When at low frequency the loss modulus G′′ is higher than the storage modulus G′, both parameters vary sharply with frequency: G′′(ω) and G′(ω²). This behavior is said to be typical of a liquid-like material. As frequency increases, G′(ω) crosses G″(ω), the response of the material beyond this cross-over frequency is said to be solid-like. Entanglements of macromolecular solutions can result in such behavior. When G′ is higher than G″ over most of the frequency range investigated, a weak gel behavior is observed due to the formation of a weak three-dimensional network of ordered chain segments.

Thus, viscoelastic structures of hydrocolloids may differ considerably. So, recent data about their behavior in mixtures with starch or model foods is discussed.

3. Starch under heating

Dynamic rheological tests allow continuous measurement of dynamic moduli during temperature and frequency sweep testing of a starch suspension. The rheological properties of starches differ, because their composition and granules' morphology is different. Thus, the identification of native starch sources is required in order to achieve the desired functionality and unique properties (Duxbury, 1989). Furthermore, concentration effects, temperature, heating rate and shear rate effects can be found and measured by rheological properties.

Concerning starch the following changes under heating can be measured using oscillatory experiments.

- Gelatinization
- Pasting
- Gelling and
- Retrogradation, which can be distinguished in short-term by network formation and long-term retrogradation that lasts several weeks

3.1. Dynamic rheology and gelatinization

During this first stage of heating, starch granules swell during the process of gelatinization. Soluble polymer molecules leach from the swollen granules and the rheological properties, such as storage modulus (G′) and loss modulus (G″) of the starch increase to a maximum. A sharp increase in G′ may occur between 60-80°C (Ahmed et al., 2008) caused by the formation of three-dimensional (3D) gel network developed by leached out amylose and reinforced by strong interactions among swollen starch particles (Fig. 1). Similar changes can occur when viscosity is measured.

The swelling of the granules is important for both viscosity increase and viscoeleasticity of the produced dispersions. Granules' morphology and rigidity, complexes with other components (e.g. lipid-amylose), amylose content, protein content are some factors that determine both peak values of the viscoelastic parameters and their breakdown thereafter.

Concerning their botanical source, among native starches (corn, rice, wheat and potato), potato starches exhibit the highest swelling power and final viscoelastic values. Their shape and size differs with respect to starches of other botanical sources. Starch granules of potato are smooth-surfaced and of different shapes form oval and irregular to cube-shaped. Starch granules of corn are angular-shaped, while those of rice are pentagonal and angular-shaped. Finally, wheat starch granules are spherical (B-granules) and lenticular-shaped (A-granules).

Moreover, potato starch granules are the largest (<110μm) in size followed by wheat (<30μm), corn (<25μm) and rice (<20μm) starches. The granule size of potato starch is variable and ranges form 1 to 20μm for small and from 20 to 110 μm for large potato granules, whereas rice starch granules commonly range from 3 to 5μm in size (Hoover, 2001; Singh et al., 2003). Large and cubical or irregularly shaped granules in potato starch exhibit higher storage and loss modulus and lower tanδ than the small and oval granules (Singh & Singh, 2001). Thus, potato starch shows higher G′, G′′ and lower tanδ than corn, rice and wheat starches during the heating cycle. Furthermore, starch dispersions may exhibit significantly higher G′ values (~100 times) as compared to flour dispersions at the first period of heating (40-60 °C) (Ahmed et al., 2008).

Amylose amount is also quite important for controlling the viscoelastic properties of starch dispersions. Amylose results in higher G′ indicating a well-cross-linked nature. Specifically G′ can increase exponentially as a function of amylose content (Biliaderis & Juliano, 1993). Concentration effects are also linear for wheat and maize starch in the range of 6-30% (Ring, 1985) and follow a power law in the case of rice starches (8-40%).

Starch is a complicated viscoelastic structure. Under heating it can be described as a composite system, in which gelatinization may be regarded as an example of a phase-separated composite gel, primarily governed by the volume fraction occupied by the swollen particles, whereas the continuous phase makes an additional contribution due to its own viscoelastic properties (Alloncle & Doublier, 1991; Dickinson,1992).

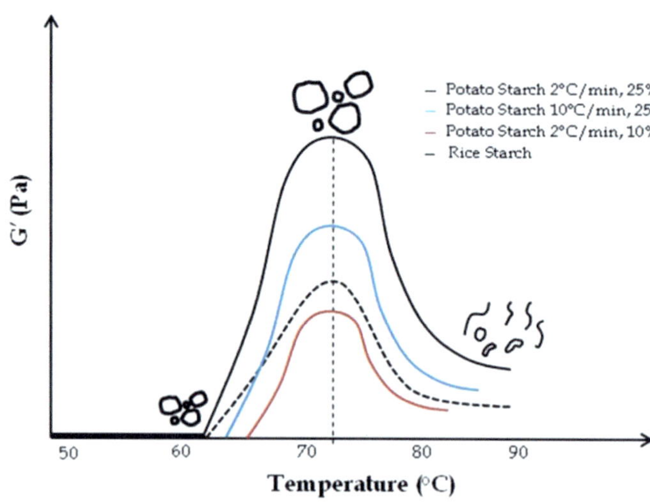

Figure 1. Viscoelastic changes of starch suspensions under heating and further cooling. Curves can shift to both axes accordingly to the factors that are mentioned. Changes in starch granules are also shown. Peak and plateau values can be seen at maximum starch swelling. Thereafter network breakdown and paste formation is shown. (Modification according to data from Ahmed et al., 2008, Food Hydrocolloids 22, pp 278-287 and Singh et al, 2007a, Starch/Stärke, 59, pp. 10-20)

3.2. Pasting and viscoelastic properties

After gelatinization the process of pasting follows. Under continuing heating, granules are further swollen and finally disrupted (Fig.1). A hot paste is then created consisting of swollen granules, granule fragments, and soluble materials. The botanical source of the starch, water content, temperature and shearing during heating determine the consistency of this paste. The network created consists of dissolved starch polymers (amylose and amylopectin) and a discontinuous phase of swollen granules, empty (ghost) ones and fragments.

Rheologically, a peak value of both viscosity and G′, G′′ is reached resulting mainly from maximum swelling. Furthermore a plateau e.g. constant values may occur (from 80-85°C) due to irreversible swelling and solubilisation of amylose (Ahmed et al., 2008) followed by a sudden drop of G′ under extensive heating and shear and time. Granules disintegrate. At this point a hot paste is created (Fig. 1). The height of the peak at a given concentration reflects the ability of the granules to swell freely before their physical breakdown. A sudden drop after the maximum indicates the breakdown on cooking as well as a great ability to swell (Adebowale & Lawal, 2003).

G′ decrease indicates the gel structure disruption due to the ′′melting′′ of the crystalline regions or disentanglements of the amylopectin molecules in the swollen particles that softens the particles (Tsai et al., 1997). The network collapses due to the loss of interactions between the particles (Ahmend et al., 2008).

3.3. Retrogradation and viscoelastic properties

The distinction between a paste and a gel is not always evident. A paste usually refers to the hot freshly cooked system and gel is formed after cooling. Both are viscoelastic materials. As the hot pastes, especially of amylose –containing starches, begin to cool they become more elastic and develop solid properties. The transition from a viscous to an elastic gel can be determined by storage and loss moduli thus the setback can be found as a transition point from viscous to solid one (BeMiller, 2011). This setback is known as retrogradation (Atwell et al., 1988). At this critical gel point the system is wall-to-wall connected (percolation threshold) and is characterized by a critical behavior with $G'(\omega)$ and $G''(\omega)$ obeying the same power law: $G'(\omega) \sim G''(\omega) \sim \omega^n$ (Doublier & Cuvelier, 2006).

The first phase of retrogradation begins as the paste cools and a formation of entanglements and/or junction zones is created between amylose molecules resulting in an elastic gel. This phase may last up to 48 h. The second phase of retrogradation involves amylopectin changes, which is a much slower process that may proceed for several weeks depending on the storage temperature. Both G′ and G′′ increase upon cooling and during short-storage, G′ and G′′ increase indicating that the gels become firmer.

4. Modified starches

Briefly, chemical modification leads to a considerable change in the rheological and pasting properties of starches. Storage (G′) and loss modulus (G′′) of acetylated, hydroxypropylated and cross-linked starches from different sources increase to a maximum and then drop during heating following the same general rheological pattern as native starches (Singh et al., 2007b).

The temperature of maximum G′ drops significantly on acetylation or hydroxypropylation, while it increases after cross- linking (Kaur et al., 2004, 2006; Singh et al., 2004). Acetylated corn and potato starches showed greater values of G′ and G′′ under heating but lower compared to their native starch gels upon cooling of heated starch gels, confirming their lower tendency to retro gradate.

Strengthening bonding between starch chains by cross-linking will increase resistance of the granules towards swelling resulting in lower G′ values in a high degree of cross-linking. Cross-link concentration and location could lead to different rheology. Botanical source also influences cross-linking and potato starches show a higher susceptibility towards cross-linking than other kinds of starches (Kaur et al., 2004).

5. Interactions with other hydrocolloids

According to an excellent review of BeMiller (2011) twenty-one different native starches in combination with thirty two different hydrocolloids have been investigated in different studies. Thus there is an increased interest in starch-hydrocolloid systems as well as a significant amount of scientific work in this area. New research works are mainly presented in this section, enhancing the knowledge about such systems.

5.1. Hydrocolloids influence in starch pastes

Starch pastes have typical biopolymer gel behavior. In typical biopolymer gel behavior greater G′ values than G′′ along the frequency sweep are observed, however in starch pastes both moduli are frequency dependent. Hydrocolloids modify the dynamic spectra of starch, although different trends can be observed.

First assumption: Hydrocolloids lead to weaker structures with less gel-like character.

With their addition, starch network shifts from an elastic-like to a more viscous-like one (Rosell et al., 2011). Starch-hydrocolloid systems can be considered as biphasic systems. When starch granules are swollen, the hydrocolloid is located entirely in the continuous phase. The concentration of hydrocolloid will then increase as the volume of the phase accessible to the hydrocolloid is reduced. This fact changes the viscoelasticity of the starch. Thus, cellulose derivatives and carrageenans can lead to less solid-like pastes than the control paste (Techawipharat et al., 2008). This assumption depends on starch type as well. In waxy starches, due to an absence of amylose, short-term retrogradation does not occur and therefore, the addition of hydrocolloids could not alter the viscoelastic characteristics of these starch pastes.

Second assumption: Hydrocolloids addition leads to associations with starches resulting in increased G′, G′′ values.

Several hydrocolloids can promote associations with starches and as a result, when they are added in starch pastes, an increase in G′, G″ is often observed, see examples: Mandala et al.(2004a); Achayuthakan & Suphantharika, (2008); Wang et al., (2008). In such systems the question is which ingredient predominates in the overall rheology, starch or hydrocolloid. In wheat starch-hydrocolloid systems, it is the hydrocolloid that predominates in the whole system, according to the shift factors found.

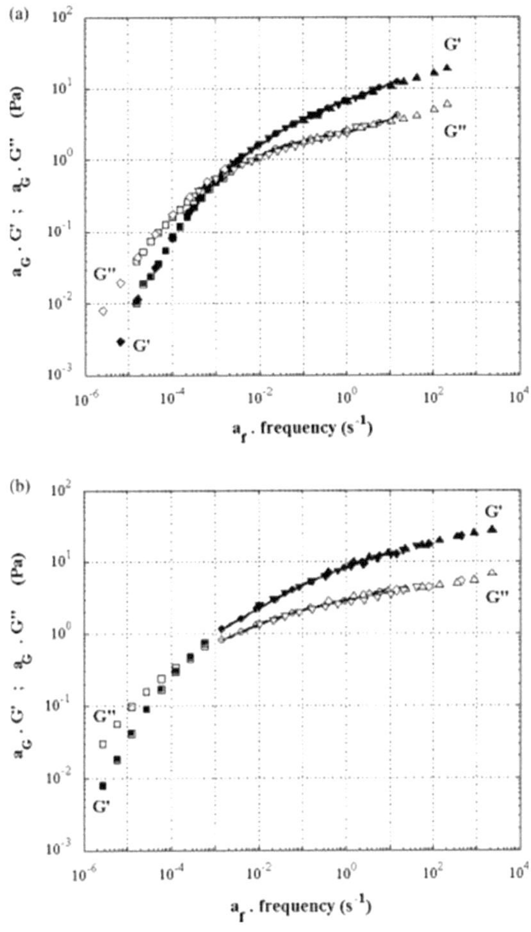

(a) In water of xanthan 0.1 wt% (□, ■) and 0.5 wt% (○, ●), starch 2 wt%/xanthan 0.5 wt% (Δ, ▲), amylose 0.3 wt%/xanthan 0.1 wt% (◊, ♦), amylose 0.3 wt%/xanthan 0.5 wt% (∇, ▼).
(b) In 0.1 M NaCl of xanthan 0.1 wt% (□, ■), 0.5 wt% (○, ●) and 0.8% (◊, ♦), starch 2 wt%/xanthan 0.5 wt% (Δ, ▲), amylose 0.3 wt%/xanthan 0.5 wt%,(∇, ▼). (From Mandala et al, 2004a. Carbohydrate Polymers 58, pp 285–292, with permission).

Figure 2. Superimposed shifted spectra.

Furthermore, except G′, G″ values, tanδ may be important in interpretations of the behavior of starch-hydrocolloids interactions. Thus, although some hydrocolloids promote an increase in G′, G″ by their addition, in fact they lead to a less solid-like system (higher tanδ values) as described above. On the contrary, in some cases, hydrocolloid addition may lead to a more solid-like system as noticed in systems of maize starch with flaxseed gum (Wang et al., 2008).

In this research work, the variation of the G′ with frequency for the maize starch alone and the flaxseed gum-maize starch mixtures with different flaxseed gum concentrations was not significant. This suggests that both the maize starch and its mixture with flaxseed gum have a typical biopolymer gel network, but flaxseed gum helps the formation of stronger gels. Concerning temperature effects, at a temperature range of 25-75°C, flaxseed gum addition shows more significant temperature dependence compared to that of maize starch alone. An increase in temperature results in a decrease in G′ of the mixture, indicating that the addition of flaxseed gum affects the thermal stability of the mixture (Wang et al., 2008).

5.2. Influence of hydrocolloids during storage

Gelation and short- or long-term retrogradation of starch can be influenced by hydrocolloids. The addition of a hydrocolloid can accelerate gelation and reduce retrogradation (Kim & Yoo, 2006, Lee et al., 2002; Mandala & Palogou, 2003; Fumami et al., 2005, 2008) but this depends on many parameters, some of which are discussed extensively in the following text.

Concerning gelation, starch-hydrocolloid mixtures may display weak gel-like behavior (Funami et al., 2008; Kim & Yoo, 2006; Lee at al., 2002). According to time-dependent curves of the mixtures of rice starch-xanthan gum, G′ values increase rapidly during the first few hours at low temperature aging (5°C) and remained steady afterwards. Gelation could be considerably shortened by the presence of xanthan gum. Increasing xanthan gum concentration increased G′ values during aging, indicating that the elastic character of xanthan gum influences the reinforcement of the overall gel properties during aging. A rapid increase and subsequent plateau of G′ can be shown by xanthan gum addition (Fig. 3). This is due to the rapid aggregation of amylose chains at the early stage and the slow aggregation of amylopectin chains at the late stage respectively (Kim & Yoo, 2006).

First-order kinetics for structure development of starch-xanthan mixtures during aging (recrystallization) and further retarding during longer storage can be developed. The rate of G′ increase (structure development) due to the retrogradation of rice starch during cold storage is apparently affected by the presence of xanthan gum and greatly dependent on the xanthan gum concentration. The same was observed for wheat starch-xanthan mixtures aging (Mandala & Palogou, 2003).

In the long-term retrogradation of different hydrocolloid-starch mixtures (Funami et al., 2008; Kim & Yoo, 2006; Lee at al., 2002) molecular associations between the gum and the amylopectin fraction inhibit the formation of crystalline structures during storage. As another factor to inhibit long-term retrogradation, gums can stabilize water molecules, therefore they can act as water binder effectively depriving amylose or amylopectin of usable water for crystallization as described in the case of corn starch fenugreek gum mixtures (Funami et al., 2008).

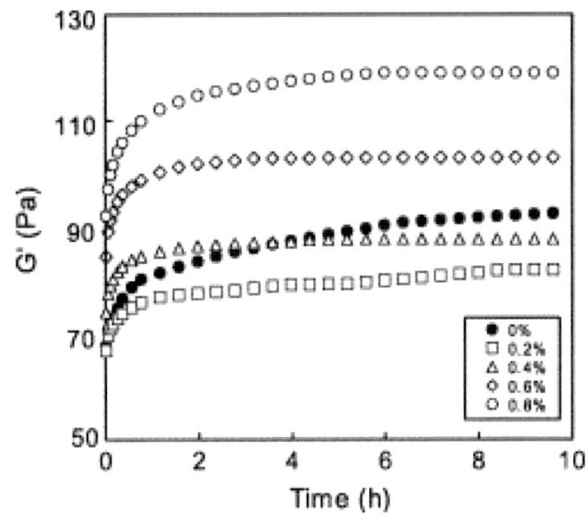

Figure 3. Changes in storage modulus (G') during aging at 4°C for 10h. Close symbol rice starch. Open symbols rice starch-xanthan gum mixtures (XG:0.2-0.8%). (From Kim & Yoo, 2006, Journal of Food Engineering 75, pp. 120-128, with permission).

Gelation and retrogradation can be also influenced by the molecular size of the hydrocolloid in a starch-gum mixture. Thus, the molecular mass and size of guar gum influences gelatinization and retrogradation behaviour of corn starch according to Funami et al. (2005a, 2005b). Viscosity and viscoelastic properties can be measured. Molecular interactions between guar gum and amylose are responsible for an earlier onset of viscosity increase for the composite system of starch-guar gum, whereas molecular interactions between guar gum and amylopectin are responsible for the increase in peak viscosity of the composite system. Moreover, the addition of guar gum accelerates the gelation of starch, in particular when the amylose fraction increases. Concerning the control of retrogradation by adding guar gum, storage modulus (G') for starch systems increases rapidly at very early stage of storage at 4°C.

Short-term retardation of retrogradation is also suggested, because the gelled fraction in the system is reduced with the addition of guar gum (loss targent increase). This happens due to the decrease in the amount of amylose leached out of the starch granules during gelatinization. There is a critical Mw up to which the amount of leached amylose can be influenced, which is $15.0x10^5$ g/mol. The effect of guar gum on the inhibition of short-term retrogradation becomes less Mw-dependent at above this Mw value. On the other hand, the higher the Mw of guar gum, the easier the guar interacts with amylopectin.

G' becomes less-frequency dependent with decreasing Mw of guar gum. These results suggest that the interactions between guar gum and amylose should hardly contribute to forming a gelled or ordered structure (Funami, 2005). Furthermore, the ability of guar gum to inhibit long-term retrogradation is enhanced markedly when the Mw of the guar is over $30.0x10^5$ g/ml. Thus, above this molecular weight guar gum can act easily on either amylose or amylopectin to retard starch crystallization.

Concluding:

- Hydrocolloid addition may decrease or increase the gel-like character of starch pastes depending on hydrocolloid and starch type as well as on gum concentration. The most common observation was the increase in both viscous and elastic character with more pronounced effects on the viscous one.

- Regarding gelation, it is induced by adding hydrocolloids. Short-term retrogradation is also related to this gelation acceleration, as well as to amylose amount.
- Gelation acceleration does not mean retrogradation acceleration as well, since a more viscous character is maintained by hydrocolloid addition in many cases.
- Concerning long-term retrogradation, it is clearly reduced by hydrocolloid addition. Factors that may contribute are the associations of hydrocolloid-amylopectin, the stabilization of water molecules and last but not least the molecular weight of the gum.

6. Applications to foodstuffs

In a food system, many other ingredients influence the process of gelatinization, pasting, breakdown and retrogradation/setback of starch pastes (BeMiller, 2011). Recent data about starch pastes and gels as governed by their ingredients and the interactions with other hydrocolloids is presented. Furthermore, the role of these interactions in controlling rheology in model food is discussed. Examples given include stability issues of ready to eat white sauces, soups and caramel sauces.

6.1. White sauces

Béchamel sauce or ''white sauce'' is used in a lot of preparations or as a basis for other more complex sauces (Heyman et al., 2010). Sauces often exhibit stability problems during prolonged storage either caused by emulsion instability or by changing polymer interactions (Mandala et al., 2004b; Mc Clemments, 2006).

Non-starch hydrocolloids added in starch paste can alter the continuous phase of the system which contains them. In a complex system like that of a sauce, apart from rheology, they can also alter water holding capacity of the sauces. The effects of partially replacing modified starch by hydrocolloids (guar gum, xanthan gum and carboxymethylcellulose) on the rheological behavior and the physicochemical stability of the sauces is discussed.

Oscillatory spectra of all sauces are very similar to those of starch gels (example model sauce and guar gum at different concentrations, (Fig. 4)). All samples exhibit a dominant elastic behavior since G′ is larger than G′′ over the studied frequency range.

When different hydrocolloids such as xanthan, guar or CMC gum are added, xanthan gum causes the greatest increase in G′ compared to the model system. Guar gum and CMC also shift the G′ curve to higher values (Fig.4), but in a less pronounced way than xanthan gum does. White sauces present similarities between their values and those of starch gels alone supporting the hypothesis of a strong influence of both native and modified starches present in the system (Caisawang & Suphantharika 2006). Same conclusions are reached by Mandala et al. (2004b). During preservation at 7°C for 30 days the overall profile of the frequency curves does not change. Slight reductions in both G′ and G′′ are noticed with the most significant changes to occur the first 2 weeks.

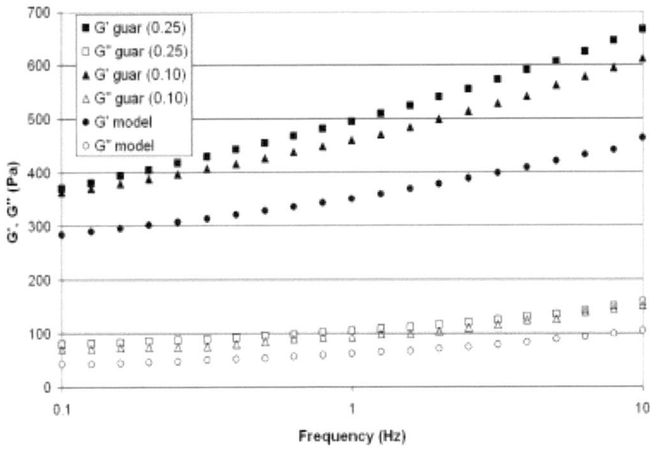

Figure 4. Effect of guar addition on oscillatory measurement of béchamel sauces. (From Heyman et al. (2010). Journal of Food Engineering 99, 115-120, with permission).

6.1.1. White sauce and freeze-thaw stability

Starches combined with different hydrocolloids are used in white sauces and the freeze/thaw stability of the produced samples is investigated. In a typical white sauce, after a freeze/thaw cycle, an increase in the viscoelastic functions is observed as a consequence of extensive starch retrogradation. By adding hydrocolloids this increase is reduced, leading to a less structured system. This can be justified by hydrocolloid interaction with solubilised amylose that reduces amylose - amylose interactions, preventing also structure ordering and hence reducing the extent of retrogradation (Arocas et al., 2009).

6.1.1.1. Ambient conditions' thawing

Specifically, the viscoelastic properties of fresh and thawed white sauces containing different corn starches (native waxy corn starch (NWS), native corn starch (NS), hydroxypropyl distarch phosphate waxy corn starch (HPS) and pregelatinized acetylated distarch adipate waxy corn starch (AAS)) are compared. Samples are frozen at -18°C and thawed at room temperature until 20°C.

A different behavior is found among the modified and the native starch sauces (Fig. 5).

The fresh modified starch sauces show higher G′ and G′′ values than the fresh native starch sauces, HPS being the one with the highest capacity and NWS the one with the lowest capacity. A high thickening capacity is ascribed to the fact that modified starches present high starch granule stability in comparison to the native starches and their granules do not break down in the thermal and shear conditions.

Moreover, a temperature increase from 20 to 80°C does not affect the values of G′ and G′′ either in the fresh or freeze/thaw samples. On the contrary, in native starch sauces a slight decrease in the values of the viscoelastic moduli is observed after 50°C, particularly pronounced after the freeze/thaw cycle. Furthermore, the values of the G′, G′′ of samples prepared with native starches after freeze/thaw cycle are much greater than those of the fresh samples due to retrogradation phenomena occurring during the freezing process.

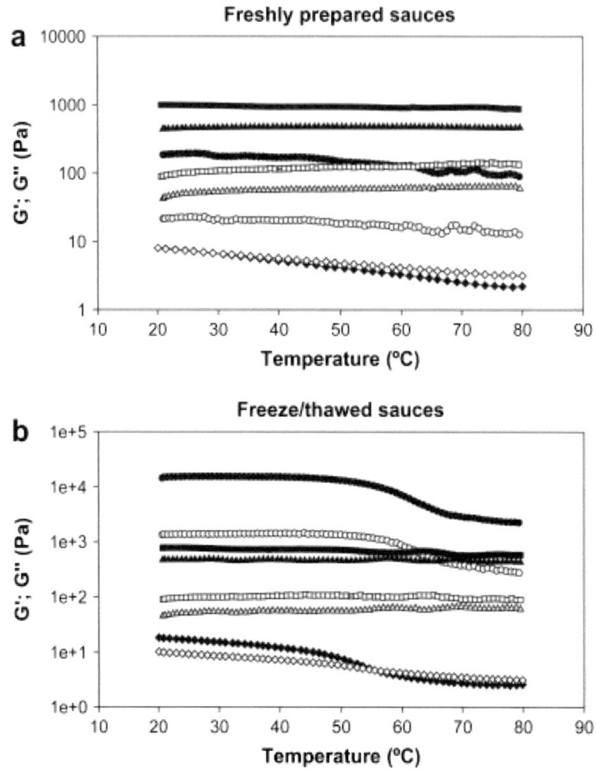

Figure 5. *G′* and *G″* as a function of increasing temperature for : a) freshly prepared sauces and b) freeze/thawed sauces. NS sauce (G′:●, G″:○), NWS sauce (G′:♦, G″: ◊), AAS sauce (G′:(▲, G″:Δ) and HPS sauce (G′:■, G″:□). Frequency: 1 Hz. γ: 0.001. Heating rate: 1.5 °C/min. (From Arocas et al., 2009, Food Hydrocolloids, 23, pp 901-907, with permission).

Concerning mechanical spectra, all samples behave as soft gels with values of G′ higher than values of G′′. A weak dependence on frequency is observed, as well as in starch-hydrocolloid mixtures. After freeze/thaw, as it was expected, structural changes occur mainly in sauces containing native starches. As a consequence, great values of G′ and G′ are observed and these samples presented a spongy structure depending also on freezing rate (the lower the freezing rate, the more pronounced the spongy structure). Thus, chemical modification is effective in providing freezing and thermal structure stability.

6.1.1.2. Microwave and water bath thawing

Differences between conductive heating and MW heating of starch dispersions are found not in the mechanism of gelatinization but on the starch crystallinity, which disappears at a higher rate when samples were heated by microwaves. Furthermore, the attainment of a certain viscosity required longer time in the conduction-heated samples. Microwaving can cause incomplete gelatinization of the starch in comparison to convection heating and potato starch can be influenced much more than corn starches (An et al., 2008).

The effects of microwave thawing and water bath thawing on white sauces prepared with two different native starches (potato and corn) and a modified waxy maize starch are compared. Starch retrogradation is strongly affected by freezing and thawing. Thus, possible reduction of starch retrogradation upon different thawing methods could be beneficial for the quality characteristics of the final product.

Furthermore, microwave treated samples are quite similar to the freshly prepared sauces compared to the water bath-thawed ones. In this regard, microwaving could be considered more suitable than the water bath for diminishing the loss of quality associated with the freezing step (Table 1).

The differences in the viscoelastic properties of the microwave-and water bath-thawed native starch sauce can be explained because of the shorter heating time required in the microwave, which reduces the time available for the retrogradation occurring from -5 to 20°C. Furthermore, the big local temperature and the differences occurring during microwave heating can lead to an improved localized melting of amylopectin and eventual melting of amylose (at temperatures near 100°C).

Starch type	Treatment	G′ (Pa)	G* (Pa)	tanδ
Corn	WB	326.0 A	328.0 A	0.114 C
	MW	265.2 B	267.0 B	0.118 C
	Fresh	80.4 E	81.9 E	0.189 B
Potato	WB	177.6 C	179.8 C	0.158 BC
	MW	114.7 D	116.7 D	0.183 B
	Fresh	97.6 DE	101.0 DE	0.265 A
Modified waxy corn	WB	270.7 B	272.5 B	0.118 C
	MW	264.7 B	267.7 B	0.149 BC
	Fresh	206.6 C	210.5 C	0.194B

Table 1. Influence of starch type and treatment in the rheological parameters G′, G* and tan δ. Frequency: 1 Hz. ABCDE Means with the same letter are not significantly different ($p < 0.05$) according to the Tukey's multiple range test. WB water bath thawed samples, MW microwave thawed samples. (From Arocas et al, 2011, Food Hydrocolloids 25, pp 1554- 1562, with permission).

This can lead to new starch/water interactions and consequent water adsorption. Generally, heating during thawing improves the quality of frozen sauces as amylose bonds formed during retrogradation are broken accompanied with re-absorption of the previously released water. Furthermore, the modified starch resists the heating applied during the sauce preparation as well as during the thawing process. Although the modified starch granules swell, little release of the starch components occurs. Thus, the differences between thawing techniques are related to their effect on structure changes related to starch retrogradation (Arocas et al., 2011).

6.2. Chilli sauce

Chilli sauces investigated presented a dominant elastic behavior compared to the viscous behavior typically observed in suspensions with network-like structure (Gamonpilas et al., 2011). Weak gel-like characteristics are found in chilli sauces containing starch and hydrocolloids, as well as in white sauces mentioned above. The presence of starch/xanthan mixture in the commercial chilli sauces promotes their elastic properties. Furthermore, the sauce with low solid content and without xanthan gum has weak network structure and inferior flow properties. The addition of xanthan gum and/or modified starch can provide a network-like characteristic of the sauce.

6.3. Soup formulation

In a soup formulation corn starch was gradually replaced with fenugreek gum (FG) till 0.9% wt. Pure systems of corn starch and fenugreek gum dispersions are also tested. Pure corn starch-FG mixtures and soups are subjected to rheological measurements after preparation. (Matia-Merino & Ravindran, 2009).

Final viscosities increase with increasing FG replacement levels. Soup containing FG (without starch) exhibit a dose–related pasting pattern and different pasting profiles than starch-FG soup formulations. Combined addition of starch-FG results in a synergistic effect of starch and FG. Thus, the viscosity increase is much greater than when the two ingredients are added individually. The viscoelasticity of the starch-FG systems and soups containing them is shown at the frequency of 1 Hz (Fig 7.a, Fig 7b). The viscous or loss modulus shows slightly more dependency with the elastic component being frequently independent and always over the viscous modulus through the entire frequency range. The main differences between starch-FG systems and soups containing them is: a) the higher viscoelasticity developed in all the soups with both G′ and G′′ being greater in value than in the pure starch/fenugreek mixtures, measured under similar conditions and b) both elastic and viscous moduli gradually increase with the substitution of starch by FG for all formulations.

6.4. Caramel sauces

Caramel sauces can have tailored sensory, rheological and textural characteristics by using potato starch and xanthan gum combinations (Krystyjan et al., 2012). Potato starch and xanthan gum are selected among other starch-hydrocolloid combinations taking into account transparency and clear appearance for further use in caramel sauces as thickeners. Commercial sauce without any stabilisers received the lowest score in sensory evaluation. Samples were stored and among other rheological characteristics, storage and loss moduli

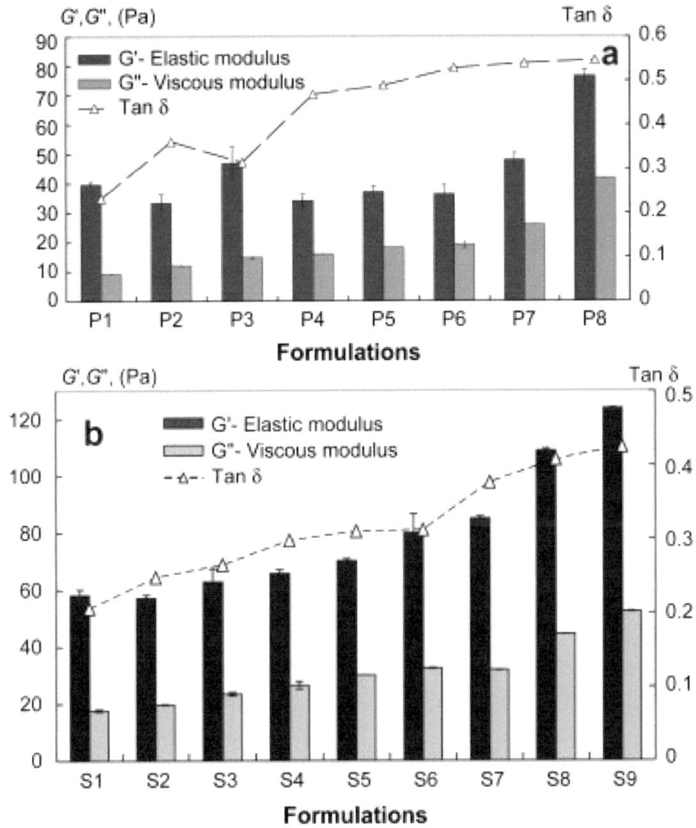

(a) 5% standard corn starch (P1), gradually replaced with fenugreek at 0.1% (P2), 0.15% (P3), 0.2% (P4), 0.25% (P5), 0.3% (P6), 0.5% (P7), 0.7% (P8), 0.9% (P9).
(b) soup formulations with 5% standard corn starch (S1), gradually replaced with fenugreek at 0.1% (S2), 0.15% (S3), 0.2% (S4), 0.25% (S5) and 0.3% (S6), 0.5% (S7), 0.7% (S8), 0.9% (S9). All measurements were carried out at 25 °C (From Matia-Merino & Ravindran, 2009, Food Hydrocolloids, 23(3), pp 1047-1053, with permission).

Figure 6. Storage modulus (G'), loss modulus (G''), and tan δ measured at 1 Hz and 1% strain

were measured. On the contrary to the assumptions in previous examples, the storage moduli of all sauces are much lower than their loss moduli. Sauces have very weak elastic and very strong plastic properties. This feature is considered disadvantageous, because the sauce can very easily flow down from the surface of the glazed products. An increase in xanthan gum amount results in moduli increase, both before and after storage. Controlled rheology can be achieved improving caramel sauces performance.

Concluding:

- Sauces and soups containing starch and hydrocolloids behave like weak gel-like products with good network structure.
- Caramel sauces containing starch and hydrocolloids are fluid-like with viscous character much more pronounced than the elastic one.

- Viscocelastic behavior of soups is similar to that of simple hydrocolloid-starch systems. However, viscous–like behavior is more pronounced in final soups than in starch-hydrocolloids' systems.
- Thawing under heating of frozen ready to eat starch white sauces results in improved rheological characteristics of the final samples. In particular, thawing under microwaves leads to better products than those being thawed in a water bath.

7. Conclusions

In this chapter oscillatory rheological data are presented to highlight the structural changes of starch suspensions during heating and further storage. Furthermore, the viscoelastic behavior of different starch types with selected hydrocolloids is presented in order to understand and control rheology in food processing. Moreover, the rheological behavior of starch-hydrocolloid mixtures is described in complex systems, they contain them that aim at successful food product development.

Starches undergo significant changes under heating that are related to structural changes and can be determined using small amplitude oscillatory shear (SAOS) tests. In many research works these tests are also combined with rapid visco analyser (RVA measurements), viscosity values, texture analyses, microscopic observations, DSC thermographs.

Hydrocolloids alter the viscoelastic character of starch pastes and different mechanisms are proposed, some of them based on their interactions with starch polymers. Hydrocolloids are found in the continuous phase of the starch pastes changing the effective concentration of starch in the final mixtures. They can also immobilize the water available, or they can interact with starch polymers. Pastes containing hydrocolloids may be less structured and less solid-like than the control pastes. Specific interest presents their role in gelation and retrogradation, since they are considered anti-staling agents and can prohibit retrogradation effects, specifically those of amylopectin.

Controlled rheological properies can be achieved by changing hydrocolloid-starch mixtures in selected products such as sauces, soups and caramel sauces, e.g. the final quality of native starches white sauces can be maintained when appropriate thawing process under fast heating (microwaves) is used.

New kinds of hydrocolloids are going to be used and starch role especially that of native starches is re-defined. The interest about such systems is going to increase and research on this field will be continued.

Author details

Ioanna G. Mandala
Agricultural University Athens, Dept. Food Science and Technology, Greece

8. References

Adebowale, K. O., & Lawal, O. S. (2003). Functional properties and retrogradation behaviour of native and chemically modified starch of mucuna bean (Mucuna pruriens). Journal of the Science of Food and Agriculture, 83, pp. 1541–1546.

Achayuthakan P. & Suphantharika M. (2008). Pasting and rheological properties of waxy corn starch as affected by guar and xanthan gum. Carbohydrate Polymers, 71, pp. 9-17

Arocas A., Sanz T. & Fiszman S.M. (2009). Influence of corn starch type in the rheological properties of a white sauce after heating and freezing. Food Hydrocolloids, 23, pp. 901-907

Acoras A., Sanz T., Hernando M.-I. & Fiszman S.M. (2011). Influence of corn starch type in the rheological properties of a white sauce after heating and freezing. Food Hydrocolloids, 23, pp. 901-907

Ahmed J., Ramaswamy H. S., Ayad A. & Alli I. (2008). Thermal and dynamic rheology of insoluble starch from basmati rice. Food Hydrocolloids 22, pp. 278-287

An, H.J., Yang H.S., Liu Z.D. & Zhang Z.Z. (2008). Effects of heating modes and sources on nanostructure of gelatinized starch molecules using atomic force microscopy. LWT-Food Science &Technology 41, pp. 1466-1471

Alloncle M. & Doublier J.-L. (1991). Viscoeleastic properties of maize starch/hydrocolloid pastes and gels. Food Hydrololloids, 5(5), pp. 455-467

Atwell, W. A., Hood, L. F., Lineback, d. R., Varriano-marston, E., & Zobel, H. F. (1988). The terminology and methodology associated with basic starch phenomena. Cereal Foods World, 33, pp. 306–311.

BeMiller J.N. (2011). Pasting, paste, and gel properties of starch-hydrocolloid combinations. Carbohydrate Polymers 86, pp. 386-423

Bilbao-Sainz C., Burtler M., Weaver T. & Bent J. (2007). Wheat and starch gelatinization undermicrowave irradiation and conduction heating. Carbohydrate Polymers 69, pp 224-232

Biliaderis C., Juliano B. (1993). Thermal and mechanical properties of concentrated rice starch gels of varying composition. Food Chemistry 48, pp. 243–250

Bowler P., Williams M.R. &AngoldR.E. (1980). A hypothesis for the morphological changes which occur on heating lenticular wheat starches in water. Starch/Stärke, 34, pp. 149

Chaisawang M. & Suphantharika M. (2006). Pasting and rheological properties if native and anionic tapioca starches as modified by guar gum and xanthan gum. Food Hydrocolloids 20 (5), pp. 641-649

Dickinson E. 1992. *Rheology In: An introduction to food colloids*, Oxford University Press, pp 59-62.

Doublier J.-L. & Cuvelier G. (2006). *Gums and Hydrocolloids: functional aspects in Carbohydrates in food*, ed. A.-C. Eliasson, CRC: Taylor &Francis, London

Duxbury D.D. (1989). Modified starch functionalities—no chemicals or enzymes. Food Processing, 50 , pp. 35–37

Ferry J.D. (1980). *Viscoelastic properties of polymers (3rd ed.)*. J. Wiley and Sons, New York

Funami T., Kataoka Y., Omoto T., Goto Y., Asai I. & Nishinari K. (2005). Food hydrocolloids control the gelatinization and retrogradation behavior of starch. 2a: Function of guar gum with different molecular weights on the gelatinization behavior of corn starch. Food Hydrocolloids 19, pp. 15-24

Funami, T., Kataoka, Y., Noda, S., Hiroe, M., Ishihara, S., Asai, I., Takahashi R.& Nishimari K. (2008). Functions of fenugreek gum with various molecular weights on the gelatinization and retrogradation behaviors of corn starch-1: Characterizations of fenugreek gum and investigations of corn starch/fenugreek gum composite system at a relatively high starch concentration; 15w/v%. Food Hydrocolloids, 22, pp. 763–776.

Funami T. (2011). Next target for food hydrocolloid studies: Texture design of foods using hydrocolloid technology. Food Hydrocolloids, 25(8), pp. 1904-191

Gamonpillas C., Pongjaruvat W., Fuonfuchat A., Methacanon P., Seetapan N. & Thamjedsada N. (2011). Physicochemical and rheological characteristics of commercial chilli sauces as thickened by modified starch or modified starch/xanthan mixture. Journal of Food Engineering 105, pp. 233-240

Heyman B., Depypere F., Delbaere C. & Dewettinck K. (2010). Effects of non-starch hydrocolloids on the physicochemical properties and stability of a commercial béchamel sauce. Journal of Food Engineering 99 ,pp. 115-120

Hoover R. (2001). Composition, molecular structure, and physicochemical properties of tuber and root starches: a review. Carbohydrate Polymers 45, pp. 253-267

Kaur, L., Singh, N., & Singh, J. (2004). Factors influencing the properties of hydroxypropylated potato starches. Carbohydrate Polymers, 55, pp. 211–223

Kaur, L., Singh, J., & Singh, N. (2006). Effect of cross-linking on some properties of potato starches. Journal of the Science of Food and Agriculture 86, pp. 1945–1954

Kim C. & Yoo B. (2006). Rheological properties of rice starch-xanthan gum mixtures. Journal of Food Engineering 75, pp. 120-128

Krystyjan M., Sikora M., Asamczyk G. & Tomasik P. (2012). Caramel sauces thickened with combinations of potato starch andxanthan gum. Journal ofFoodEngineering 112, pp 22-28

Lee M.H., Baek M.H., Cha D.S., Park H.J. & Lim, S.T. (2002). Freeze-thaw stabilization of sweet potato starch gel be polysaccharide gums. Food Hydrocolloids, 16, pp. 345-352

Mandala I. & Palogou E. (2003). Effect of Preparation Conditions and Starch/Xanthan Concentration on Gelation Process of Potato Starch Systems. International Journal of Food Properties 6, pp. 311-328

Mandala I. G., Michon C. & Launay B. (2004a). Phase and rheological behaviors of xanthan/amylose and xanthan/starch mixed systems. Carbohydrate Polymers 58, pp. 285–292

Mandala I.G., Savvas T.P., Kostaropoulos A.E. (2004b). Xanthan and locust bean gum influence on the rheology and structure of a white model-sauce. Journal of Food Engineering 64(3), pp. 335-342

Mc Clemments D.J. (2006). Non-covalent interactions between proteins and polysaccharides. Biotechnology Advances 24(6), pp. 621-625

Matia-Merino L. & Ravindran G. (2009). Starch–fenugreek (Trigonella foenum-graecum L.) polysaccharide interactions in pure and soup systems, Food Hydrocolloids, 23(3), pp. 1047-1053

Rosell C.M., YokoyamaW. & Shoemaker C. (2011). Rheology of different hydrocolloids-rice starch blends. Effect of successive heating-cooling cycles. Carbohydrate Polymers 84, pp. 373-382

Ross-Murphy S.B. (1984). *Rheological methods, Biophysical Methods in Food Research.* vol.5 (H.W.-S. Chan, ed.) SCI Critical Reports on Applied Chemistry, Blackwell, Oxford, pp. 137-139

Singh, J., & Singh, N. (2001). Studies on the morphological, thermal andrheological properties of starch from some Indian potato cultivars. Food Chemistry, 75, pp. 67–77

Singh N., Singh J., Kaur L., Singh S. N. & Singh G. B. (2003). Morphological, thermal and rheological properties of starches from different botanical sources. Food Chemistry 81 (2), pp. 219-231

Singh, N., Chawla, D., & Singh, J. (2004). Influence of acetic anhydride on physicochemical, morphological and thermal properties of corn and potato starch. Food Chemistry, 86, pp. 601–608.

Singh N., Nakaura Y., Inouchi N. & Nishinari K. (2007a). Fine structure, thermal and viscoelastic properties of starches separated from Indica rice Cultivars. Starch/Stärke, 59, pp. 10-20

Singh J., Kaur L. & McCarthy O.J. (2007b). Factors influencing the physic-chemical, morphological, thermal and rheological properties of some chemically modified starches for food applications- A review. Food Hydrocolloids 21, pp. 1-22

Stanley D.W., Stone A.P. & Tung M.A. (1996*). Mechanical properties of food. In Handbook of Food Analysis*, Volume I, Ch.4. L.M.L. Nollet (ed.), New York: Marcel Dekker, Inc., 93-136

Techawipharat J., Suphantharika M., BeMiller J. N. (2008). Effects of cellulose derivatives and carrageenans on the pasting, paste, andgel properties ofrice starches. Carbohydrate Polymers 73, pp. 417-426

Tsai, M. L., Li, C. F., & Lii, C. Y. (1997). Effects of granular structure on the pasting behavior of starches. Cereal Chemistry, 74, 750–757.

Wang Y., Wang L.-J., Li D., Ozkan N., Chen X. D. & Mao Z.-H. (2008). Effect of flaxseed gum addition on rheological properties of native maize starch. Journal of Food Engineering 89, pp. 87-92

Whistler R.L. & BeMiller J. N. (1997). Carbohydrate Chemistry for Food Scientists. St. Paul, MN: Eagan Press ,pp. 117-164

Viscoelastic and Textural Characteristics of Masa and Tortilla from Extruded Corn Flours with Xanthan Gum

Luis Carlos Platt-Lucero, Benjamín Ramírez-Wong,
Patricia Isabel Torres-Chávez and Ignacio Morales-Rosas

Additional information is available at the end of the chapter

http://dx.doi.org/10.5772/49975

1. Introduction

Corn plays an important role in the diet of many countries, especially México (Serna-Saldivar et al 1990), where corn tortilla is the principal ancestral food, and its consumption per capita is approximately 120 kg (Martínez et al. 2004). Corn tortilla can be produced with fresh masa using the traditional nixtamalization process, or with instant nixtamalized corn flour (Arámbula et al. 1999). Both processes result in industrial effluents known as nejayote (Serna-Saldivar 1996). An alternative process to avoid such contamination is extrusion which is defined as a continous process in which mechanical cutting is combined with heat to obtain gelatinization of the starch and denaturation of the proteins. It yields a plastified and restructured product with new shapes and textures. Some of the characteristics for which extrusion has gained popularity are versatility, reduced costs, high productivity, high-quality products, different product shapes, energy efficiency, and generation of new products and absence of effluents (Harper 1989).

Extrusion has been utilized as a continuous process of nixtamalization to produce instant flour, then to make corn tortillas, giving good product such as that of fresh masa obtained from the traditional process (Arámbula et al. 1998; Arámbula et al. 2002; Galicia 2005; González 2006; Martínez-Flores et al. 1998; Milán-Carrillo et al. 2006; Reyes-Moreno et al. 2003). However, during its storage, tortilla became hard and diminishes in flexibility. This can be due to fact that after its preparation starch reorganizes, and as a consequence produce crystalline structures, which is known as retrogradation, altering the texture and nutritional characteristics of tortillas (Aguirre-Cruz et al. 2005).

To improve and preserve the quality of tortillas made from nixtamalized corn flour during their storage, new alternative have been studied. Among these alternatives is the use of

INTECH
open science | open minds

© 2012 The Author(s). Licensee InTech. This chapter is distributed under the terms of the Creative Commons Attribution License http://creativecommons.org/licenses/by/3.0), which permits unrestricted use, distribution, and reproduction in any medium, provided the original work is properly cited.

some additives such as hydrocolloids which are water-soluble, high-molecular weight heteropolysaccharides. These compounds vary in shape and function and add flexibility to the tortillas produced, acting as a fat replacement, water binder, texturizer and adhesive. Among the main hydrocolloids utilized in tortillas is carboxymethylcelullose (CMC), guar gum, alginates, carrageenans and xanthan gum (Gurkin 2002).

Roman-Brito et al (2007) studied the effect of xanthan gum on nixtamalized corn flour with 0.25%(w/w), 0.50%(w/w), and 0.75%(w/w) of xanthan gum to make tortillas. These authors observed a decrease in hardness and increase in flexibility in the tortillas during their storage at 4°C. Yau et al. (1994) also extended the stability of tortillas made from corn flour during their storage at 25°C with xanthan gum at 1% (w/w), along with other additives. Arämbula et al. (1999), prepared tortillas with extruded corn flour with the addition of hydrocolloids: CMC, arabic gum, guar gum and xanthan gum at 0.5% (w/w) with different concentrations of lime. Tortillas were obtained with good results regarding to their textural characteristics (rollability, extensibility and shear force) with masa containing 0.2 %(w/w) of lime and xanthan gum added before extrusion. The unique structure and properties of xanthan gum account for its potential in versatile applications in the food industry (Hanna et al. 1997). The effect of hydrocolloids has not been studied at any concentration or mixture of them in the production of tortillas from extruded corn flour. It could help to keep a soft texture of tortillas during storage. The addition of hydrocolloids can modify the rheological (viscoelasticity) properties of corn masa, which are important in the tortilla making process. The understanding of these parameters is crucial for the design of equipment and definition of operation parameters in the tortilla industry (Aguirre-Cruz et al. 2005).

On the other hand, one of the methods most frequently utilized for the study of viscoelastic properties of masa is the dynamic method. In this test, results obtained refer to parameters that help to characterize appropriately the materials whose rheological behavior is complex, such as dynamic moduli of storage and loss, respectively the energy stored elastically and that dissipated as heat during a cycle of deformation (Faubion and Hoseney 1990).

The storage modulus G' is an indicator of the elastic component of the material, and the loss modulus G'' is an indicator of the viscous component. The storage and loss moduli are usually reported as a function of frequency. The phase angle represents a simple mean of elastic and viscous natures of the material. On some occasions, this property is expressed as the tangent of the phase angle (Tan δ), that is, the ratio between the loss and storage moduli (G''/G'). Moisture content is an important element in the determination of viscoelastic properties of cereals, which are reduced proportionally with moisture content (Masi et al. Measurement of viscoelastic characteristics have been utilized in dispersions in masas of nixtamalized corn dehydrated with hydrocolloids (Aguirre-Cruz et al. 2005), and of commercial corn flour at different moisture content: 35%, 40% and 50%, respectively (Broulliet-Fourmann et al. 2003).

The aim of this research was to determine the effect of xanthan gum on viscoelastic and textural characteristics of masa and tortilla from extruded nixtamalized corn flour, and to find the best combination of extrusion process factors to produce corn flour with xanthan gum to make tortillas with the best texture. We use response surface methodology (RSM).

2. Experimental procedure

2.1. Raw material

White corn (Dekalb variety) from Sinaloa de Leyva, Sinaloa, México was used. Corn was cleaned using a vibrating cleaner (Blount, model M2BC. Bluffon, IN, USA), and stored at 5 °C until used.

2.2. Extrusion process

Samples of corn (2 kg) were ground in a mill (Pulvex, model 200, Mexico, D.F.), with a 0.8-mm sieve. The ground corn was mixed with commercial grade lime ($Ca(OH_2)$) at 0.3 % (w/w) and xanthan gum (Spectrum Chemical, Gardena, CA, USA) at different concentrations (0.3-0.7 % of corn weigth) and using an industrial mixer (Hobart, model AS200T. Troy, Ohio, USA). Next, distilled water was added to this blend up to reach the moisture content appropriated (range 25-35 % of the corn weight). To obain a complete hydration of the ground corn particles, samples were packed in a polyethylene bag and stored for 12 h at 5 °C. Before extrusion, samples were tempered at 25°C during 4 h.

The extrusion process was carried out in a single-screw laboratory extruder (Brabender, model 837416. Duisburg, Germany) with a 19 mm screw-diameter, length-to-diameter 25:1, nominal compression ratio 1:1, and a die opening of 3 mm. The first three zones of the extruder were maintained at 60 ºC, and the fourth zone was varied (110-130 °C), with a screw speed of 112 rpm. A screw-operated feed hopper fed the extruders at 45 rpm. Extrudates were dried at 60 °C for 1 h in a tunnel type dryer (no brand), and cooled at room temperature (25°C). To obtain the extruded nixtamalized corn flour (ENCF), the extrudates were ground in a mill (Pulvex, model 200, Mexico, D.F.) with a sieve 0.8-mm diameter, and packed in plastic bag at 5°C.

2.3. Corn flours evaluation

The ENCFs were analyzed for water absorption index (WAI) and subjective water absorption capacity (WAC). These response variables are the most critical for making tortillas at commercial level. WAI was measured using the method of Anderson et al. (1969) with a modification: The distilled water temperature was 25 °C instead of 30 °C, and theWAI was expressed as g of gel / g of dry matter. WAC was determined using the method described by Flores-Farías et al. (2002). The quantity of water added was recorded as the capacity for water absorption of the flour in mL of water / 100 g of flour.

2.4. Corn masa viscoelasticity

Corn masa was prepared from each ENCF using 100 g sample and adding distilled water. The quantity of water utilized corresponded to the WAC. Once prepared masas, they were allowed to stand for 30 min in a plastic bag at room temperature (25 °C). Samples of 2 g of masa each were weighed out to be utilized in the rheometer.

The oscillatory dynamic scanning test was performed utilizing a dynamic mechanical spectrometer (Rheometrics Scientific, model RSF III. Piscataway, NJ, USA) equipped with parallel plates of 25 mm diameter and a chamber for temperature control (peltier). A sample was placed between the plates separated by a gap of 2.5 mm. The excess of masa was cut off using a plastic instrument. Next, petroleum jelly was applied where the sample was air exposed to prevent loss of moisture. The frequency sweep test was carried out using a software (RSI Orchestrator, Rheometrics Scientific). Each test was run to a deformation of 0.04% and at 25°C, which gave a minimum of structure disorder and with sufficient assurance of the level of torsion (Broulliet-Fourmann et al. 2003). The deformation used was previously determined to work in the viscoelastic linear region in a frequency range from 0.1 to 100 rad/s. The viscoelastic parameters obtained in the frequency range used were the storage modulus (G') and loss modulus (G'') in kPa, and the tangent of the phase angle (Tan δ).

2.5. Corn tortilla preparation

Two kg of masa from each ENCF were mixed with distilled water to obtain masa. The amount of water was based on the WAC of each extruded treatment. Masas from ENCF were transported to a commercial factory. A roller machine (Rodotec, model RT-100, Guadalupe, N.L., México) was used with a mold of 14 cm diameter, and was adjusted to a masa weight of 25 g. Tortillas were baked in an oven (integrated to the roller machine) with 3 temperature zones: zone 1, 270 ± 10 °C; zone 2, 320 ± 30 °C; and zone 3, 300 ± 25°C, with a residence time of 60 s. The baked tortillas were cooled at room temperature (25 °C).

2.6. Corn tortilla textural evaluation

To determine firmmess and rollability, tortillas packaged in plastic bags were placed at room temperature (25°C). Firmness and rollability were measured at 2 h, 24 h and 48 h after baking. The firmness test was carried out according to the procedure reported by Ramírez-Wong et al. (2007), modifying the cross head speed of the texturometer (Instron, model 4465. Canton, MA, USA) to 50 cm/min. Firmness was expressed as maximal force (MF) in kPa. Regarding tortilla rollability, three strips 2 cm wide were cut from each tortilla and individually tested (Waniska 1976). Each strip of tortilla was rolled up in a wooden cylinder 2 cm in diameter, and was examined for degree of rupture, which was established on a scale of 1 to 5, where 5 indicated no tear of the tortilla (maximum rollability), 3 partial tear, and 1 complete tearing. Five tortillas were used for each treatment in each test.

2.7. Experimental design and statistical analysis

Response surface methodology (RSM) was used, and the process variables were: temperature of the fourth zone of the extruder (T, 110-130 °C), moisture content of conditioned ground corn (MC, 25-35 % of the corn weight), and concentration of xanthan gum (XG, 0.3-0.7 % of corn weigth). A central composite rotatable design was utilized (Table 1), with three factors and 5 levels (Montgomery 2001; Myers 1971). In order to observe the difference between specific treatments, Tukey's test was used at a level of significance of p= 0.05 . To find the best variables combination of the process to obtain the extruded corn flour, the conventional graphical method was used and maximizing WAI, Tan δ; minimizing G', G" and MF. To obtain contour plots for the visualization and selection of the best combination of T, MC and XG to prepare the extruded nixtamalized corn flour, the contour plots of each of the response variables were utilized, by means of the method of superimposition of surfaces. Data analysis and the elaboration of surface response and contour plots were performed using Design Expert version 6.0.7 software (Design Expert, 2002).

3. Results and discussion

3.1. Corn flours evaluation

WAI is a parameter that gives an idea of the absorption of water of corn flour, and is an indicator of yield of fresh masa (Molina et al. 1977). The highest value of WAI in ENCFs was of 3.6 g of gel/g of dry matter, and was obtained at high concentration (0.84%) of xanthan gum (treatment 14, Table 1), which would indicate the capacity of the gum to form gels. This could be due to the high affinity of hydrocolloids for water, because of its branched structure. During hydration, water molecules hydrogen bond with hydroxyl (or carboxyl) groups found in the unit components (sugars) of hydrocolloid molecules, inducing this association with increased capacity for water retention (Dickinson 2003). Aguirre-Crus et al. (2005) observed in their research an increase in the capacity for water retention in suspensions of masa of corn dehydrated with hydrocolloids at different temperatures.

WAC is the quantity of water that is absorbed by the flour to obtain a masa of appropriate consistency for the preparation of tortillas and is a subjective test. WAC was affected very significantly ($p < 0.01$) By the treatment. The WAC range in the ENCFs was between 74.8 - 89 mL water/100 g flour (Table 1). Arámbula et al. (1999) reported in extruded flours with xanthan gum a high WAC value of 88.5 mL/100g. However, the value for extruded flour reported by González (2006) was 72 mL water/100 g flour, which was low; probably due to that none type of gum was added. Arámbula et al. (2002) found in extruded flour a WAC of 70 mL/100g with 0%, and 80 mL/100g with 3% of addition of corn pericarp, respectively.

Treatment[a]	Process Factors[b]			Response Variables[c,d]					
	T(X 1)	MC (X2)	XG(X3)	WAC	WAI	G'	G"	Tan δ	MF
1	110 (-1)	25 (-1)	0.30 (-1)	76.9	3.0	274.3	58.3	0.21	53.3
2	130 (+1)	25 (-1)	0.30 (-1)	77.0	2.8	269.3	59.8	0.22	59.0
3	110 (-1)	35 (+1)	0.30 (-1)	76.3	3.2	223.0	54.8	0.25	47.0
4	130 (+1)	35 (+1)	0.30 (-1)	78.6	3.3	302.0	60.0	0.20	48.5
5	110 (+1)	25 (-1)	0.70 (+1)	74.8	2.6	231.0	51.6	0.22	58.5
6	130 (-1)	25 (-1)	0.70 (+1)	79.6	2.2	270.3	63.3	0.23	58.6
7	110 (-1)	35 (+1)	0.70 (+1)	79.0	3.4	213.3	48.2	0.23	54.3
8	130 (+1)	35 (+1)	0.70 (+1)	79.6	3.2	234.0	51.2	0.22	41.8
9	103.18 (-1.681)	30 (0)	0.50 (0)	82.8	3.5	231.0	50.7	0.22	44.9
10	136.82 (+1.681)	30 (0)	0.50 (0)	76.6	3.4	219.3	52.7	0.24	51.0
11	120 (0)	21.59 (-1.681)	0.50 (0)	80.1	2.5	257.0	61.6	0.24	61.3
12	120 (0)	38.41 (+1.681)	0.50 (0)	79.0	2.9	294.7	72.1	0.24	55.9
13	120 (0)	30 (0)	0.16 (-1.681)	83.1	3.0	243.7	50.7	0.21	49.7
14	120 (0)	30 (0)	0.84 (+1.681)	89.0	3.6	211.0	47.8	0.23	43.7
15	120 (0)	30 (0)	0.50 (0)	87.3	3.4	218.7	54.7	0.25	46.0
16	120 (0)	30 (0)	0.50 (0)	87.0	3.2	190.3	45.7	0.24	43.6
17	120 (0)	30 (0)	0.50 (0)	87.3	3.5	225.3	55.8	0.25	46.5
18	120 (0)	30 (0)	0.50 (0)	87.0	3.5	218.7	49.3	0.23	49.1
19	120 (0)	30 (0)	0.50 (0)	87.3	3.2	176.6	46.5	0.26	41.2
20	120 (0)	30 (0)	0.50 (0)	87.0	3.1	210.3	53.5	0.25	41.2

[a]Numbers do not correspond to the order of processing. [b]T, extrusion temperature (°C); MC, moisture content (%, w/w); XG, xanthan gum (%, w/w); values in parentheses are the coded levels. [c]WAC, subjetive water absorption capacity (mL water/100/g flour; WAI, water absorption index (g gel/g dry matter); G', storage modulus (kPa), G", loss modulus (Kpa); Tan δ, tangent of the phase angle; MF, maximal force (kPa). [d]Mean of three replicates.

Table 1. Experimental designa used to obtein different combinations of extrusion temperature/moisture content/xanthan gum for production of extrusion-nixtamalized corn flour.

3.2. Masa viscoelasticity evaluation

In Figures 1(a, b), 2(a, b) and 3(a, b) are presented the storage modulus (G'), loss modulus (G") and tangent of the phase angle (Tan δ), respectively. It is observed that all of these viscoelastic parameters increased with frequency. The range 0.1-10 rad/s was the most susceptible to structural changes (Broulliet-Fourmann et al. 2003). In this frequency range, there was a considerable increase in the viscoelastic parameters G', G" and Tan δ, whereas small gradual changes occurred at high frequencies. Hence, the values of the viscoelastic parameters of 10 rad/s were selected for the optimization of the variables of the process for making tortillas from ENCFs. Values of G' were higher than those of G" (Table 1), indicating that the elastic behavior predominated over the viscous behavior of masas. Similar trend was found by Aguirre-Cruz et al. (2005) in samples of suspension of corn masa at 10% (w/w) of solids in their heating/cooling kinetics, and by Broulliet-Fourmann et al. (2003) in corn flours at different moisture contents (35%, 40% and 50%).

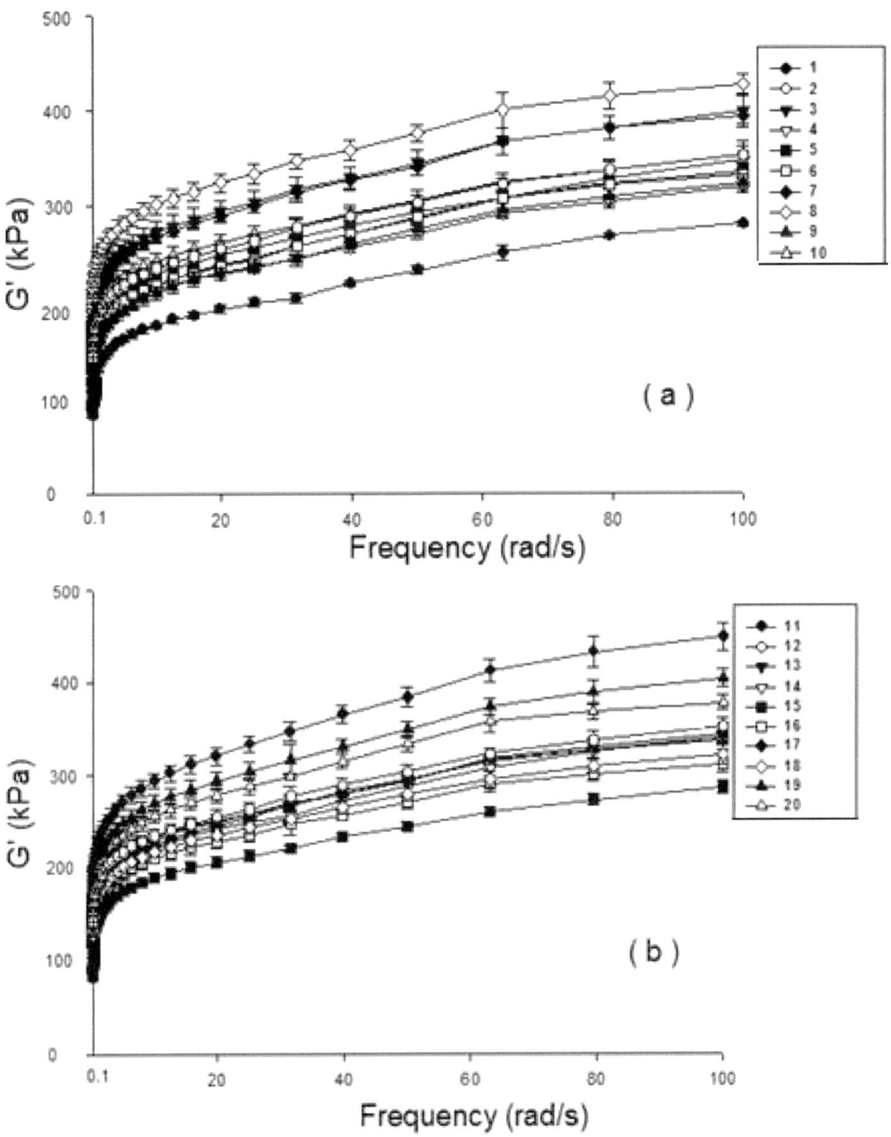

Figure 1. Storage modulus (G') vs frequency for extruded nixtamalized corn flours in treatments: (a) 1-10 and (b) 11-20. Error bars indicate standard error of means.

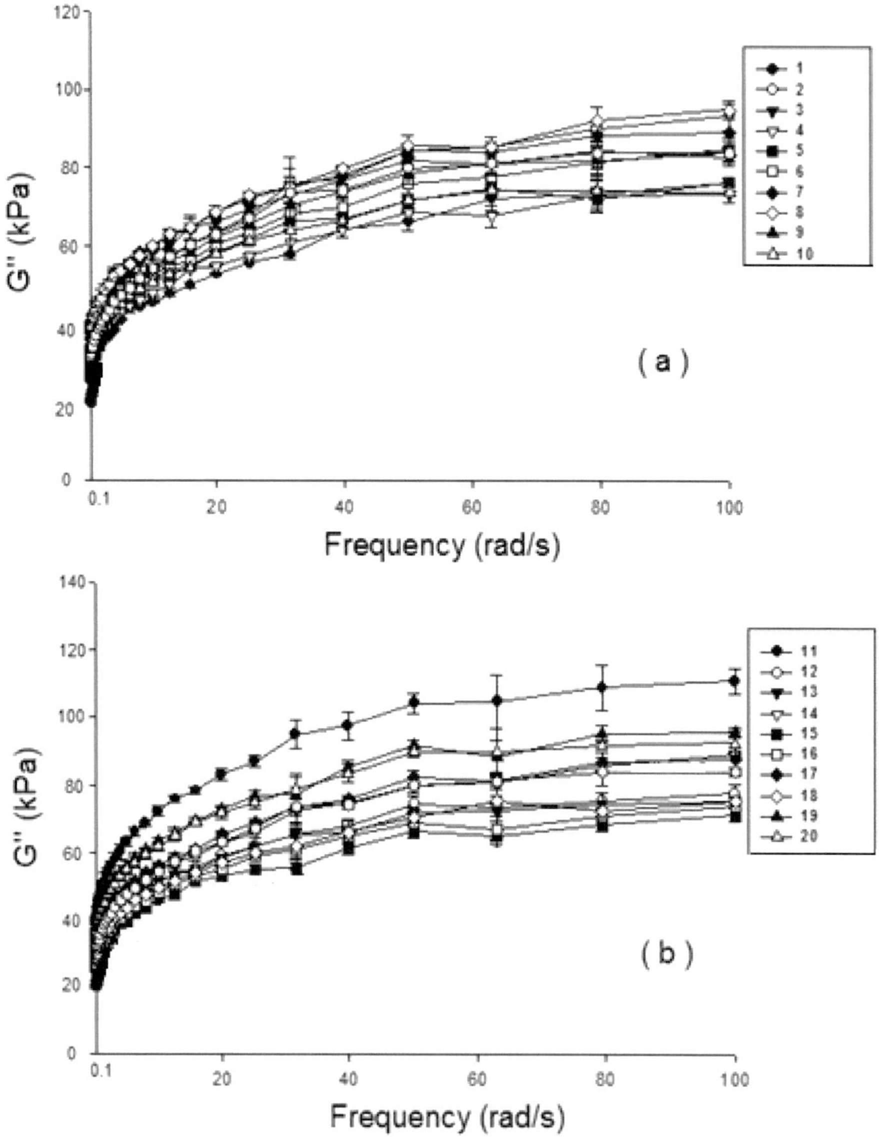

Figure 2. Loss modulus (G'') vs frequency for extruded nixtamalized corn flours in treatments: (a) 1-10 and (b) 11-20. Error bars indicate standard error of means.

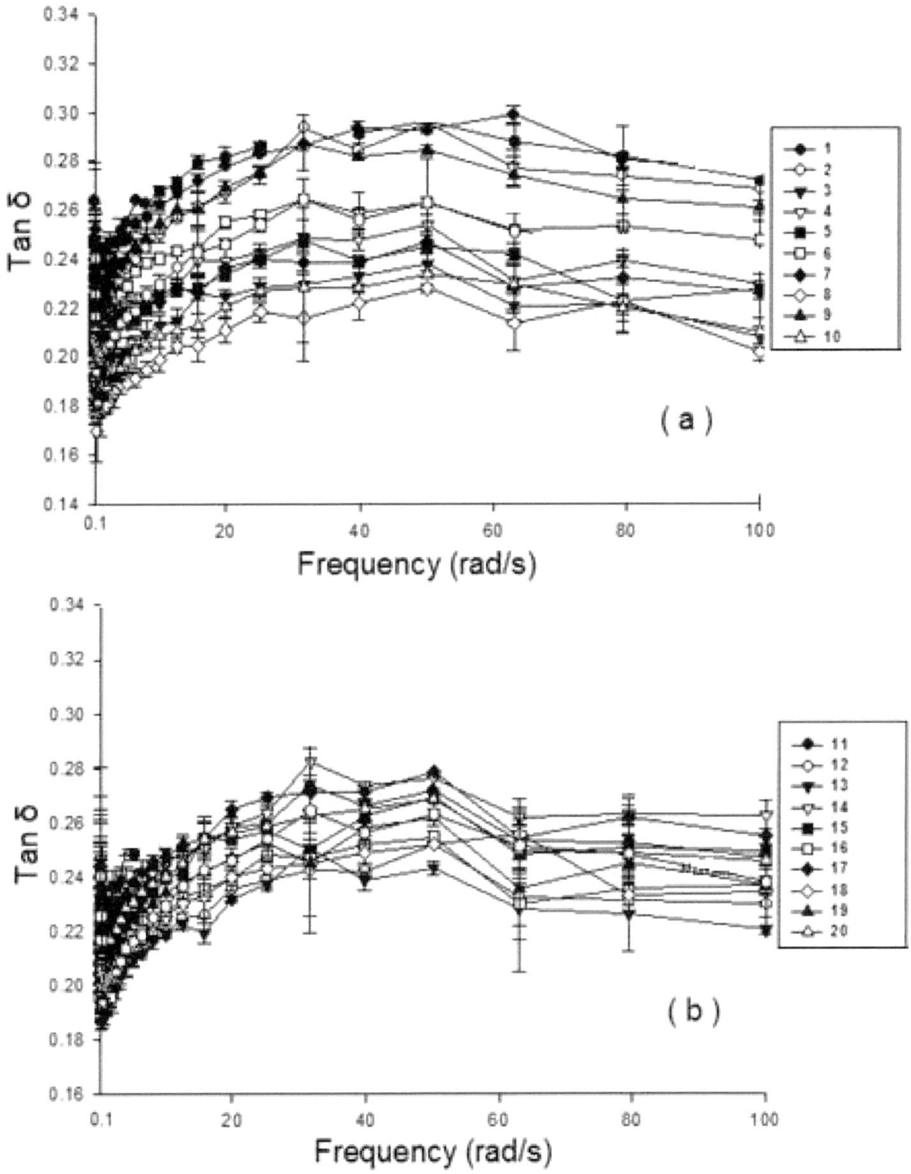

Figure 3. Tangent of the phase angle (Tan δ) vs frequency for extruded nixtamalized corn flours in treatments: (a) 1-10 and (b) 11-20. Error bars indicate standard error of means.

In the storage modulus (G'), the lowest value (176.6 kPa) was with treatment 19 (Table 1) with no significant difference ($p < 0.05$) with treatment 16, whose conditions were the same. When the concentration of xanthan gum increased, at the same conditions of temperature and moisture content, G' decreased. It probably was due to that the water molecules and bound components (amylose-hydrocolloid) of masas with hydrocolloids formed a gel with a softer structure, which affected G' values (Aguirre-Cruz et al. 2005). Regarding to the loss modulus (G''), the lowest value (45.7 kPa) was obtained for treatment 16 (Table 1), however, there were not significant diferences ($p < 0.05$) with treatments 7, 14 and 19. The highest value of Tan δ was observed in treatment 19 with 0.26 (Table 1), which differed significantly ($p < 0.05$) to the rest of the treatments. Since Tan δ values for all masas were in the range of 0.2-0.3, it is corresponding to that of an amorphous polymer (Ferry 1980). Similar values were obtained by Aguirre-Cruz et al. (2005) during cooling of diluted corn masa.

3.3. Corn tortilla evaluations

Values for the physical characteristics of tortillas from ENCFs after two hours they were made, time at which they are eaten, were similar to those of commercial tortillas. The physical characteristics evaluated were: weight, (range 21.3-25.2 g); diameter (range 12.2-12-7 cm); and thickness (range of 1.4-1.9 cm). Similar results were reported by Ramirez-Wong (1989) who evaluated corn tortillas obtained by the traditional process of nixtamalization.

Figure 4 presents the tortilla firmness as maximum force (MF) to rupture the tortillas made from the different ENCFs after 2 h, 24 h and 48 h of storage at room temperature (25°C). In general, for all the treatments (Table 1), as time progressed tortilla firmness increased. At 2 h of storage time, the lowest MF (41.2 kPa), which is the best value for tortilla firmness (softer) were observed in treatments 8, 9, 14, 15, 16, 17, 19 and 20 (Table 1), and they were not significantly different ($p > 0.05$). At 24 h of storage time the lowest MF (62.8 kPa) was observed for treatment 20. At 48 h of the storage time the lowest value (62.3 kPa) was similar to that of 24 h, corresponding to the same treatment (20), where there was a significant difference ($p<.05$) compared to all other treatments. The lowest values of firmness in tortillas during storage corresponded to high concentrations of xanthan gum (0.5% to 0.84%). Treatments with high xanthan gum concentrations retained the tortilla moisture and improving its textural characteristics. This finding is similar to that reported by Arámbula et al. (1999), who obtained better results when utilizing xanthan gum compared to other addtitives such as guar gum, CMC and Arabic gum.

Tortillas made with extruded nixtamalized corn flours showed an increase in their MF during their storage at 25°C. Regarding to texture, the most important changes during storage occurred the first 24 h (Fernández et al. 1999; Ramírez-Wong 1989). Since some starch crystals are retained after baking of the tortilla, they serve as nuclei which facilitate the rapid association of starch molecules, and structural changes occur during the initial 24 h following baking, which in turn leads to rapid retrogradation or increasing the texture of this product (Fernández et al. 1999). A similar tendency for MF was obtained by Roman-Brito et al. (2007) for tortillas made with nixtamalized flour containing xanthan gum and stored for 3 days at 4°C; however, the increase was less due to the lower storage temperature. Results obtained for tortillas made from ENCFs are similar to those reported by Galicia (2005) and Gonzalez (2006) after 2 h of storage. Neverthaless, at 24 h and 48 h of storage the MF values obtained in the present study were lower than those reported for these authors because they did not use gum in masas preparation.

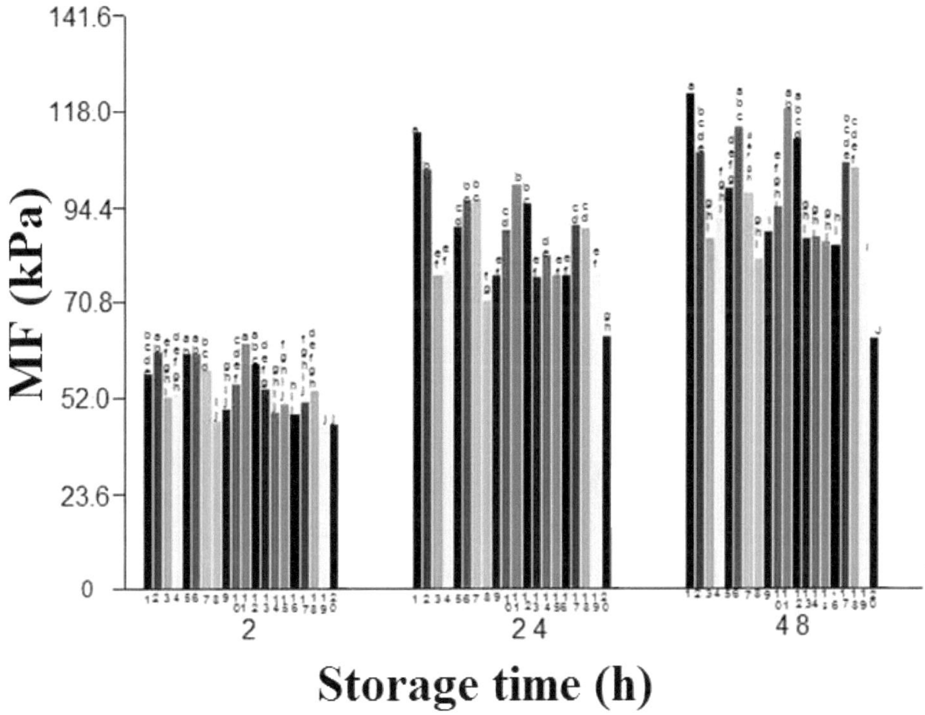

In each storage time, MF values with the same letter are not statistically significant ($p > 0.05$).

Figure 4. Effect of storage time on the maximum force (MF) to rupture for tortillas made from extruded nixtamalized corn flours.

Figure 5 presents tortilla rollability for all ENCF treatments after 2 h, 24 h and 48h of storage at room temperature (25°C). In general, for all the treatments (Table 1), as time progressed tortilla rollability decreased. For all ENCF treatments, the best tortilla rollability was obtained at the 2 h of storage. The most rapid loss of tortilla rollability occureed within 24 h of storage time, such as in the study of tortillas made from corn flour with xanthan gum added and stored at 4°C by Roman-Brito et al. (2007). It could be due to the very rapid association (retrogradation) of amylose and of amylose and amylopectin in starch (Fernández et al. 1999). At 2 h of storage time, the highest rollability (a value of 5) was obtained for all treatments, except for 11 and 12 (Table 1). At 24 h of storage, treatments 15, 19 and 20 showed the highest rollability values (3.6-3.8). For 48 h of storage, treatment 15 showed the highest value (3.6). The highest values for rollability during storage corresponded to the xanthan gum concentration of 0.5%, which offers greater flexibility to the tortillas as in the study made by Arámbula et al. (1999), who used the same concentration and the same gum.

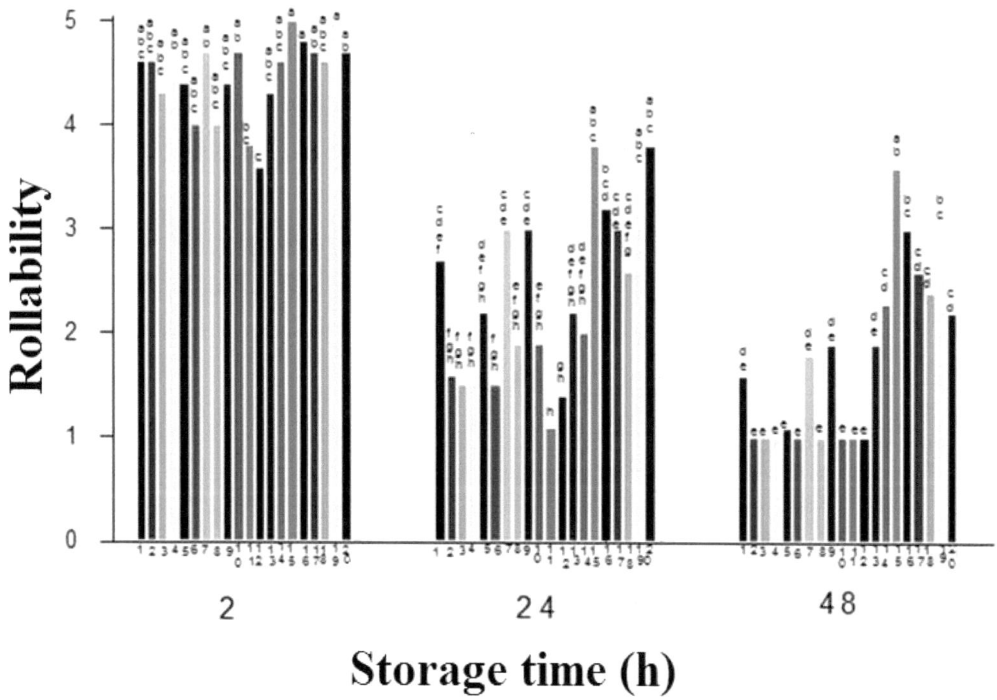

In each storage time, rollability values with the same letter are not statistically significant (p > 0.05).

Figure 5. Effect of storage time on the rollability of tortillas made from extruded nixtamalized corn flours.

3.4. Extrusion process optimization

To find the best variables combination of the process (T, MC and XG) to obtain the extruded nixtamalized corn flour, response surface methodology (RSM) was used. The evaluations to optimize the extrusion process were: WAI, G′, G′′, Tan δ and MF.

Water absorption index. Analysis of variance showed that WAI of extruded flours depended on the linear term of the conditioning moisture content (MC, $p < 0.01$), quadratic term of MC [$(MC)^2$, $p < 0.01$] and combined term of moisture content with xanthan gum [(MC)(XG), $p < 0.10$]. The prediction model in terms of original variables for WAI was:

$$Y_{WAI} = -4.37 + 0.583MC + 0.137(MC)(GX) - 0.01(MC)^2 \quad (1)$$

The regression model explained 75% of the total variation ($p<0.01$) in WAI of extruded nixtamalized corn flours. Figs. 6 (a,b,c) show the effect of WAI extruded flours as a function of T, MC and XG, noting that an increase conditioning moisture content of flour increase the water absorption index, as it does the increase in xanthan gum concentration at high levels of moisture content. Vargas-López (1987) mentioned that extruded grits of corn-sorghum and corn starch exhibit a maximum WAI at high temperatures and high moisture content.

The highest value of WAI (3.34 g gel/g dry matter) was observed at T=110-120.94°C/MC=29.68-35% (w/w)/XG=0.48-0.70% (w/w) (Figs. 6 (a,b,c)). Similar values of WAI for extruded flours were obtained by Reyes-Moreno et al. (2003)

Storage modulus. The linear term of xanthan gum (XG, $p<0.01$) and quadratic term of moisture content [$(MC)^2$, $p < 0.01$) affected significantly G′ in masa of ENCFs. The prediction model in terms of original variables for G′ was:

$$Y_{G'} = 1035 - 64XG + 0.97(MC)^2 \quad (2)$$

The regression model explained 70% of the total variation ($p < 0.05$) in G′ of extruded nixtamalized corn flours. Figs. 7 (a,b,c) show the effect of G′ in masa of ENCFs as a function of T, MC and XG, observing that with any value of moisture content from approximately the central point (30%), G′ increaces, due to its quadratic effect, and that increasing the concentration of XG decreases G', because of its linear effect. The lowest value (214.58 kPa) was observed at T=110-130°C/MC=28-32.1% (w/w)/GX=0.57-0.70% (w/w) (Figs. 7 (a,b,c).

Loss modulus. The viscous modulus G" in masa of ENCFs depended on the quadratic term of moisture content [$(MC)^2$, $p < 0.01$)]. The prediction model in terms of original variables for G″ was:

$$Y_{G''} = 235 + 0.22(MC)^2 \quad (3)$$

The regression model explained 72% of the total variation ($p < 0.01$) in G″ of masa from extruded nixtamalized corn flours. Figs. 8 (a,b,c) show the effect of G″ in masa from ENCFs as a function ot T, MC and XG, noting that for any value of moisture content starting at the central point (30%), G" increased due to the quadratic effect. The lowest value (49.35 kPa) was observed at T=110-128.32°C/MC=27.4-32.72% (w/w)/XG=0.52-0.70% (w/w)/ (Figs. 8 (a,b,c)).

Tangent of the phase angle. Analysis of variance showed that Tan δ of masa from ENCFs depended on the quadratic terms of temperature (T^2, $p < 0.05$) and xanthan gum [$(XG)^2$, $p < 0.01$], and on the interaction temperature-moisture content [(T)(MC), $p < 0.05$)]. The prediction model in terms of orginal variables for Tan δ was:

$$Y_{Tan\delta} = -1.59 - 0.0002(T)(MC) - 0.00007(T)^2 - 0.31(XG)^2 \quad (4)$$

The regression model explained 71% of the total variation ($p < 0.05$) in Tan δ of masa from extruded nixtamalized corn flours. Figs. 9 (a,b,c) show the effect of Tan δ of masa from ENCFs as a function of T, MC and XG, observing that at any concentration of xanthan gum from approximately the central point (0.5%), Tan δ decreases due to its quadratic effect. The highest value (0.239) was shown at T=116.61-124.55°C/MC=0.25-0.35% (w/w)/XG=0.47-0.59% (w/w) (Figs. 9(a,b,c)).

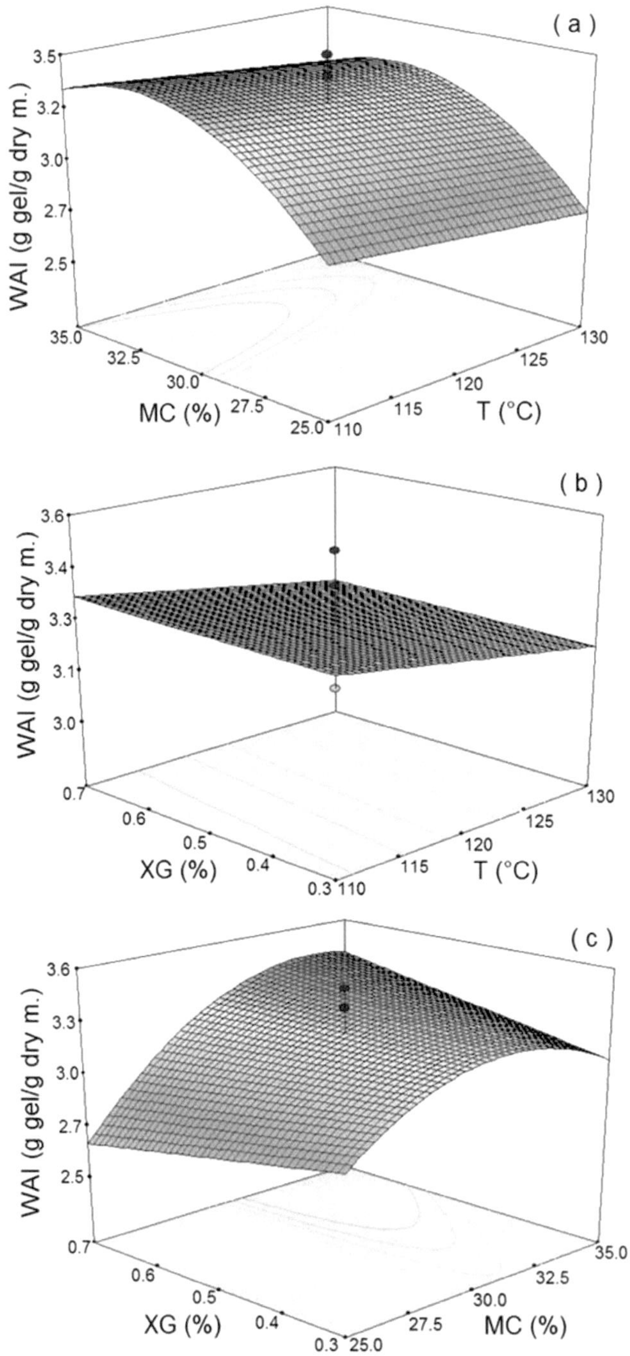

Figure 6. Response surface and contour plots for the effect of extrusion process factors on water absorption index (WAI) from extruded nixtamalized corn flour. (a) Effect of moisture content (MC, %) and extrusion temperature (T, °C) on WAI; (b) Effect of xanthan gum (XG, %) and T on WAI; (c) Effect of XG and MC on WAI.

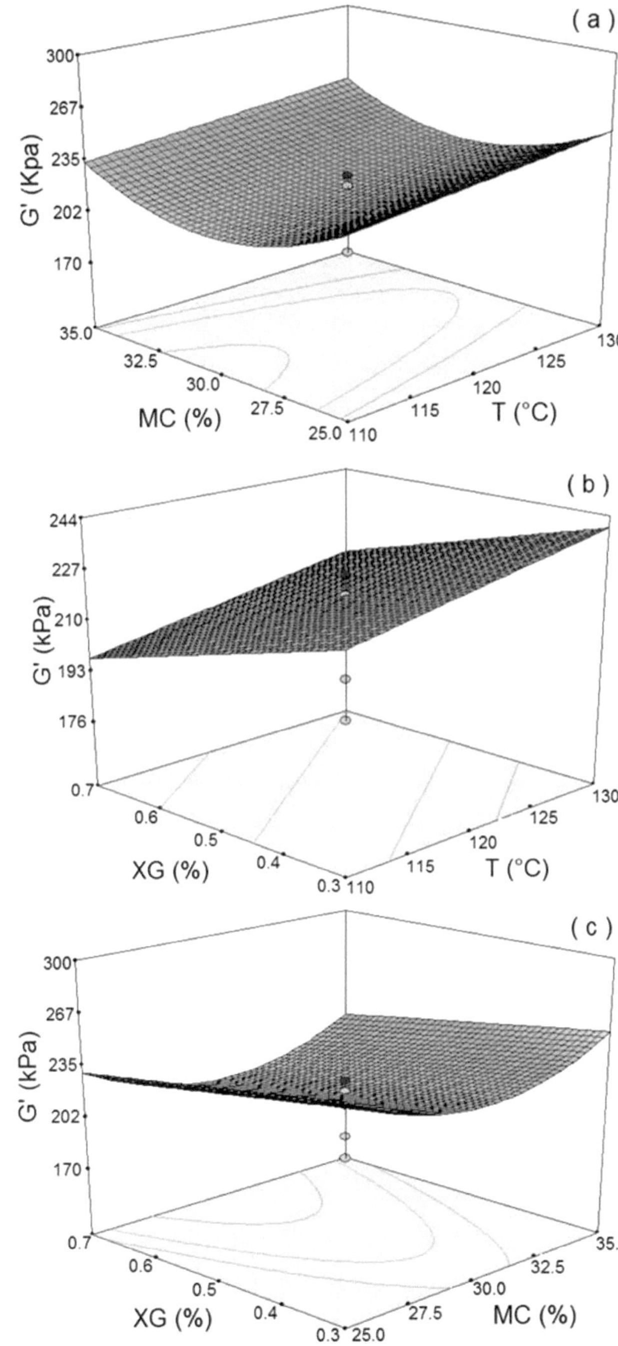

Figure 7. Response surface and contour plots for the effect of extrusion process factors on storage modulus (G') of masa from extruded nixtamalized corn flours. (a) Effect of moisture content (MC, %) and extrusion temperature (T, °C) on G'; (b) Effect of xanthan gum (XG, %) and T on G'; (c) Effect of XG and MC on G'.

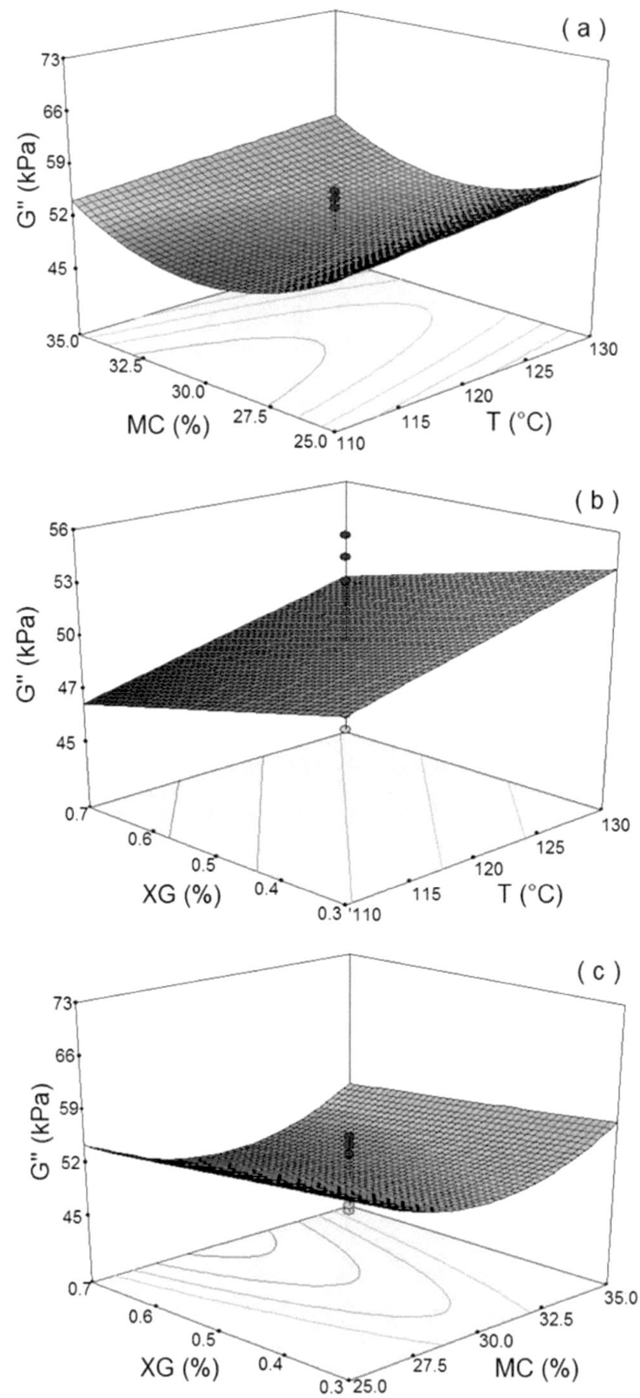

Figure 8. Response surface and contour plots for the effect of extrusion process factors on loss modulus (G'') of masa from extruded nixtamalized corn flour. (a) Effect of moisture content (MC, %) and extrusion temperature (T, °C) on G''; (b) Effect of xanthan gum (XG, %) and T on G''; (c) Effect of XG and MC on G''.

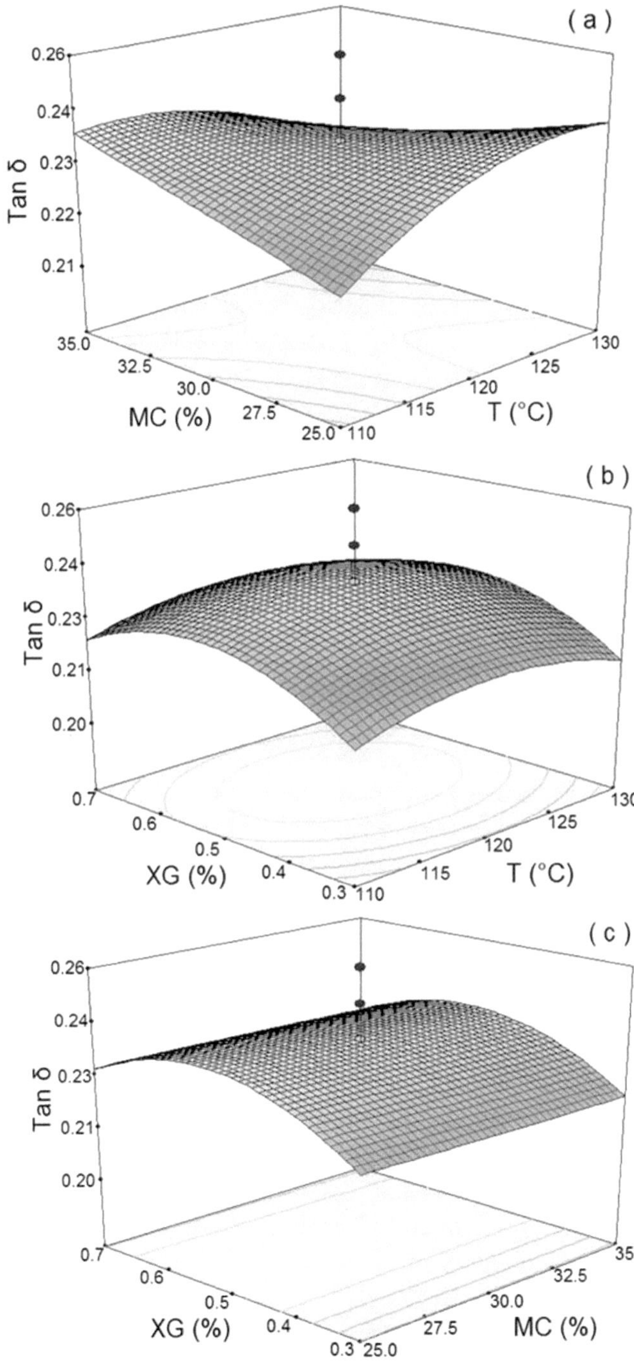

Figure 9. Response surface and contour plots for the effect of extrusion process factors on tangent of the phase angle (Tan δ) of masa from extruded nixtamalized corn flour. (a) Effect of moisture content (MC, %) and extrusion temperature (T, °C) on Tan δ; (b) Effect of xanthan gum (XG, %) and T on Tan δ; (c) Effect of XG and MC on Tan δ.

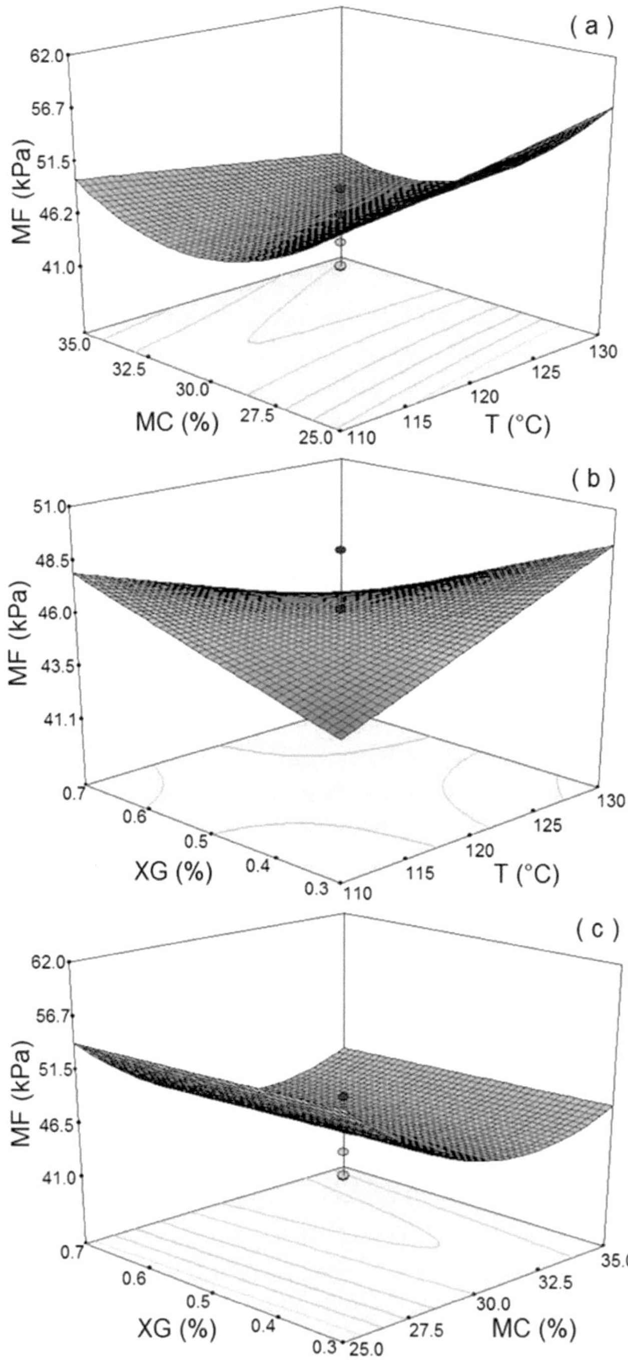

Figure 10. Response surface and contour plots for the effect of extrusion process factors on maximum force (MF) for tortilla made from extruded nixtamalized corn flour. (a) Effect of moisture content (MC, %) and extrusion temperature (T, °C) on MF; (b) Effect of xanthan gum (XG, %) and T on MF; (c) Effect of XG and MC on MF.

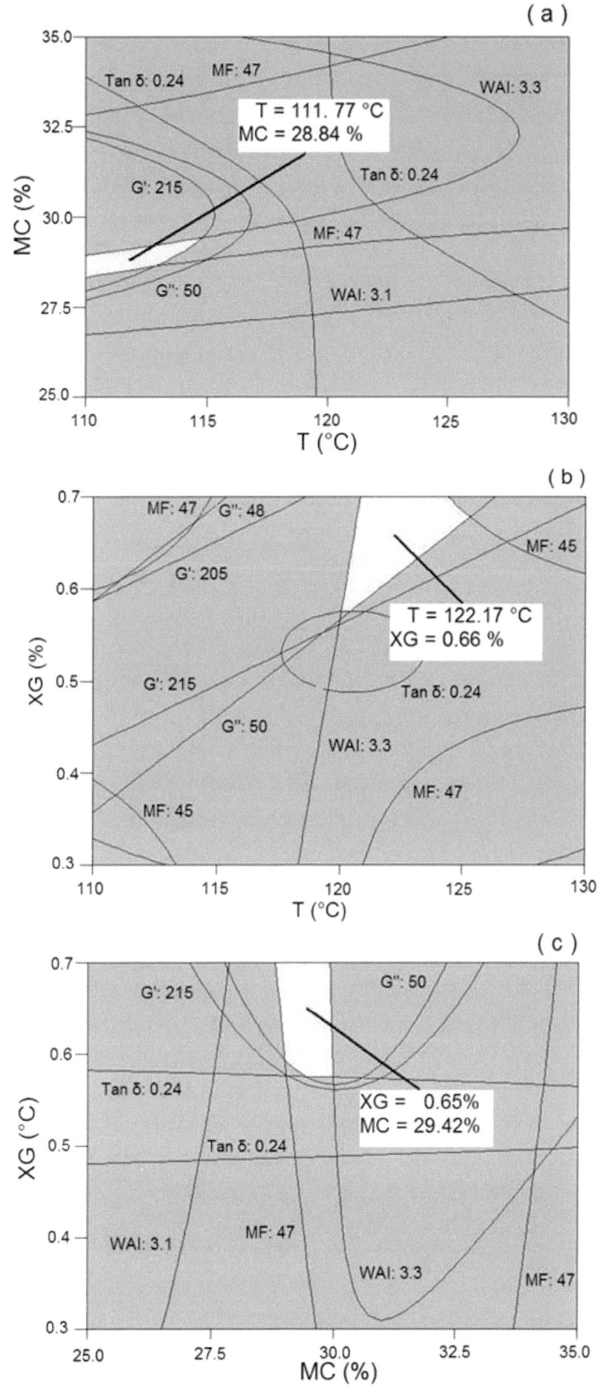

Figure 11. Regions of best combinations of process factors for producing optimized extruded nixtamalized corn flour, using a single-screw extruder. (a) Effect of moisture content (MC, %) and extrusion temperature (T, °C); (b) Effect of xanthan gum (XG, %) and T; (c) Effect of XG and MC.

Maximum force. Changes in MF of tortilla made from ENCFs were affected by the linear term of moisture content (MC, $p < 0.01$), quadratic of moisture content $(MC)^2$, $p < 0.01$) and the combinations of temperature-moisture content [(T)(MC), $p < 0.10$] and temperature-gum [(T)(XG), $p < 0.01$]. The prediction model in terms of original variables for MF was:

$$Y_{MF} = 14.75 - 7.36MC - 0.04(T)(MC) - 1.17(T)(XG) + 0.2(MC)^2 \tag{5}$$

The regression model explained 82% of the total variation ($p < 0.01$) of MF of tortilla made from ENCFs. Figs. 10(a,b,c) show the effect of MF in tortilla made from ENCFs as a function of T, MC and XG, observing that at any value of moisture content for conditioning starting at approximately the midpoint of the matrix, the maximum force is increased due to the quadratic effect. The lowest value (46.16 kPa) was shown at T=115 - 130°C/MC=30.69-34.87% (w/w)/GX=0.42-0.7% (w/w) (Figs. 10 (a,b,c).

The superimposition of contour plots of the effect of variables of the extrusion process (T, MC and XG) on WAI of flour, G', G'' and Tan δ of masa, and MF of tortilla made from ENCFs, was used to obtain Figs. 11 (a,b,c), which in turn was utilized to determine the best combinations of the extrusion process variables. The central points of the regions of optimization in Figs. 11 (a,b,c) correspond to the values of the process variables of T=111.77°C/MC=28.84%, XG=0.66%/T=122.17°C and XG=0.65%/MC=29.42%, respectively. The optimal combination for the operation conditions of the single-screw extruder derived from the averages of those values were: T=116.67°C/MC= 29.13%/XG = 0.65%. The optimal conditions were validated using experiments, which were similar to those values predicted by RSM. These values can be used to obtain ENCF with the highest WAI and WAC, and tortillas with less firmness (softer) and more flexibility (more rollable).

4. Conclusions

The viscoelastic parameters (G', G'', and Tan δ) were affected by xanthan gum concentration, inducing a weakening of the existing structure of masas from extruded nixtamalized corn flour. Tortillas from masas with high values of Tan δ improved their textural characteristics. The optimal combination of variables to obtain ENCF using a single-screw extruder was: T=116.67°C/MC= 29.13% / XG= 0.65%. The optimal conditions were validated experimentally, and results were similar to those values predicted by RSM. Besides the information of basic nature that can be obtained with the dynamic method, it can be considered a practical rheological tool capable of differentiating masa from extruded nixtamalized corn flours.

Author details

Luis Carlos Platt-Lucero, Benjamín Ramírez-Wong*,
Patricia Isabel Torres-Chávez and Ignacio Morales-Rosas
Universidad de Sonora, Departamento de Investigación y Posgrado en Alimento, Hermosillo, Sonora, México

* Corresponding Author

Acknowledgement

The Authors want to thank to SEP-PROMEP for the financial support of the Project "Aplicación de Métodos Físicos, Reológicos y Biológicos en el Procesamiento de Maíz", Red: Tecnologías Convencionales y Alternativas para el Procesamiento de Cereales.

5. References

Aguirre-Cruz, A., Méndez-Montealvo, G., Solorza-Feria, J., and Bello-Pérez, L.A. 2005. Effect of carboxymethylcellulose and xanthan gum on the thermal, functional and rheological properties of dried nixtamalised maize masa. Carbohydrate Polymers 62: 222-231.

Anderson, R.A., Conway, H.F., Pteifer, V.F. and Griffin E.L. Jr. 1969. Gelatinization of corn grits by roll and extrusion cooking. Cereal Sci Today. 14: 1-12.

Arámbula- Villa, G., Figueroa, J.D.C., Martinez Bustos, F., Ordorica F.C.A., and González-Hernandez. 1998. Milling and processing parameters for corn tortillas from extruded Instant dry masa flour. J. Food Sci. 63:338-341.

Arámbula, V.G., Mauricio, S.R.A, Figueroa, C.J.D., González-Hernández, J., and Ordorica, F.C.A. 1999. Corn masa and tortillas from extruded instant masa and tortillas from extruded instant corn fluor containing hidrocollids and lime. J. Food Sci. 6:120-124.

Arámbula, V.G., González-Hernández, J,Moreno, M.E., and Ordorica, F.C.A. 2002. Characteristics of tortillas prepared from dry extruded masa flour added with maize pericarp. J. Food Sci. 67: 1444-1448..

Brouillet-Fourmann, S., Carrot, C., and Mignard, N. 2003. Gelatinization and gelation of corn starch followed by dynamic mechanical spectroscopy analysis. Rheol. Acta 42: 110-117.

Design Expert 2002. Version 6.0.7 by STAT-EASE. Inc, MN. USA.

Dickinson, E. 2003. Hydrocolloids at interfaces and the influence on the properties of dispersed systems. Food Hydrocolloids. 17, 25-39.

Faubion, J. M and Hoseney, R.C. 1990. The Viscoelastic Properties of Wheat Flour Doughs En Dough Rheology and Baked Product Texture, editado por H. Faridi y J.M Faubion. New York. Van Nostrand Reinhold. pp. 42-62.

Fernández, D. A., Waniska, R. D., and Rooney, L. W. 1999. Changes in starch properties of corn tortillas during storage. Starch/stärke. 51(4):136-140.

Ferry, J.R. 1980. Viscoelastic properties of polymers (3rd ed.). New York: Wiley.

Flores-Farias, R., Martínez-Bustos, F., Salinas-Moreno, Y., and Ríos, E. 2002. Characterization of Commercial Nixtamalized Maize Flours. Agrociencia 36:557-567.

Galicia G.T. 2005. Producción y evaluación de la calidad de harinas nixtamalizadas obtenidas de maíz de calidad proteica. Utilizando el proceso de extrusión y técnicas de optimización. Tesis de Maestria. Universidad de Sonora.

González, V.I. 2006. Evaluación de cambios fisicoquímicos que sufre el almidón en el proceso de elaboración de tortillas, utilizando harinas nixtamalizadas por extrusión de maíz de alta calidad proteica. Tesis de Maestria. Universidad de Sonora.

Gurkin, S. 2002. Hydrocolloids-ingredients that add flexibility to tortilla processing. Cereal Foods World. 47:41-43.

Hanna, M.A., Chinnaswamy, R., Gray, D.R., and Miladinov, V.D. 1997. Extrudates of starch-xanthan gum mixtures as affected by chemical agents and irradiation. J. Food Sci. 62(4): 816-820.

Harper. J.M. 1989 Foods extruders and their applications. In "Extrusion Cooking". C. Mercier, P. Linko y J.M. Harper (ED). American Association of Cereal Chemists. St. Paul, MN, EUA. Pp. 1-15.

Martínez-Flores, H., Martínez-Bustos F., Figueroa J., and González-Hernández, J. 1998. Tortilla from extruded masa as related to corn genotype and milling process. J. Food Sci. 63: 130-133.

Martínez, F.H.E., Gayton, M.M., Figueroa, C.J., Martínez, B.F., Reyes, V.M., y Rodríguez, V.A. 2004. Efecto de algunos conservadores sobre la vida util de tortillas de maiz obtenidas a partir de masa extrudida. Agrociencia 38: 285 - 292.

Masi, P., Cavella, S. and Sepe, M. 1998. Characterization of dynamic viscoelastic behavior of wheat flour doughs at different moisture contents. Ceral Chem. 75:428-432.

Milán-Carrillo, J., Gutiérrez-Dorado,R., Perales-Sánchez, J.X.K., Cuevas-Rodríguez, E.O., Ramírez-Wong, B., and Reyes-Moreno, C. 2006. The optimization of the extrusion process when using maize flour with a modified amino acid profile for making tortillas. Int. J. Food Sci. and Tech. 41:727-736.

Molina, M., R, Letona M. and Bressani R. 1977. Drum drying for the improved production of instant tortilla flour. J. Food Science 42:1432.

Montogomery, D. C. 1991. Design and Analysis of Experiments. 3ra Ed. John Wiley and Sons, Inc. New York, NY, EUA. p.278.

Myers, R. H. 1971. Response Surface Methodology. Allyn and Bacon Publisher, Boston. MA, EUA. Pp. 26-106.

Ramírez-Wong, B. 1989. Measurement and evolution of textural and rheological properties of fresh corn masa. Ph.D. Dissertation. Texas A&M University, College Station, TX.

Ramírez-Wong,B., Walker, C.E., Ledesma-Osuna, A. I, Torres, P.I., Medina-Rodríguez,C.L, López-Ahumada,G.A., Salazar-García, M.G., Ortega-Ramírez, R., Johnson,A.M. and Flores, R.A. 2007. Effect of flour extraction rate on white and red winter wheat flour compositions and tortilla texture. Cereal Chem. 84(3):207–213.

Reyes-Moreno, C., Milán-Carrillo, J., Gutierrez-Dorado, R., Paredes-López, O., Cuevas-Rodriguez, E.O., and Garzón-Tiznado, J.A. 2003. Instant flour from quality protein maize (Zea mays L). Optimization of extrusión proces. Lebensm-Wiss. U.-Technol. 36:685-695.

Román-Brito, J.A., Agama-Acevedo, E., Méndez-Montealvo, G., and Bello-Pérez, L.A. 2007. Textural studies of stored corn tortillas with added xanthan gum. Cereal Chem. 84(5):502-505.

Serna-Saldivar, S. O. 1996. Química, almacenamiento e industrialización de los cereales. Ed. Tecnológico de Monterrey. México. Pp. 275-305.

Serna- Saldivar, S. O., Gómez, M. H., and Rooney, L. W. 1990. Technology, chemistry, and nutritional value of alkaline-cooked corn products. In "Advances in cereal science and technology. Vol. X. ACC. Pomeranz. Y. Pp.243-307.

Vargas-López, J.M. 1987. Producción y evaluación de una harina para elaborar tortillas a partir de una mezcla maíz / sorgo en un extrusor-cocedor de bajo costo. Tesis de Maestría. Universidad de Sonora.

Waniska, R.D. 1976. Methods to assess quality of boiled sorghum, gruel and chapaties from sorghum with different characteristics. M.S. Thesis. Texas A&M University, College Station, TX.

Yau. J.C., R.D., and Rooney, L.W. 1994. Effects of Food Additives on Storage Stablility of Corn Tortillas. Cereal Foods World 39:396-401.

Chapter 12

Use of the Stress-Relaxation and Dynamic Tests to Evaluate the Viscoelastic Properties of Dough from Soft Wheat Cultivars

Elisa Magaña-Barajas, Benjamín Ramírez-Wong,
Patricia I. Torres-Chávez and I. Morales-Rosas

Additional information is available at the end of the chapter

http://dx.doi.org/10.5772/45993

1. Introduction

Wheat is the only cereal capable of produce strong and cohesive dough, qualities responsible for the wide variety of food products made from it. In Mexico, wheat is classified into five groups based on functionality; those wheat varieties in group three are called soft, and they are used in products such as cakes and cookies.

Wheat dough is classified as viscoelastic material (Faubion & Hoseney, 1990) which has elastic and viscous characteristics conferred by gluten. Gluten network is formed by hydrophobic interactions between water and protein polymers of the flour (mainly glutenins and gliadins). The nature of the glutenin mainly influences the functional properties of strength and elasticity, while the gliadin fraction provides extensibility and viscosity to the dough (Lu & Grant, 1999). Studies have been conducted in search of those factors that influence dough's viscoelastic properties to better understand the behavior of dough, so it can be handled and utilized properly.

Smith et al. (1970) utilized the dynamic test in a gluten-starch-water system and found that high protein content is reflected in high values of the storage module (G′) and loss module (G′′). Besides protein content, it has been found that the soluble protein fraction is a determinant factor for the rheological properties of dough by acting as a lubricating agent (Rouille et al., 2005). The importance of the total gliadin fraction in the structure of gluten was observed by Lee & Mulvaney (2003). Gupta et al. (1993) utilized an extensograph to evaluate doughs and found a high correlation between unextractable proteins and maximum resistance (Rmax), which led them to believe that a high molecular weight fraction of glutenin contributes to the resistance of the dough.

INTECH
open science | open minds

© 2012 The Author(s). Licensee InTech. This chapter is distributed under the terms of the Creative Commons Attribution License http://creativecommons.org/licenses/by/3.0), which permits unrestricted use, distribution, and reproduction in any medium, provided the original work is properly cited.

Another factor affecting the rheological properties of dough is starch. It has been seen that damage to starch and the presence of other minor components affect rheological properties (Dexter et al., 1994; Lynn & Stark, 1995). Campos et al. (1997) and Chiotelli et al. (2004) suggested that the rheological changes observed in dough subjected to low heating tests may be due to starch molecules present. Zeng et al. (1997) found that 80% of the variation in the viscosity of starch paste of 13 cultivars of wheat was due to the concentration of amylose and amylopectin, and Morris et al. (1997) observed that this variation is due to the presence of isoenzymes synthesized in relation to the starch granule. Georgopulus (2006) observed that the rheological properties of dough are affected more by the removal of the native lipids of the flour than for its gluten.

Rheological properties of wheat flour are determinants of the desired characteristics in the final product, as well as for the design of equipment and processes. Therefore, it is necessary to find reliable rheological tests that are a useful tool for characterizing dough of wheat flours (Safari-Ardi & Phan-Thien, 1998). Rheology of wheat dough is broad, and the tests utilized are diverse. Initially, empirical tests were more used than the fundamental tests, but disadvantages were seen such as: dependence on the instrument, the form and quantity of the sample utilized and lack of theoretical basis (Faubion & Hoseney, 1990), which led to the development of fundamental tests. The most important fundamental viscoelastic tests utilized for wheat dough are the force-deformation ratio, creep test, the dynamic test and stress-relaxation test (Mohsenin, 1978).

Recently, the dynamic test and the stress relaxation test have been used to characterize the dough viscoelasticity. Safari-Ardi & Phan-Thien (1998), utilizing the dynamic test observed that there were no differences among dough elaborated with different flours, probably due to the small deformation used (<1%). Therefore, it was decided to evaluate viscoelastic properties of dough applying a stress-relaxation test. In this technique, the dough is rapidly deformed at a predetermined level and the stress is measured over time, where deformations are >1% (Faubion & Hoseney, 1990; Mohsenin, 1978; Rao et al., 2000).

Several researches have utilized and recommended the stress-relaxation test (Lee and Mulvaney, 2003; Rao et al., 2000; Uthayakumuran et al., 2002; Wikström & Eliasson,, 1998; Yadav et al., 2005). Safari-Ardi et al. (1997) found when using the stress-relaxation test, it was possible to differentiate dough from distinct wheat flours, and demonstrated that besides the method being consistent, the data of the stress-relaxation test obtained at high amplitudes (1-15% deformation in $3x10^3$ s) were very precise (Safari-Ardi & Phan-Thien, 1998). Utilizing the stress-relaxation test, it has been found that the distribution of the quantity of protein (high molecular weight glutenin) and its molecular weights are related to the relaxation time (Rao et al., 2000; Uthayakumaran et al., 2002). In addition, it has been shown that moisture content and strain affect the relaxation characteristics of dough by doing them no linear (Yadav et al., 2005). Similarly, Smith et al. (1970) observed that the relaxation time increases with mixing, and Rao et al., (2000) concluded that this parameter is less for strong dough than that moderately strong.

The study of viscoelastic properties of flour from different wheat cultivars has been reported for different authors (Larsson & Eliasson, 1996a, 1996b; Wikström & Eliasson, 1998; Safari-Ardi & Phan-Thien, 1997, 1998; Van Bockstaele et al., 2008), nevertheless, none of them specifically study the viscoelastic characteristics of soft wheat, and this could be important for the cookies and baked industry. The purpose of this research was to evaluate the rheological properties of soft wheat dough from some cultivars in order to determine in the most appropriate manner its final use and in particular its quality. The viscoelastic properties of soft wheat dough were evaluated by applying the stress-relaxation and the dynamic test, and relate them with empirical rheological measurements and physicochemical characteristics.

2. Experimental procedure

2.1. Raw material

Samples of soft wheat from four cultivars were used: Barcenas, Cortazar, Salamanca and Saturno. Samples were obtained in the Central part of Mexico (El Bajío Zone), and they were sent to the Departamento de Investigación y Posgrado en Alimentos from the University of Sonora at Hermosillo Sonora, México, where was carried out this study.

2.2. Flours elaboration

Wheat samples of the four cultivars were cleaned (Blount/Ferrell-Ross, model M2BC), and placed in plastic bags, which were stored at refrigeration temperature (4°C) until use. Wheat samples were conditioned based on the international approved method number 26-95 (American Asociation of Cereal Chemists, [A.A.C.C], 2000), at a moisture content of 16%, utilizing a conditioner (Chopin Instruments, Villeneuve-La-Garenne, France). Samples were allowed to stand for a period of 24 h before preparation of flours. The conditioned samples were milled using the approved method number 26-10 (A.A.C.C., 2000), and an experimental mill (Brabender, model Quadrumat Senior, South Hackensack, NJ). For maturation, flours were allowed to stand for a period of 15 days.

2.3. Flours quality analysis

For proximate chemical analysis of flours, the official methods of the A.A.C.C. International (2000) were used, and the following determinations were made: protein content (approved method number 46-13) using a nitrogen analyzer (LECO, model FP-528, MI, USA), and the protein factor was N x 5.7; ash content (approved method number 08-03); and moisture content (approved method number 44-40).

Wet gluten content was determined utilizing the approved method number 38-11 (A.A.C.C., 2000) and the apparatus Glutomatic (Falling Number, model 2100, Huddinge, Sweden). Sedimentation volume was determined using the approved method number 56-61A (A.A.C.C., 2000). The falling number was measured utilizing approved method number 56-81b (A.A.C.C., 2000) and the apparatus Falling Number (model 1400, Huddinge, Sweden).

2.4. Rheological tests

2.4.1. *Empirical test*

The farinographic parameters of water absorption, stability and time of development were obtained based on the farinographic approved method number 54-21 (A.A.C.C., 2000), and utilizing a farinograph (Brabender Instruments, type 810143, South Hackensack, NJ).

To characterized flour-water dough for extensibility and resistance to extension, it was used approved method number 54-30A (A.A.C.C., 2000), and the Alveograph (Chopin Instruments, Villeneuve-La-Garenne, France).

2.4.2. *Fundamental test*

Sample preparation. Dough was prepared with 100 g of flour of each cultivar, and adding distilled water corresponding to the absorption obtained from the farinograph. Flour was placed and mixed in a mixer with 300 g capacity (National MFG, Lincoln, NE, USA) without the distilled water. Then it was mixed for 1 min with one afterward; the dough was allowed to stand for 30 min in a proofing chamber (National MFG, Lincoln, NE, USA) with a controlled environment (30°C, 95% relative humidity). Next, samples of dough weighing individually 2.7 g were placed in plastic bag, hermetically sealed, and left at room temperature (25°C) until use in the rheometer.

Dynamic method. Oscillatory measurements were conducted at 0.1% strain within the linear viscoelastic regime (previously carried out), over a frequency range of 0.1 to 100 rad/s. Frequency sweep experiments were conducted in a Rheometer (Rheometrics Scientific, model RSF III, Piscataway, NJ, USA). The rheometer was equipped with 25-mm diameter parallel plates that were maintained at 25° C. A dough sample of 2.7 g was round by hand and placed between the plates of the rheometer. Sample was gently flattened to fit the plate geometry using a Teflon-coated spatula. The upper plate was lowered to a fixed gap between plates of 2 mm. The exposed edges of dough were trimmed, and to avoid drying, petroleum jelly was injected around the plates. Sample was allowed to rest for a period of 20 min before the test. The rheometer was run in the frequency scanning in oscillatory mode, and at a controlled temperature. The viscoelastic parameters determined were: the storage modulus (G', Pa), the loss modulus (G'', Pa) and tangent of phase angle ($\delta = G''/G'$). These parameters were obtained using the software analysis program of the rheometer (RSI Orchestrator, Rheometrics Scientific) (Magaña-Barajas et al., 2011).

Stress-relaxation test. Sample of dough was prepared and placed in the rheometer in manner similar to that utilized in the dynamic test. After placing the sample in the rheometer, it was allowed to rest for a period of 20 min before the test. Based on preliminary tests and some researches (Li et al., 2003; Rao et al., 2000; and Safari-Ardi & Phan-Thien 1998), stress relxation was measured at intervals of 0.1 s applying a 15% shear strain for 30 min at a controlled temperature of 25ºC operating in shear stress mode. The parameters determined were maximum stress at 15% strain (G_0) and the relaxation time (τ).

2.5. Experimental design and statistical analysis

A completely randomized experimental design was performed, where the factor was the wheat cultivar: (Barcenas, Cortazar, Salamanca and Saturno). For the purpose of determining the effect of the factor on the different determinations, analysis of variance (ANOVA) was used with a 95% confidence interval. In addition, simple correlations (r) between different determinations were carried out. ANOVA was carried out with the Statistical Analytical System Software (SAS Institute, Inc. Cary, NC, 2002).

3. Results and discussion

3.1. Flours quality

Analysis of variance showed that the wheat cultivar affected significantly ($p<0.05$) the proximate chemical analysis of flours. Table 1 presents mean values for protein, moisture, ash and wet gluten content, sedimentation volume, and falling number each flour from the soft wheat cultivars. Cultivar Barcenas showed the highest values of protein content, ash content, wet gluten, sedimentation volume and falling number. Moisture content was similar for all the cultivars.

In general, it has been observed that wheat cultivar affected the physical and chemical properties of flours (Lin & Czuchajowska, 1997). In some investigations (Carrillo et al., 1990; Chiotelli et al., 2004; Yamamoto et al., 1996), marked physicochemical differences have been observed in cultivars of soft wheat, and that falling number has been seen to influence the rheological properties of dough. Sedimentation volume indicates the quality of the proteins presents; in hard wheat the close values to 70 mL indicates high baking quality (A.A.C.C., 2000; Pomeranz, 1988). Falling number values greater than 400 s have been reported that the amylolytic activity is low. Therefore, wheat is considered that has been exposed to low rainfall during growth up to harvest time, consequently resulting in healthy grains (Carcea et al., 2006; Pomeranz, 1988). In this study, for all the flours, the Falling number was greater than 400 s, which indicate that flours were obtained from healthy soft wheat cultivars.

3.2. Rheological tests

3.2.1. Empirical test

Table 2 shows the mean values for water absorption, stability and developing time each flour. Salamanca had the highest values of water absorption and developing time, while cultivar Barcenas had the greatest stability. Values for water absorption, stability and developing time are directly proportional with protein content. In addition, water absorption is directly proportional to the diameter of the cookie (Bloksma, 1990; Bloksma & Bushuk, 1988; Farrand, 1969; Yamamoto et al., 1996).

Table 2 shows also the alveographic parameters for flours from the soft wheat cultivars. Cultivar Salamanca showed the highest values of W, P and P/L. These high values probably were due to its protein content, a factor that also results in high water absorption (Unbehend et al., 2004; Yamamoto et al., 1996).

Cultivar	Moisture Content (%)	Protein Content[a] (%)	Ash Content[a] (%)	Wet Gluten[a] (%)	Sedimentati on Volume[a] (mL)	Falling Number (s)
Barcenas	14.08 ± 0.33[b]	11.81 ± 0.50	0.84 ± 0.18	34.87 ± 4.82	32.03 ± 1.62	476.60 ± 31.73
Cortazar	14.06 ± 0.23	10.58 ± 0.31	0.69 ± 0.21	31.77 ± 2.47	22.04 ± 1.18	422.51 ± 9.51
Salamanca	14.10 ± 0.45	11.27 ± 0.49	0.76 ± 0.18	33.00 ± 1.02	28.92 ± 8.66	449.67 ± 27.54
Saturno	14.05 ± 0.61	10.32 ± 0.41	0.55 ± 0.04	30.86 ± 2.46	25.46 ± 1.94	409.98 ± 13.85

[a] Dry basis [b] Standard Deviation

Table 1. Physicochemical characteristics of flour from soft wheat cultivars

Cultivar	Farinograph			Alveograph		
	Water Absorption (%)	Stability (min)	Developing Time (min)	W[b] (10^{-4}J)	P[c] (mm H_20)	P/L[d]
Barcenas	54.04 ± 3.84[a]	4.28 ± 2.01	3.70 ± 0.66	114.75 ± 14.25	49.76 ± 5.04	0.46 ± 0.11
Cortazar	56.16 ± 1.48	1.96 ± 0.59	3.39 ± 0.39	85.50 ± 10.22	52.63 ± 7.54	0.57 ± 0.05
Salamanca	56.39 ± 1.70	3.75 ± 2.14	4.54 ± 1.67	139.62 ± 66.88	67.01 ± 16.41	0.70 ± 0.15
Saturno	53.28 ± 1.51	3.51 ± 0.68	3.14 ± 0.54	106.33 ± 17.58	48.69 ± 4.13	0.52 ± 0.10

[a] Standard Deviation; [b]W= Deformation energy of dough; [c]P= Maximum overpressure, tenacity; [d]P/L= Curve configuration ratio, extensibility

Table 2. Rheological characteristics of flour from soft wheat cultivars

3.2.2. Fundamental test

Two techniques were used to characterize the viscoelasticity of dough from soft wheat cultivars: a) the dynamic method; and b) the stress-relaxation test, which were performed in a rheometer.

Dynamich method. The viscoelastic properties of dough obtained from the soft wheat cultivars were the storage modulus (G′), loss modulus (G′′) and tangent of angle phase (Tan δ) (Figures 1, 2 y 3, respectively). To evaluate if there were statistically differences among the viscoelastic properties of dough from the different wheat cultivars, an analysis of variance was performed on the G′, G′′ and Tan δ values at a frequency of 5 rad/s, where the dough viscoelasticity was in the linear viscoelastic domain (Figures 1, 2 and 3). According to the ANOVA, G′, G′′ and Tan δ obtained with the dynamic test at a frequency of 5 rad/s had similar values ($p>0.05$), indicating that dough from the soft wheat cultivars were in the same values of viscoelasticity (Table 3). For all the soft wheat cultivars, the G′ values were greater than those for G′′, and Tan δ was greater than 0.50.

This indicates that in soft wheat cultivars dough prevailed the viscous behavior over the elastic behavior.

Figures 1 and 2 show the dependence of G′ and G′′ with frequency (Hibber & Wallace, 1966). In both figures, values of G′ and G′′ are increasing, which could indicate that dough hardness increase, and it could be due to content of gluten (Chiotelli et al., 2004) and protein quality, resulting in the gluten of wheat dough not as weak as expected (since soft wheat flour contains a weak gluten). Several authors (Huebner et al., 1999; Yamamoto et al., 1996) agreed that content and quality of protein and gluten content are factors affecting positively these viscoelastic parameters.

In Figure 3 is presented the behavior of Tan δ with frequency for dough from all soft wheat cultivars. Initially, Tan δ shows a slight decrease; at intermediate levels (1-10 rad/s) values at a given frequency are similar; and at high frequencies this parameter shows the same initial phenomenon that at low frequency values. Tan δ for all the soft wheat cultivars is higher than 0.5, which indicates that viscous behavior is higher than the elastic behavior.

Stress-relaxation test. Dough was subjected to a deformations much greater than that in the dynamic test (>1%). The parameters evaluated from this test were the initial (maximum) stress at 15% strain (G_0), and the relaxation time (τ). This last parameter is related to the process of flow occurring when dough is relaxed, and is defined as the time required for the force to fall 1/e times, or by 36.8% of its original value (Smith et al., 1970).

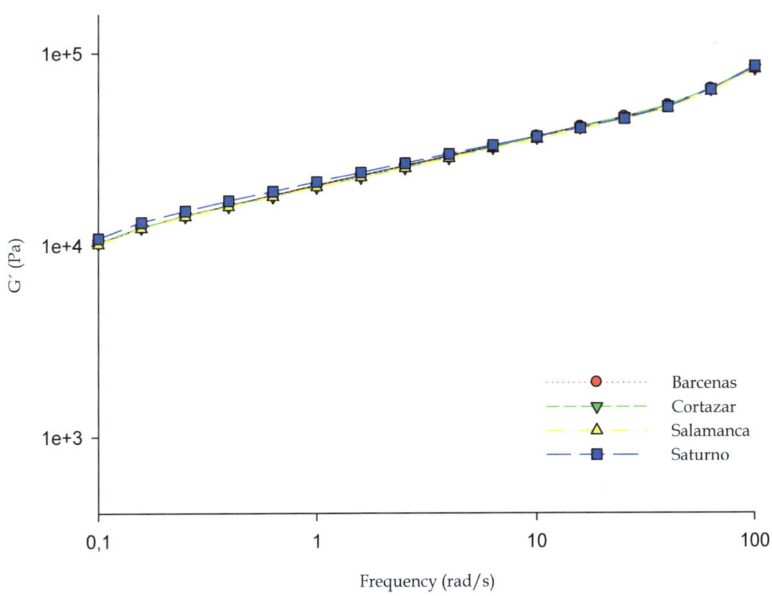

Figure 1. Storage modulus (G′) evaluated with the dynamic test for dough made of soft wheat cultivars.

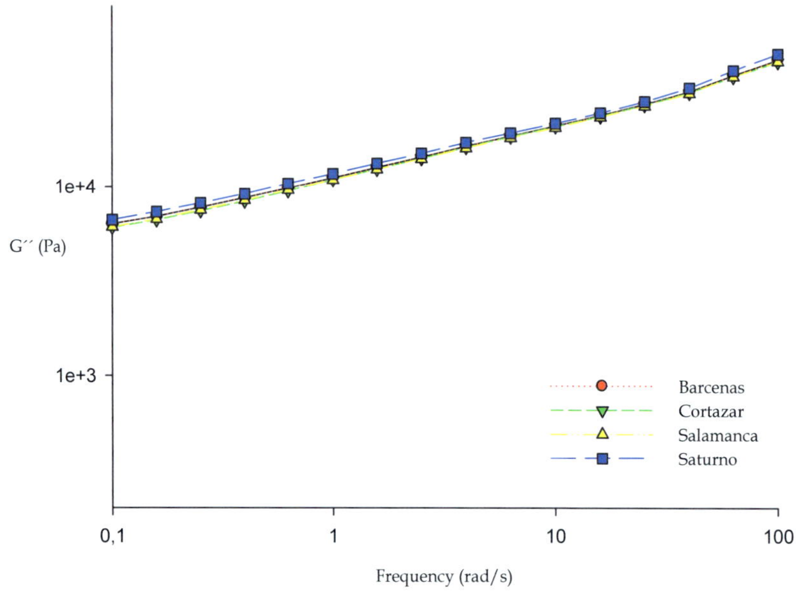

Figure 2. Storage modulus (G´) evaluated with the dynamic test for dough made of soft wheat cultivars.

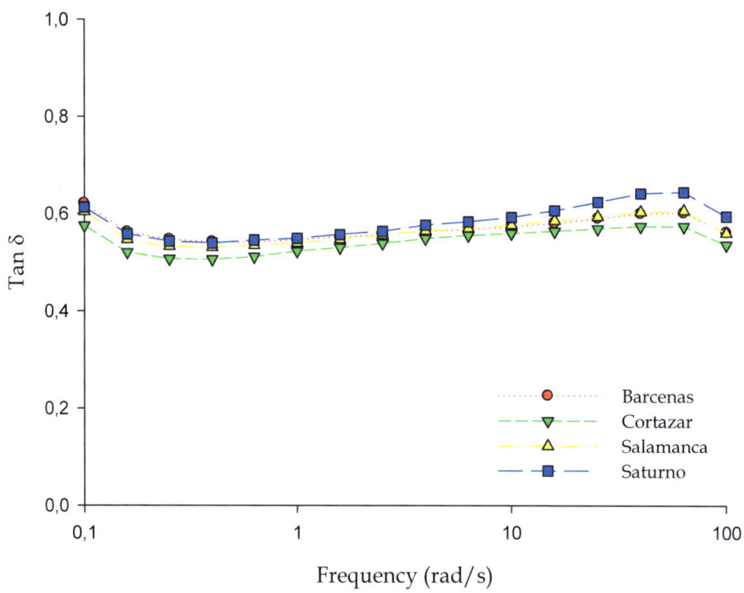

Figure 3. Tangent of angle phase (Tan δ) evaluated with the dynamic test for dough made of soft wheat cultivars.

An ANOVA was carried out to see if there were differences between the viscoelastic properties of the different soft wheat cultivars determined with the stress-relaxation test. It was shown that the soft wheat cultivar affected only to τ ($p<0.05$).

Figure 4 shows the behavior and mean values for the viscoelastic parameters of τ and initial maximum stress (G_0) of dough from the soft wheat cultivars, obtained in the stress-relaxation test. Initially, dough was subjected a high deformation (15%) yielding in response a high initial stress, and as time progress, dough relaxed and stress diminished. It is observed that all curves present a single maximum stress coinciding with Rao et al. (2000). Li et al (2003) found that relaxation process occur in two steps: the first, correspond at the increase of distribution of protein polymers short chain (gliadin and glutenins of low molecular weight). The second peak corresponds at the protein polymers of large chain like glutenins of high molecular weight. In this case, the soft wheat cultivars evaluated had slow protein content (9.9%) and the content of protein polymers short chain (44.22% of gliandins, of total) is highest than the content of protein polymers short chain (data not showed), changing the ratio glutenins/gliadins.

Cultivar	Dynamic Test[g]			Stress Relaxation Test[h]	
	G′[b] (Pa)	G′′[c] (Pa)	Tan δ[d]	G_0[e] (Pa)	τ[f] (s)
Barcenas	29080 ± 3384[a]	16378 ± 1926	0.56 ± 0.03	510.42 ± 119.61	0.36 ± 0.07
Cortazar	28445 ± 11489	15621 ± 1114	0.54 ± 0.02	466.00 ± 55.61	0.38 ± 0.02
Salamanca	28246 ± 4174	16007 ± 2714	0.56 ± 0.01	484.28 ± 130.50	0.33 ± 0.07
Saturno	29750 ± 2967	17121 ± 1524	0.057 ± 0.02	462.45 ± 56.50	0.41 ± 0.02

[a] Standard Deviation; [b] G′: Storage modulus; [c] G′′: Loss modulus; [d] Tan δ: Tangent of phase angle; [e] G_0: Initial maximum stress; τ^f: Relaxation time; [g]: the viscoelastic parameters were evaluated at frequency of 5 rad/s; [h]: the viscoelastic parameters were measured at 15% shear strain

Table 3. Viscoelastic characteristics of dough made from soft wheat cultivars evaluated with the dynamic and stress relaxation tests

In Table 3 is presented G_0 and τ values for all flours. The wheat cultivar Saturno had the highest value of τ (0.41 s). The high values of τ indicates that the recovery of deformed dough structure is slow, reflecting a weak gluten. That wheat cultivar presents the lowest values of moisture, protein content, ash content, wet gluten and falling number, which probably caused a lesser strength in gluten. This could explain the strength of the dough. Dough of cultivars Salamanca and Barcenas had the lowest τ values (0.33 y 0.36 s, respectively). These varieties have high values in almost all the physicochemical properties. Low values of τ correspond at strong gluten. For this reason when dough is subject to a deformation quickly recovers its original form. In general, variation of values of τ (range 0.33 s to 0.41 s) was probably attributed at the content and type of protein, and moisture content of the wheat cultivar. This has been observed in several investigations, where it has been found that moisture and gluten properties (such as gluten strength) affect the parameters of this test (Fu et al., 1997; Larsson & Eliasson, 1996a; Li et al., 2003; Yadav et al., 2005).

Results obtained show an inverse relationship between the values of G_0 and τ. Applying a high strain to dough with three-dimensional network structure of strong gluten, values of G_0 are obtained almost immediately after deformation.

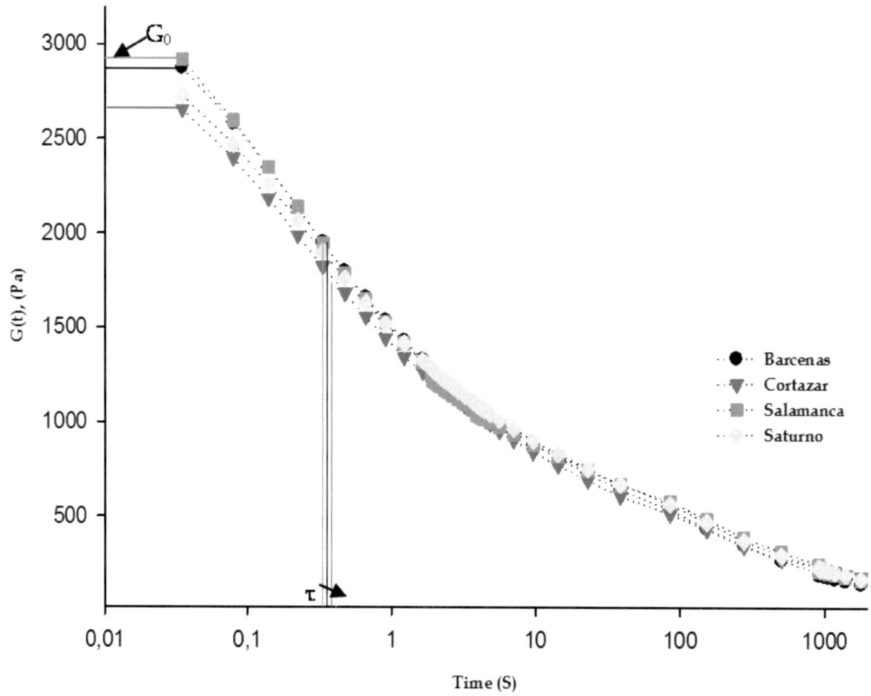

Figure 4. Relaxation modulus, G (t) vs time of dough from soft wheat cultivars, at 15% of deformation. τ: relaxation time.

The relaxation time was the parameter showing the greatest difference among the soft wheat cultivars. However, the differences obtained in τ among dough from the soft wheat cultivars were less than estimated. This despite to the range of low protein content of the soft wheat cultivars (10 to 12%, approximately), and besides, the fraction of gliadin and glutenin were of low molecular weight and approximately similar in all the samples (70% gliadin and 30% glutenin of low molecular weight, data not shown). This could indicate that tests used to evaluate the viscoelasticity in dough were high sensibility being capable to characterize dough from different soft wheat cultivars. Characterization and differentiation of dough was better with the stress-relaxation test than with the dynamic test, coinciding with others authors (Lee and Mulvaney, 2003; Rao et al., 2000; Uthayakumuran et al., 2002; Wikström & Eliasson,, 1998; Yadav et al., 2005; Safari-Ardi et al., 1997).

3.2.3. Correlations between viscoelastic characteristics and other determinations

To determine if there were significant relations among the viscoelastic properties of dough from soft wheat cultivars, and physicochemical and rheological evaluations in flours, simple correlations *(r)* were performed.

Five significant correlations were found among the viscoelastic characteristics and physicochemical determinations. Regarding to the stress-relaxation test, there were four highly significant correlations ($p<0.01$). The relaxation time (τ) showed a strong correlation with developing time (farinograph) (r=-0.959), which indicates indirectly a high protein content and strong gluten, resulting in a rapid response upon deformation, as a consequence lower τ values. G_0 showed a strong correlation with protein (r=0.984), which could be reflected in the greater strength of gluten, thereby a high resistance upon deformation. A second correlation of G_0 was with wet gluten (r=0.987), which probably is due to the fact that a greater amount of gluten causes greater resistance to deformation, and this is in agreement the study carried out by (Li et al., (2003). Finally, the last correlation of G_0 was with falling number (r=0.986).

The dynamic test only yielded a highly significant negative correlation ($p<0.01$) for the viscoelastic parameter of G′ (storage modulus) with water absorption (r=-0.975), which could be explained by the fact that high water absorption can give indirect evidence of a high moisture content, which in turn has a negative effect on G′ (Faubion & Hoseney, 1990). No correlations were found between the stress-relaxation test and dynamic test utilized to determine the viscoelastic properties of the dough of soft wheat cultivars.

4. Conclusions

The viscoelastic characterization of dough made from flours of soft wheat cultivars was achieved by fundamental tests. The soft wheat cultivar showed to be a determinant factor in all characteristics of the flour. Apparently, the strength of the gluten network is critical in the rheological and viscoelastic characteristics of dough. All the cultivars presents more viscous behavior than elastic behavior (Tan $\delta>0.5$). Wheat varieties evaluated showed a range τ of between 0.33 to 0.41 s.

The stress-relaxation test was better than the dynamic test in differentiating the viscoelastic characteristics of the soft wheat dough. This is confirmed by the highest number of correlations relative to the stress relaxation test, compared with the dynamic method. In our case, τ offers more information than that of G_0, making it an important parameter for characterizing dough. Parameters most affecting τ was content and quality of the protein moisture content and falling number in flours. The stress-relaxation test is a simple and rapid method to perform viscoelastic evaluations, and it is believed to be a technique suitable for the characterization and differentiation of viscoelastic characteristic among dough made from soft wheat cultivars, which is important for determining the final use.

Author details

Elisa Magaña-Barajas, Benjamín Ramírez-Wong,
Patricia I. Torres-Chávez and I. Morales-Rosas
Universidad de Sonora, Hermosillo, Sonora, México

Acknowledgement

We thank to CONACyT for financial support of the project Integral Study of the Quality of Mexican Wheat and its Potential Use, project no. G35201-B,and the project Study of Water

Re-distribution in Gluten from Frozen Dough, and its Effect on Protein and Viscoelasticity Changes no. 154090

5. References

A.A.C.C. (2000). In: Approved Methods of the Association, Methods. The Association, St. Paul, MN.

Bloksma, A. H. & Bushuk, W. (1988). Rheology and chemistry of dough. In wheat: chemistry and technology. Ed. Pomeranz, Y., Ed., Am. Assoc. Cereal Chemistry: St Paul, MN.

Bloksma, A. H., (1990). Dough structure, dough rheology, and baking quality. *Cereal Foods World*, 35, 237-245, ISSN 0146-6283.

Campos, D. T., Steffe, J. F. & Ng, P. K. W. (1997). Rheological behavior of undeveloped and developed wheat dough. *Cereal Chemistry*, 74, 489-494, ISSN 0009-0352.

Carcea, M., Salvatorelli, S., Turfany V. & Mellara F. (2006). Influence of growing conditions on the technological performance of bread wheat (Triticum estivum L). *International Journal of Food Science and Technology*, 41, 102-107, ISSN 0168-1605.

Carrillo, J. M., Rousset, M., Qualest, C. O. & Kasarda, D. D. (1990). Use of recombinant inbred lines of wheat for study of associations of high-molecular-weight glutenin subunit alleles to quantitative traits. *Theorical and Applied Genetetics*, 79, 321-330, ISSN 0040-5752.

Chiotelli, E. Rolee, A. & Le Meste, M. (2004). Rheological properties of soft wheat flour doughs: effect of salt and triglycerides. *Cereal Chemistry*, 81, 459-468, ISSN 0009-0352.

Dexter, J. E., Preston, K. R., Martin, D. G. & Gander, E. J. (1994). The effects of protein content and starch damage on the physical dough properties and bread-making quality of canadian durum wheat. *Journal of Cereal Science*, 20, 139-151, ISSN 0733-5210.

Farrand, E. A. (1969). Starch damage and α-amylase as bases for mathematical models relating to flour water-absorption. *Cereal Chemistry*, 46, 103-116, ISSN 0009-0352.

Faubion, J. M. & Hoseney, R. C. (1990). The viscoelastic properties of wheat flour doughs. In dough rheology and baked product texture, H. Faridi and J. M. Faubion, Eds. (pp. 42-62). New York. Van Nostrand Reinhold.

Fu, J., Mulvaney, S. J. & Cohen, C. (1997). Effect of added fat on the rheological properties of wheat flour doughs. *Cereal Chemistry*, 74, 304-311, ISSN 0009-0352.

Georgopoulos, T., Larsson, H. & Eliasson, A. C. (2006). Influence of native lipids on the rheological properties of wheat flour dough and gluten. *Journal of texture studies*, 37,49, ISSNN 1745-4603.

Gupta, R. B., Khan, K. & MacRitchie, F. (1993). Biochemical basis of flour properties in bread wheats. I. Effects of variation in quantity and size distribution of polymeric proteins. *Journal of Cereal Science*, 18, 23-44, ISSN 0733-5210.

Hibberd, G. E. & Wallace, W. J. (1966). Dynamic viscoelastic behavior of wheat flour doughs. part I: linear aspects. *Rheological Acta*, 5, 193-198, 0035-4511.

Huebner, F. R., Bietz, J. A., Nelson, T., Bains, G. S. & Finney, P. L. (1999). Soft wheat quality as related to protein composition. *Cereal Chemistry*, 76, 650-655, ISSN 0009-0352.

Larsson, H. & Eliasson, C. (1996a). Phase separation of wheat flour dough studied by ultracentrifugation and stress relaxation. I. Influence of water content. *Cereal Chemistry*, 73, 18-24, ISSN 0009-0352.

Larsson, H. & Eliasson, C. (1996b). Phase separation of wheat flour dough studied by ultracentrifugation and stress relaxation. II. Influence of mixing time, ascorbic acid, and lipids. *Cereal Chemistry*, 73, 25-31, ISSN 0009-0352.

Lee, C. C. & Mulvaney S. J. (2003). Dynamic viscoelastic and tensile properties of gluten and glutenin gels of common wheats of different strength. *Journal of Agriculture and Food Chemistry*, 51, 2317 -2327, ISSN 0021-8561.

Li, W., Dobraszczyk, B. J. & Schofield, J. D. (2003). Stress relaxation behavior of wheat dough, gluten, and gluten protein fractions. *Cereal Chemistry*, 80, 333-338, ISSN 0009-0352.

Lin, P.-Y. & Czuchajowska, Z. (1997). General characteristics and milling performance of club wheat vs soft white winter wheat. *Cereal Foods World*, 42, 861-867, ISSN 0146-6283.

Lu W. & Grant, L.A. (1999). Role of flour fractions in breadmaking quality of frozen dough. *Cereal Chemistry*, 76, 663-667, ISSN 0009-0352.

Lynn, A. & Stark, J. R. (1995). Effect of mechanical damage on starch granules-some new observations. *Carbohydrate Letters*, 1, 165-171, ISSN 1073-5070.

Magaña-Barajas, E., Ramírez-Wong, B., Torres, P.I., Sánchez-Machado, D.I. & López-Cervantes, J. (2011). Efecto del contenido de proteína, grasa y levadura en las propiedades viscoelásticas de la masa y la calidad del pan tipo francés. *Interciencia*, 36, 248-255, ISSN 0378-1844.

Mohsenin, N. N. (1978). Physical properties of plant and animals materials: structure, physical characteristics and mechanical properties. Gordon and Breach Science Publishers. (pp. 99, 114-116, 123-143). New York.

Morris, C. F., Shackley, B. J., King, G. E. & Kidwell, K. K. (1997). Genotypic and environmental variation for flour swelling volume in wheat. *Cereal Chemistry*, 74, 16-21, ISSN 0009-0352.

Pomeranz, Y. (1988). Wheat chemistry and technology american association of cereal chemists, Inc. St. Paul, Minnesota, USA.

Rao, V. K., Mulvaney, S. J. & Dexter, J. E. (2000). Rheological characterization of long and short-mixing flour based on stress relaxation. *Journal of Cereal Science*, 39, 159-171, ISSN 0733-5210.

Rouillé, J., Della Valle, G., Lefebvre, J., Sliwinski, E., & Van Vliet, T. (2005). Shear and extensional properties of bread doughs affected by their minor components. *Journal of Cereal Science*, 42, 45-57, ISSN 0733-5210.

Safari-Ardi, M., Phan-Thien, N. & Oliver, J. (1997). Stress relaxation measurements of wheat flour doughs at varing sheat strains. In: Safari-Ardi, M. & Phan-Thien, N. (1998). Stress relaxation an oscillatory test to distinguish between dough prepared from wheat flours of different varietal origin. *Cereal Chemistry*, 75, 80-84, ISSN 0009-0352.

Safari-Ardi, M. & Phan-Thien, N. (1998). Stress relaxation an oscillatory test to distinguish between dough prepared from wheat flours of different varietal origin. *Cereal Chemistry*, 75, 80-84, ISSN 0009-0352.

Smith, J. R., Smith, T. L. & Tschoegl, N. W. (1970). Rheological properties of wheat flour doughs. III. Dynamic shear modulus and its dependence on amplitude, frequency, and dough composition. *Rheological Acta*, 9, 239-252, ISSN 0035-4511.

Unbehend, L., Unbehend, G. & Kersting H. J. (2004). Rheological properties of some croatian and german wheat varieties and their relation to protein composition. CODEN ACALDI *Acta Alimentaria*, 33, 19-29, ISSN 0139-3006.

Uthayakumaran, S., Beasley, H. L., Stoddard, F. L., Keentkot, M., Patien, N., Tanner, R. I. & Bekers, F. (2002). Synergistic and Aditive Effect of Three High Molecular Weight Glutenin Subunit Loc. I. Effect on Dough Rheology. *Cereal Chemistry*, 79, 294-230, ISSN 0009-0352.

Van Bokstaele, F., De Leyn, I., Eeckhout M. & Dewettinck, K. (2008). Rheological properties of wheat flour dough and the relationship with bread volume. II Dynamic Oscillation Measurements. *Cereal Chemistry*, 85, 762-768, ISSN 0009-0352.

Wikström, K. & Eliasson Ann-Charlotte. (1998). Effects of enzymes and oxidizing agents on shear stress relaxation of wheat flour dough: additions of protease, Glucose oxidase, ascorbic acid, and potassium bromated. *Cereal Chemistry*, 73, 331-337, ISSN 0009-0352.

Yadav, N., Roopa, B. S. & Bhattacharya, S. (2005). Viscoelasticity of a Simulated Polymer and Comparison with Chickpea Flour Doughs. *Journal of Food Science*, 70, 273, ISSN 1750-3841.

Yamamoto, H., Worthington, S. T., Hou, G. & Ng, P. K. W. (1996). Rheological properties and baking qualities of selected soft wheats grown in the United States. *Cereal Chemistry*, 73, 215-221, ISSN 0009-0352.

Zeng, M., Morris, C. F., Batey, I. L. & Wrigley, C. W. (1997). Sources of variation for starch gelatinization, pasting, and gelation properties in wheat. *Cereal Chemistry*, 74,63-71, ISSN 0009-0352.

Section 4

Other Applications

Chapter 13

Micro-Rheological Study on Fully Exfoliated Organoclay Modified Thermotropic Liquid Crystalline Polymer and Its Viscosity Reduction Effect on High Molecular Mass Polyethylene

Youhong Tang and Ping Gao

Additional information is available at the end of the chapter

http://dx.doi.org/10.5772/45996

1. Introduction

Molten thermotropic liquid crystalline polymers have physical properties that are dependent on both small molecule liquid crystals and flexible chain polymers. Liquid crystalline polymers (LCPs) are typical examples of self-ordered polymeric systems, due largely to their intrinsic molecular anisotropy. The properties of LCPs are strongly influenced by flow-induced changes in the degree of molecular orientation during processing [1].

Blends containing small amounts of a thermotropic liquid crystalline polymer (TLCP) in a matrix of thermoplastic have attracted technical interest in recent years for two main reasons. Firstly, by the use of TLCP to enhance the mechanical properties of the matrix polymer through in situ formation of fibrous TLCP dispersion during processing, it may be possible to develop 'self-reinforced' composites that exploit the outstanding tensile properties of fibres made from LCPs. Secondly, it is known that TLCP can act as a flow modifier, resulting in a substantial reduction in pressure drop during melt extrusion. Previous studies by Chan et al.[2] have shown that a small amount of TLCP (1.0 wt %) added to high molecular weight polyethylene (HMMPE, Chevron Phillips Marlex® HXM TR570) caused drastic bulk viscosity reduction (> 95.0%) to occur at 190 °C, when TLCP was in its nematic phase. A significant improvement in extrudate surface smoothness has also been observed, coupled with an increase in the processing window from 34 1/s to up to 1000 1/s. Whitehouse et al. [3] blended 0.2%, 0.5% and 2.0% TLCP with high density polyethylene (HDPE, Chevron Phillips Marlex® HMN 6060), and the blends were then rheologically

INTECH
open science | open minds

© 2012 The Author(s). Licensee InTech. This chapter is distributed under the terms of the Creative Commons Attribution License http://creativecommons.org/licenses/by/3.0), which permits unrestricted use, distribution, and reproduction in any medium, provided the original work is properly cited.

characterized at 185 °C when the TLCP was in the nematic regime; substantial viscosity reductions of between 85% and 90% compared with pure HDPE were observed.

It is worth noting that even in early studies of polymer blends containing TLCP, researchers had already attempted to introduce inorganic reinforcements into such blend systems [4]. The addition of inorganic fillers not only enhanced the mechanical properties of the blends but also reduced the anisotropy of the resulting materials [5]. Rheological characterization revealed that TLCP could reduce the melt viscosity of glass-filled thermoplastics [6]. Much work has been published on TLCP systems containing different inorganic solid reinforcements, such as glass fibres [7], carbon black [8], whiskers [9] and silica [10]. Most of the studies have used high inorganic solid reinforcement content and have focused on the balance between the mechanical properties and processability of such blends.

Layered silicates have long been used as fillers in polymeric systems to improve mechanical, thermal and other properties in the resulting polymer composites. In layered silicate itself, the local interaction between layers causing the presence of domains similar to those found in studies of liquid crystalline and ordered block copolymer systems have been analyzed by many researchers [11-14].

It seems that intercalation or exfoliation of layered silicates in polymers should induce nanocomposites to exhibit similar rheological behaviour to that found in the nematic state structures in LCPs. Nanocomposites based on thermotropic liquid crystalline polymer and organically modified layered silicate have been studied by several groups with different foci. Zhang et al. [15] synthesized a series of liquid-crystalline copolyester/organically modified montmorillonite nanocomposites by intercalation polycondensation with different surfactant modified clay in LCP. X-ray diffraction and transmission electron microscope studies indicated that, after ion exchange with suitable surfactants, clay formed delaminated morphology and was well dispersed in LCP. Chang et al. [16] reported nanocomposites of TLCP with an alkoxy side-group and an organoclay (Closite 25 A) prepared by the melting intercalation method above the melt transition temperature of TLCP, with liquid crystallinity of the hybrids being lost when the organoclay content exceeded 6.0 %. Vaia et al. [17] directly annealed a powder mixture of TLCP and clay within the nematic region of TLCP under high hydraulic pressure. Reversible intercalation formed, with a loss of liquid crystallinity which was attributed to the confinement of LCP chains on the clay pseudo-2D geometry. An extensive study of a series of nanocomposites with a segmented main-chain liquid crystalline polymer having a pendent pyridyl group (PyHQ12) or a pendent phenylsulfonyl group (PSHQ12) and commercial Closite 20 A or 30B clays by examining rheological and other properties was reported by Huang and Han [18, 19] to demonstrate whether functionality of TLCP was essential to obtain highly dispersed clay in nanocomposites with TLCP as matrix. Only intercalated morphology formed when the nanocomposites were based on a TLCP without functionality. The formation of hydrogen bonds caused a very high degree of dispersion but a considerable loss in the degree of liquid crystallinity in a PyHQ12/30B nanocomposite. From the above studies, it can be seen that functionality in TLCP is necessary to obtain highly dispersed nanocomposites, but at the same time, there is the possibility of loss some degree of liquid crystallinity in the TLCP.

In this study, the organoclay-modified TLCP (TC3 white) is prepared by a method combining ultra-sonication, centrifugation, solution casting and heating-shearing separation. TC3 white has a very high degree of dispersion of organoclay but without any loss in the degree of liquid crystallinity of TLCP. The effects of the fully exfoliated organoclay on the thermal and rheological properties of TLCP are investigated in detail. Based on this material, we characterize the rheological behaviour of purified TLCP and TC3 white with 1 wt% in HMMPE matrix by a capillary rheometer at 190 °C and 230 °C, where the TLCP has its nematic and nematic-isotropic biphase structures. Schematic drawings of the conformation of organoclay, TLCP molecules and polyethylene molecules before and after yielding are shown. A binary flow pattern model with simulated results is presented and the predicted results show good consistency with experimental data.

2. Experimental

2.1. Materials and samples preparation

2.1.1. Materials

The HMMPE, Chevron Phillips Marlex® HXM TR571, with a melt flow index (MFI) of 2.5g/10 min (ASTM D1238, 190 ºC/21.6 kg) was kindly supplied by Phillips Petroleum International Inc., USA. The TLCP, a copolymer containing 30% p-hydroxybenzoic acid, 35% hydroquinone and 35% sebacic acid (HBA/HQ/SA), used here was synthesized and kindly supplied by B. P. Chemicals Ltd, UK. The as-received TLCP is a light brown powder that has been characterized previously [20]. The Organoclay, Closite 20A modified by dimethyl dihydrogenated tallow ammonium ions, was kindly supplied by Southern Clay Products.

2.1.2. Purification of as-received TLCP

The as-received dried TLCP powder was dissolved into chloroform and followed by mechanically stirred at room temperature overnight. The deionized water was added to the mixture and was mechanically stirred for several minutes. Water extraction was repeated several times to remove any water soluble content, e.g. pure sebacic acid; the subsequent solution was subjected to centrifugal separation (KUBOTA 2010, Japan) at a speed of 3,800 rpm for 1800 seconds. Three layers of solution were obtained. The bottom layer was a brownish precipitate which was believed to be a mixture of HBA, HBA rich TLCP, catalyst or other heavier components in the TLCP powder. The top layer was a white cloudy layer, which was believed to be a SA rich TLCP component. The middle colorless portion was extracted. To obtain more thorough purification, the whole centrifugal separation and extraction processes were repeated several times. The TLCP powder was precipitated out by dropping the extracted solution into methanol. The TLCP fine powder was filtered out and dried at 120 °C for 3 days to remove any residual solvent contents.

2.1.3. Organoclay-modified as-recieved TLCP (TC3 white) preparation

Initially, TLCP was dried in a vacuum oven at 120 °C for 2 days and organoclay in an oven at 100 °C overnight. The materials were dissolved and dispersed in chloroform respectively with a TLCP/organoclay weight ratio 97:3. The solutions were then stirred for about 4 hours at room temperature. After that, TLCP and organoclay were mixed together and the mixture was sonicated by ultrasonic pin vibration (Branson digital sonifier 450, USA) with 45% power for about 2 hours at room temperature. Subsequently, the mixture was centrifuged with a speed of 3,800 rpm for 45 minutes to separate the bottom layer precipitate which was believed to be redundant organoclay and unsolvable TLCP, and the top layer, if any, which was believed to be impurities in TLCP, from the middle portion; then the solvent was volatilized at room temperature. A nanocomposite was obtained, which was dried in an oven at 60 °C for 12 hours and in a vacuum oven at 120 °C for 2 days. The nanocomposite showed severe shear-induced phase separation phenomenon at 190 °C or higher temperatures and can be effectively separated by a capillary rheometer at 190 °C at a low speed (5.0 1/s) [21]. The extruded material was named TC3 white because of its white color, which will be used here for study.

2.1.4. Polyethylene blends preparation

For HMMPE blends, the dried 1.0 wt% TC3 white (or purified TLCP) in powder form and HMMPE in pellet form were mechanically pre-mixed at room temperature until macroscopically homogeneous. The mixture was then extruded using a Dr. Collin twin screw extruder (Dr. Collin GMBH, Germany) at 190 °C with two-time extrusion at different speeds (75 rad/s and 300 rad/s respectively). The extrudate was palletized and kept dry inside an oven overnight to remove moisture generated during the process.

2.2. Characterizations

2.2.1. Wide angle X-ray diffraction

Wide angle X-ray diffraction (WAXD) was conducted at room temperature on a Philips powder X-ray diffraction system (Model PW 1830, The Netherlands). TLCP and TC3 white were first dried in a vacuum oven for 2 days, then hot pressed at 170 °C to form films with thickness of 300 μm; the organoclay was characterized in powder form. WAXD was conducted with Cu $k\alpha$ radiation of wavelength 1.5406 $\mathring{A}$.

2.2.2. Fourier transform infraRed & nuclear magnetic resonance spectra

A Fourier transform infrared (FTIR) spectrometer (Bio-Rad FTS 6000, USA) was used at room temperature with a liquid cell container for solutions. Spectral resolution was maintained at 2 cm^{-1}. Dry nitrogen gas was used to purge the sample compartment to reduce the interference of water and carbon dioxide in the spectrum. The solutions were the TLCP chloroform and the TLCP/organoclay chloroform solutions before evaporation procedure. The ^{13}C nuclear magnetic resonance (NMR) spectra were measured at room temperature on a Bruker ARX 300 NMR spectrometer using chloroform-d as the solvent, and the chemical shifts were reported on the δ scale using tetramethylsilane (TMS) as the internal reference.

2.2.3. *Polarized optical microscopy*

The mesophase structures of the liquid-crystalline phase of TLCP and its nanocomposite were investigated by polarized optical microscopy (POM) using an Olympus microscope BX 50 with a Cambridge shear system CSS450 connecting a hot stage. The most outstanding feature of this setup is that it allows investigation of texture changes at different temperatures and under varying shear rates. Mesophase structure images were obtained at 185 °C after preshearing samples with a low shear rate i.e. 0.5 1/s for more than 3600 seconds to remove any shear history and anchored defects, and to give a common shear history or structure to samples before isothermal treatment in a quiescent condition for sufficient time.

2.2.4. *Thermogravimetric analysis & differential scanning calorimetry*

Thermal stability analysis was carried out by using a Hi-Res TGA 2950 thermogravimetric analysis (TGA) apparatus (TA instruments, USA). The test was carried out in air with a heating rate of 20.0 °C/min from 50.0 °C to 600.0 °C, and then isothermally treated for 30 minutes at 600.0 °C. The phase transition temperature of the nanocomposite based on TLCP was determined via differential scanning calorimetry (DSC) (PYRIS diamond DSC, Perkin-Elmer Instruments, USA), using indium as the calibration standard, with heating or cooling rate of 10.0 °C/min under nitrogen atmosphere.

2.2.5. *Advanced rheometric expansion system and capillary rheometer*

Controlled strain rheological measurements were carried out using an advanced rheometric expansion system (ARES) (TA instruments, USA) with a 200 g-cm transducer within the resolution limit of 0.02 g-cm. 50 mm cone and plate fixtures with nominal cone angle 0.04 rad and nominal gap 0.0508 mm, as well as 50 mm parallel plate fixtures were used for TLCP and its nanocomposite reported here. All measurements were performed at 185 °C in N_2 atmosphere, where TLCP had been shown to exhibit stable rheological properties under the nematic phase. Care was taken to ensure a controlled thermomechanical history as follows: the rheometer was heated to the testing temperature and allowed to reach equilibrium. Fresh samples, dried in a vacuum oven for 2 days at 120 °C, were loaded in the preheated rheometer, heated up to 185 °C then held for 10 minutes. Decreased gap to the testing gap and kept isotherm for 30 minutes to reach thermal and deformation equilibrium before measurements were started. The experiments were repeated no less than three times to check reproducibility. In each case, a fresh sample was used. For all tests reported here on HMMPE blends, 25 mm parallel plate fixtures were used. All measurements were performed at 190 °C with a 2000 g-cm transducer within the resolution limit of 0.2 g-cm and a 200 g-cm transducer within the resolution limit of 0.02 g-cm. Before testing, equipment was preheated and equilibrated at the test temperature for at least 30 minutes [22].

The rheological behaviors of the HMMPE blends were also characterized by a capillary rheometer (CR) (Göttfert Rheograph 2003A, Germany) at 190 °C and 230 °C. Here, the controlled piston speed mode was used with the round hole capillary dies (nominal L/D ratio equal to 30/1 and die entrance angle 180°). The real die diameters used here were recalibrated before use (Calibrated dies diameters D = 0.924 mm and 0.542 mm for the nominal D = 1.0 and 0.7 mm dies).

2.2.6. Scanning electron microscopy & transmission electron microscopy

The TC3 white film embedded in epoxy was ultra-microtomed with glass knives on an ultracut microtome (Leica ultracut-R ultramicrotomed, Germany) at room temperature to give sections with a nominal thickness of 75 nm. Transmission electron microscopy (TEM) images were obtained with a transmission electron microscope at 200 kV (JEOL 2010, Japan).

The morphology of the extrudates generated during the capillary rheometric experiment was examined by high resolution scanning electron microscopy (SEM) (JEOL 6700F, Japan) with the acceleration voltage 5 kv. All samples were sputter-coated with a ~200 $\mathring{A}$ layer of gold to minimize charging. The samples were quenched by compressed air from a hose placed near the die exit, providing a cooling ring. This 'froze' the structure of the TLCP droplets or fibrils before they could relax completely. Micrographs of the surfaces of these samples were taken after etching with a 10 wt% aqueous sodium hydroxide solution at 75 °C for 30 minutes.

The extrudate embedded in epoxy was ultra-microtomed with glass knives on an ultracut microtome at room temperature to produce sections with a nominal thickness of 100 nm. The sections were transferred to Cu grids. To enhance the phase contrast, the sections were stained with a ruthenium tetraoxide vapor for 2 hrs. TEM images were obtained with a transmission electron microscope at 200 KV (JEOL 2010, Japan). The 'frozen' extrudate for SEM was used for ultra-microtome. All images were obtained from sample sections microtomed along the flow direction.

3. Results and discussion

3.1. Dispersion and morphology of organoclay

The structures or morphologies of polymer-layered silicate nanocomposites depend on interaction of the polymer with the layered silicate and the dispersion of layered silicate in the polymer matrix. WAXD is a useful and simple measurement to characterize clay morphologies in nanocomposites. The (001) plane diffraction peak observed in the WAXD pattern of TC3 white disappeared, compared with the pattern of layered silicate having d-spacing 2.35 nm, as shown in Figure 1. The clay dispersion in the polymeric matrix was further examined using TEM. Typical TEM photographs for the TC3 white nanocomposite are shown in Figure 2. The dark plates, 15-25 nm in length, are the organoclays with surfaces paralleling the observed plane. The organoclays were fully exfoliated and well dispersed in the TLCP matrix without any agglomeration.

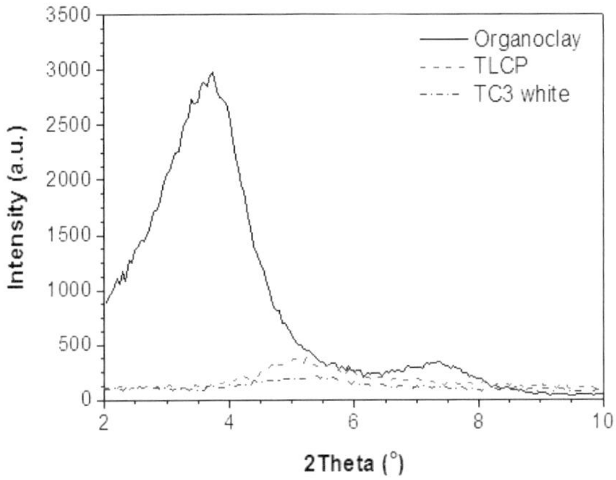

Figure 1. WAXRD patterns of organoclay, TLCP and TC3 white at room temperature.

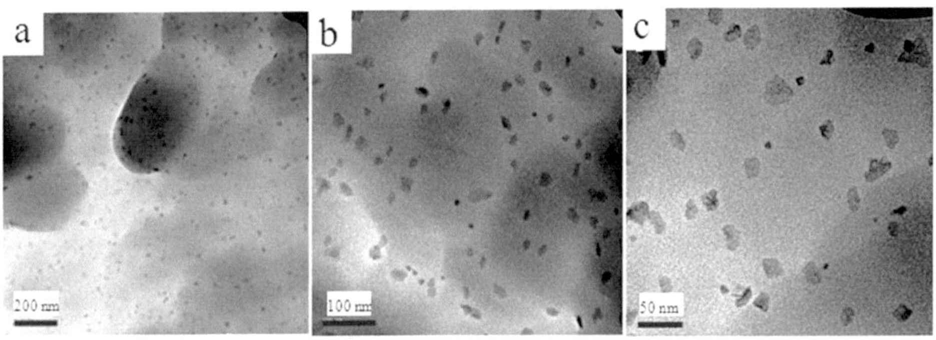

Figure 2. TC3 white TEM micrographs.

3.2. Molecular interactions in nanocomposite

IR spectroscopy is very sensitive to polymer microstructure and has been widely used in the investigation of hydrogen bonding, macromolecular orientation and crystallinity in polymer materials. FTIR with liquid cell was used with chloroform as background. Figure 3 shows the FTIR spectra for chloroform solutions with TLCP and TC3 white under ultrasonic irradiation at room temperature. It can be seen that the peak at the wavenumber of 1060 cm^{-1} in TLCP shifts to the wavenumber of 1045 cm^{-1} in TC3 white. The absorption peak at 1060 cm^{-1} is believed to represent $-C-O-$ in the TLCP molecules. The peak shift is the result of weak interaction between the positively changed N^+ ion in the surfactant 2M2HT residing on the surface of the organoclay Closite 20A with $-C-O-$ group in the TLCP molecules.

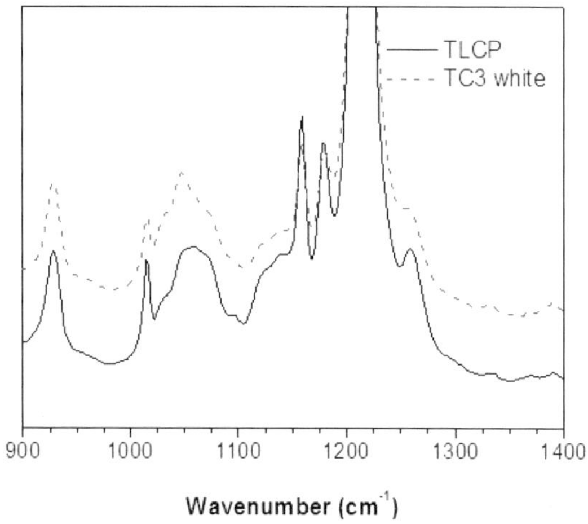

Figure 3. FTIR of TLCP and TC3 white chloroform at room temperature with liquid cell.

Figure 4 shows ^{13}C NMR spectra obtained from TLCP and TC3 white at room temperature. It is well known that a chemical shift between 0-40 pm corresponds to $-C-C-$coupling. After careful analysis of these regions, it was determined that the peaks at 25.2, 29.4 and 34.7 ppm shown in Figure 4(a), 4(b) and 4(c) belong to the spacer $-(CH_2)_8-$in TLCP. No chemical shift occurred but the peaks width increased after addition of the organoclay. The relatively broad peaks are probably due to the shielding effects of the layered silicate gallery on the alkyl chains.

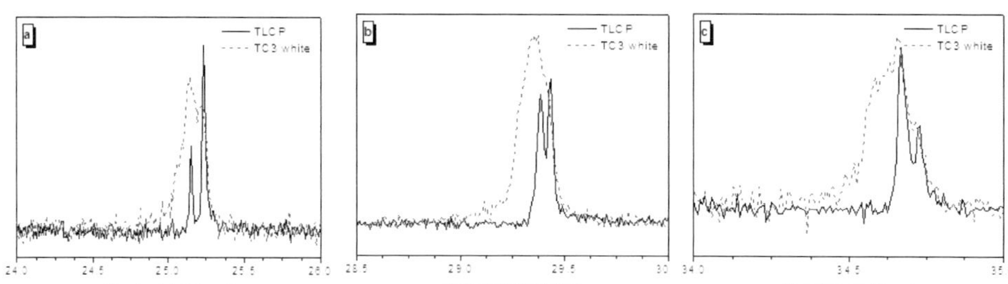

Figure 4. ^{13}C NMR spectroscopes of TLCP and TC3 white in $CDCl_3$ at room temperature.

3.3. Liquid crystalline mesophase

Figure 5 displays the in-situ microstructure evolution of TLCP and TC3 white using polarization transmission optical microscopy. The incident polarization was oriented along the flow direction and the analyzer was oriented perpendicular to this polarization. The first images are the microstructures of the melted TLCP (Figure 5 (a)) and TC3 white (Figure 5(b)) in the quiescent state at 185 °C. The TLCP micrograph shows textures surrounding different colors (blue, yellow and pink) regions, which indicate domains of different orientations. Compared with TLCP, TC3 white shows the domains (mainly darker yellow) with few other colors because of the lower light intensity. Upon startup of the shear flow, the domain textures became deformed, stretched and aligned along the shear direction. A gradual increase in the light intensity was observed. During shearing, TLCP (in Figure 5(c)) exhibited colorful textures (yellow, red and blue with some dark areas). TC3 white (Figure 5(d)) exhibited a mainly color (yellow with a few blue and red areas). With the polarized light, darkening may result from the structure becoming either completely isotropic or perfectly oriented along the principal axes. However, the darkening was not observed without the use of a polarizer; therefore isotropy can be ruled out as the primary cause for this darkening phenomenon.

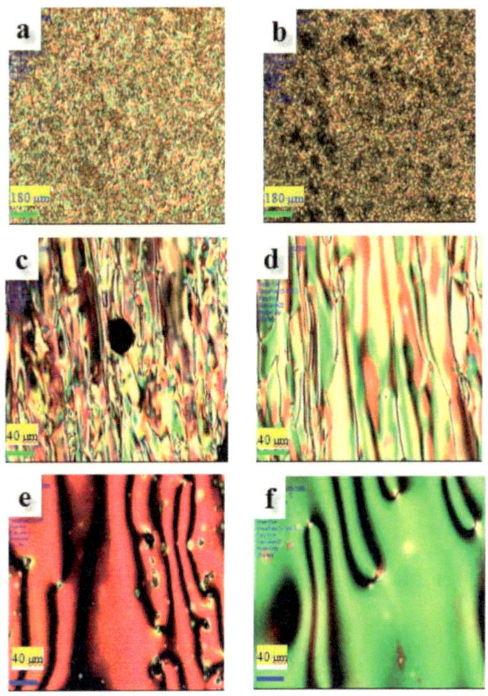

Figure 5. POM images of TLCP (a, c, e) and TC3 white (b, d, f) at 185 °C (a) and (b) before steady shear (X4); (c) and (d) during steady shear at 0.5 1/s (X20); (e) and (f) after steady shear and relax to steady state (X20).

The microstructures relaxation after the flow cessation is illustrated in Figure 5 (e) and 5(f). Once a steady flow was reached, a monodomain with a few defect structures was formed. After the same shear history, TLCP and TC3 white showed different colors. The areas of high Frank elasticity around the ±1/2 strength defects showed the highest and lowest retardation colors, whereas the bulk of the texture showed an intermediate retardation color. The static texture of the nanocomposite after steady shear is similar to that of the homopolymer. The existence of the fully exfoliated organoclays did not affect the mesophase structure of the TLCP molecules, which may due to the good dispersion of organoclays and their small sizes.

3.4. Thermal properties

The thermal stability of polymeric material is usually studied by thermogravimetric analysis. Generally, the incorporation of clay into the polymer matrix is found to enhance thermal stability, as the clay acts as a superior insulator and mass transport barrier to oxygen during oxidation in the air condition.

The TGA curves of clay, TLCP, and TC3 white in the air flow condition are shown in Figure 6. No weight loss occurred below 200 °C in any of the samples. Because of the presence of some organic molecules in interlayer spaces, the clay began to lose weight at 230 °C; at 600 °C, the clay weight loss was 30 %, similar to the result reported by Chiu et al. [23] Comparing TLCP with TC3 white, the thermal stability of TC3 white increased. For example, for TLCP, the temperature at which the weight loss started was 335 °C whereas for TC3 white, it was 349 °C. For TLCP, the temperature at which the weight loss was halved was 530 °C and for the nanocomposite, it was 549 °C.

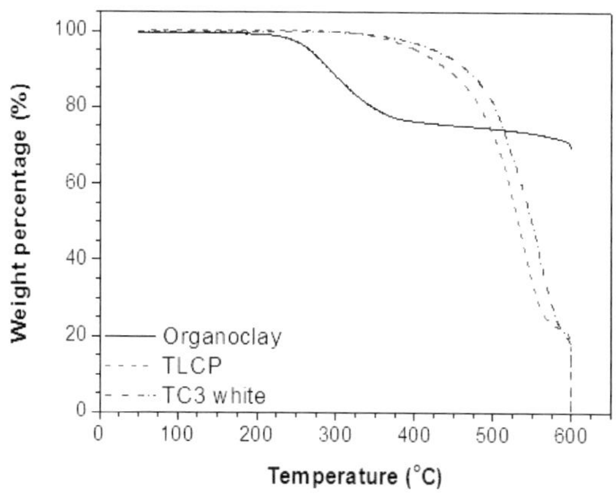

Figure 6. TGA of organoclay, TLCP and TC3 white in air.

TLCP used here was a semicrystalline nematic forming TLCP. From the DSC results, as shown in Figure 7(a), the endothermic peak appearing at ~ 125.5 °C represented the glass transition temperature of TLCP. Due to the broad sequential distribution of the HBA and SA-HQ segments in TLCP [24], two broad endothermic peaks at ~152 °C and ~164 °C represented the melting temperatures of the crystalline phase of TLCP with different segments. With the introduction of the well-dispersed nanoclay into the system, the glass transition temperature shifted to a higher temperature, peaking at ~ 142 °C, and the two melting temperatures peaked at ~ 159 °C and ~ 167 °C. The exothermic peaks are exhibited in Figure 7(b). The exothermic peak positions also changed and became sharper with the introduction of clay. For TLCP, there were two peaks, at ~166 °C and ~159 °C, corresponding to the different crystalline peaks with different molecular chain lengths as shown. With the introduction of the exfoliated layered silicates, the sharp exothermic peak was followed by a broad peak in the nanocomposite at corresponding temperatures of 163.5 °C and 169 °C respectively. With the effect of the layered structure, the arrangement of liquid-crystalline polymer chains became more ordered. There is no doubt that the transition temperatures in the heating and cooling curves increased in TC3 white.

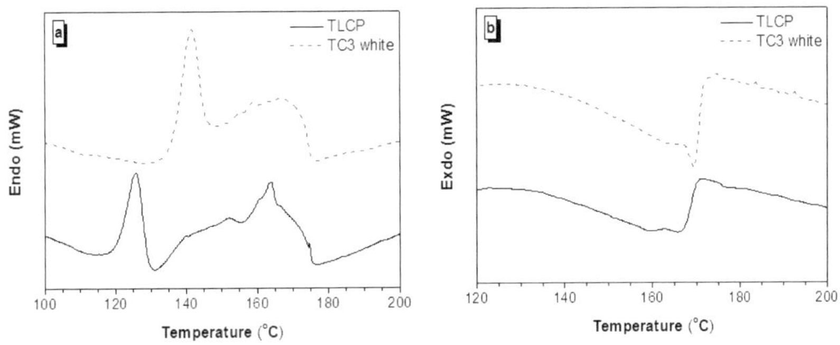

Figure 7. Second heating (a) and cooling (b) curves of TLCP and TC3 white at N_2 atmosphere.

3.5. Rheological properties

3.5.1. Organoclay modified TLCP nanocomposite

Linear viscoelasticity

Dynamic frequency sweeps at strains well within the linear viscoelastic regime of each material were performed in a range of frequencies covering 5 decades (0.01 – 1000 rad/s). The results shown in Figure 8 exhibit the storage modulus G', loss modulus G'' and complex viscosity η^* for TLCP and TC3 white. For TLCP, G', G'' and η^* became too small and scattered for accurate measurement at lower frequencies (in the region of 0.04 rad/s to 0.25 rad/s). This was due to the low viscosity of the melt at 185 °C, which led to torque values beyond the limits of the transducer. In the low frequency region (0.04 rad/s to 0.25 rad/s), TC3 white had dependable data but the TLCP data was scattered. In this region, slopes for G′ and G″ were 0.12 and 0.31 for TC3 white, and 0.30 and 0.55 for TLCP. The corresponding slope values for η^* were 0.83 and 0.58. Pseudo-solidlike behaviors existed in TLCP and TC3 white, with TC3 white exhibiting more solidlike behavior than TLCP. The polydomain structure in the bulk state of TLCP is the reason for TLCP behavior [25]. The percolated network formed by the exfoliated organoclay with the TLCP molecules enhanced those polydomain structures, causing more solidlike behavior and higher η^* in the lower frequency region in TC3 white. In the middle frequency region, both showed plateaus of nearly constant viscosity. The TC3 white curve paralleled the x-axis and TLCP had a small slope value of -0.15. At high frequencies, TC3 white showed a gradual change of slope to -0.21, whereas TLCP maintained the same slope. In this region, the slopes of G' and G'' for TLCP were 0.70 and 0.86 but for TC3 white, they were 1.83 and 0.92. There were dramatic differences between the two materials. The slopes for TC3 white approached the theoretical values for polymers with flexible chains, which are 2 and 1, respectively. With the help of exfoliated clay in high frequencies, TC3 white performed like flexible chains, whereas TLCP still showed some entanglement behavior.

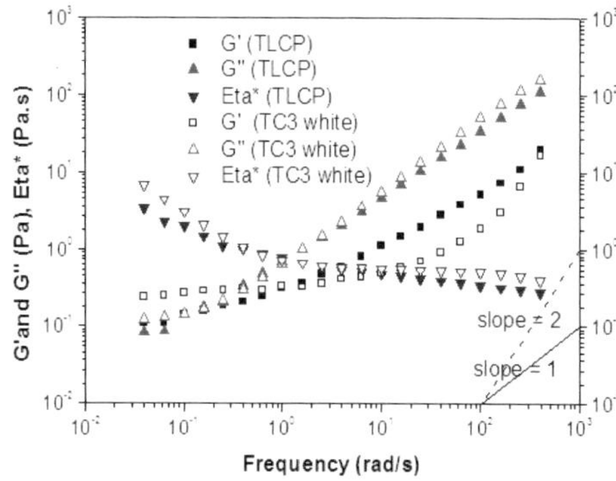

Figure 8. Storage modulus G', loss modulus G'' and complex viscosity Eta* in dynamic frequency sweep for TLCP and TC3 white at 185 °C.

It has been claimed that the Cole-Cole plot of storage modules vs. loss modulus can provide information about structures. The Cole-Cole plot for TLCP and TC3 white is shown in Figure 9. From a log-log plot of G'' vs. G', we can determine that the slope α in $G' \sim (G'')^{\alpha}$. For a single Maxwell element, $\alpha = 2$, however, complex systems with associations often deviate from the single element, and this can be seen by an exponent α that deviates from 2. For TLCP, the slope was constant with a value of 0.776 in the whole region, which has a large deviation with flexible chain polymers ($\alpha = 2$). This is due to TLCP intrinsic anisotropic properties. For TC3 white, with the loss modulus increasing, the slope changed dynamically from almost zero (pseudo-percolation behavior or solid-like behavior), gradually approaching the theatrical flexible chain value, 2. This plot showed that there were different structural responses for TC3 white in different frequency regions.

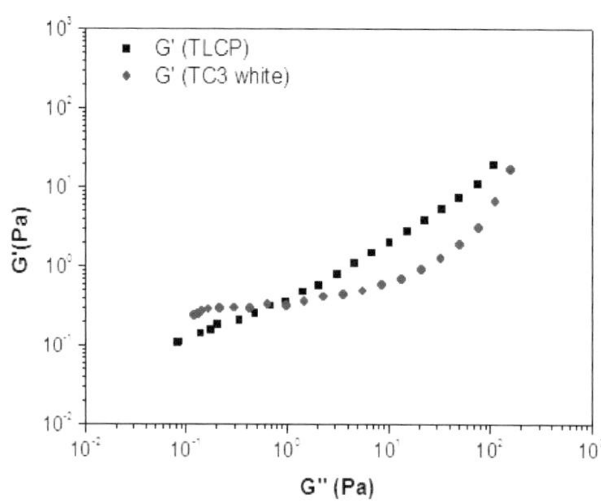

Figure 9. The cole-cole plot of loss modulus G'' vs. storage modulus G' for TLCP and TC3 white at 185 °C.

Cox-Merz rule

Figure 10 provides global view of TLCP and TC3 white viscosities as a function of frequency or steady shear rate. Unlike other organoclay based polymeric nanocomposites, the measured viscosity of TC3 white is similar to that of the TLCP matrix at the same frequency or shear rate. The curves, especially the TC3 white curve, are reminiscent of the three-region viscosity curve reported by Onogi and Asada [25] for lyotropic polymer. The phenomenon can be explained by polydomain structures. With the addition of the organoclay, the three-region viscosity phenomenon was enhanced. TC3 white performs more like lyotropic polymer, with a more rigid structure than the semi-rigid TLCP. The TC3 white curves fitted the Cox-Merz rule, i.e. $\eta(\dot{\gamma}) = |\eta^*(\omega)|$, where $\dot{\gamma} = \omega$. However, the Cox-Merz rule failed for TLCP, a finding which has also been reported by other researchers [26].

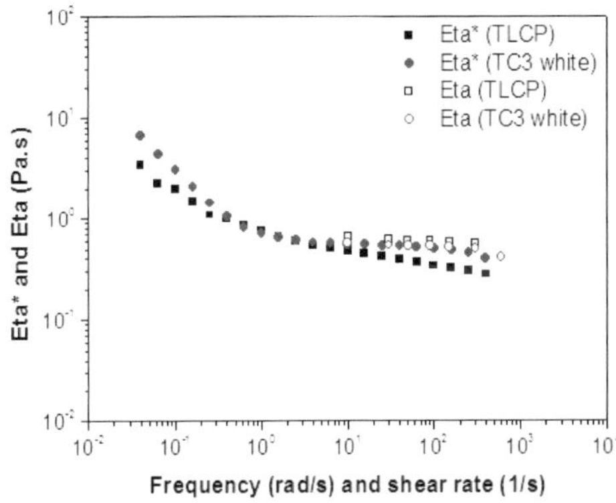

Figure 10. Cox-Merz rule on TLCP and TC3 white at 185 °C.

The first normal stress difference measurement

The cone and plate fixture, with its constant rate of shear and direct measurement of the first normal stress difference N1 by total thrust, is probably the most popular rotational geometry for studying non-Newtonian effects. Figure 11 shows the total thrust (normal force) for TLCP and TC3 white at different shear rates at 185 °C using a 50 mm diameter cone and plate fixture. The normal force evolution at a particular constant shear rate was clearly exhibited with a few peaks (or shoulders) at small strain before reaching a steady state. These peaks corresponded to the structure being broken and aligning during shear [27].

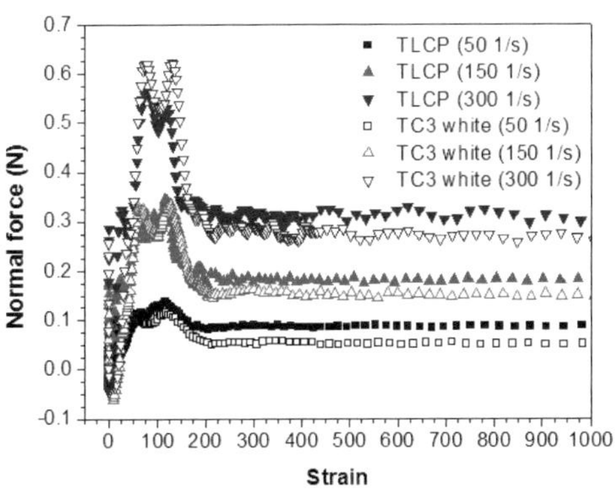

Figure 11. Normal force for TLCP and TC3 white at different shear rates with cone and plate fixture at 185 °C.

For the total thrust data measured in the cone and plate fixture, some equations can be used to calculate the N1 for TLCP and TC3 white [28].

For the cone and plate fixture:

$$N_1 = \frac{2F_z}{\pi R^2} \tag{1}$$

Where, F_z is the total thrust measured in the cone and plate fixture. R is the radius of the cone and plate fixture.

Due to the inertia and secondary flow effect, i.e. in cone and plate rheometers, inertia forces tend to pull the plates together rather than push them apart, a corrected term must be introduced to eliminate this effect:

$$(F_z)_{inert} = 0.075\pi\rho\Omega^2 R^4 \tag{2}$$

Where, ρ is the density of the material, Ω is the angular rotation rate.

After this correction the N1 can be calculated:

$$N_1 = \frac{2F_z}{\pi R^2} - 0.15\rho\Omega^2 R^2 \tag{3}$$

The steady state N1 at different shear rates at 185 °C can then be calculated. The graph is shown in Figure 12. Due to the low viscosity, the maximum shear rate that could be measured was 300 1/s for TLCP and 600 1/s for TC3 white. The sample spun out with a dramatic and continuous decrease in stress and viscosity when the shear rate was larger than the above value. In Figure 12, N1s are positive for TLCP and TC3 white with comparable values.

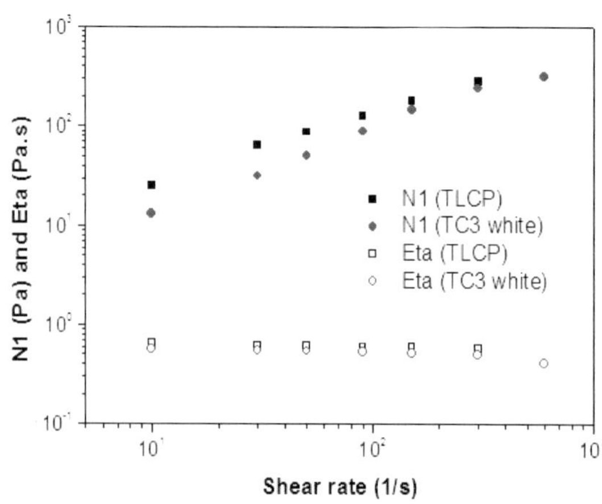

Figure 12. Shear rate dependence of shear viscosity Eta and first normal stress difference N1 for TLCP and TC3 white at 185 °C.

For N1, molecular theory, especially the Doi theory (for rod-like nematics) [29] predicts that the director (or local average orientation) will rotate or tumble with the flow. At low shear rates, N1 will be positive. At intermediate rates (when the flow rate is a little slower than the reciprocal of the longest molecular relaxation time), nonlinear effects are important, producing competition between tumbling and steady alignment of the director along the flow. As a result, the director oscillates about a steady value [30]. In this "wagging" regime, which is peculiar to tumbling, polymeric nematics and the local molecular order are significantly weakened, and N1 is negative. At very high rates of shear, the director aligns in the flow direction, and N1 is again positive. The Doi molecular theory for rod-like nematics was remarkably applicable with experimental data for PBG solutions [31]. Although a similar coupling between flow and liquid-crystalline order is conceivable in thermotropics, experiments have not yielded a corresponding agreement with theory. In fact, the expected pattern of sign changes in N1 has not been observed [32]. The N1 curves in Figure 12 for these two systems are in the high-shear-rate region, with N1>0. Figure 12 displays the shear rate dependence of η and N1 for TLCP and TC3 white at 185 °C. From comparison of the N1 values between TLCP and TC3 white, it can be observed that the values of N1 for TC3 white are a little lower than those for TLCP, but the rate of the N1 increase in TC3 white is higher. Combined with the viscosity variations with shear rate, these phenomena indicate that the presence of exfoliated clay increases elasticity in a flow-aligning state, and the exfoliated clay is oriented along the shear direction, even assisting the neighboring TLCP molecules to align in the flow direction, with the result that there is a decrease of viscosity and an increase of N1 slope, as shown in Figure 12.

The dispersion of the exfoliated clay in the TLCP matrix without and with deformation is depicted schematically in Figure 13(a) and 13(b), respectively, where the dark ellipses represent clay platelets, the wavy lines and cylinders represent TLCP chains with flexible and rigid components, and the short dashed lines represent the interaction between TLCP and clay. It should be noted that in the conformation of TLCP, the phenyls in HBA and HQ are not coplanar, and hence the SA flexible segments connected to the HQ are not in a line with HBA. In the schematic diagram, the effect of the exfoliated clay on TLCP is clearly delineated. Without deformation (Figure 13(a)), the disordered dispersion of the exfoliated clays in the TLCP matrix is presented. On the other hand, with deformation (Figure 13(b)), shear-induced molecular alignment in both TLCP and TC3 white occurs along the shear direction.

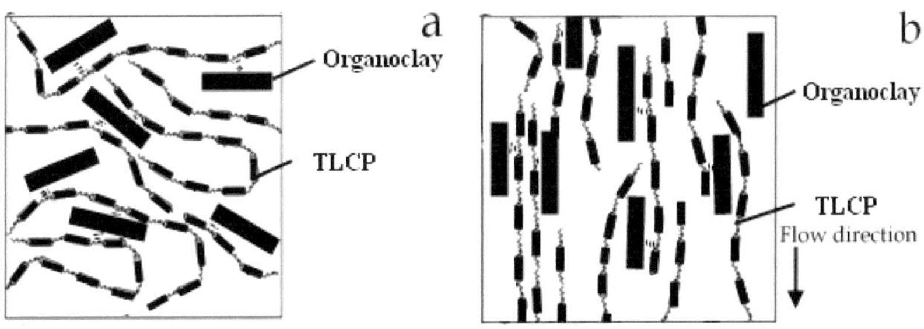

Figure 13. Schematic drawing of TLCP molecular chains influenced by the exfoliated organoclays (a) without and (b) with deformation.

3.5.2. High molecular mass polyethylene blends

Shear deformation

The linear viscoelasticity and large amplitude stress relaxation tests were performed by an ARES at 190 °C. Complex viscosities in frequency sweep and stress relaxation modulus with strain 300.0 % are exhibited in Figure 14(a) and (b) separately. In Figure 14(a), little difference is shown between the curves, indicating that the purified TLCP and TC3 white had little influence on the HMMPE matrix in the linear viscoelastic region. Large amplitude stress relaxation tests were performed in the nonlinear region. The curves in Figure 14(b) show no difference over the entire relaxation periods. All the information demonstrates that the purified TLCP and TC3 white had little effect on the shear deformation of the HMMPE at the above particular conditions.

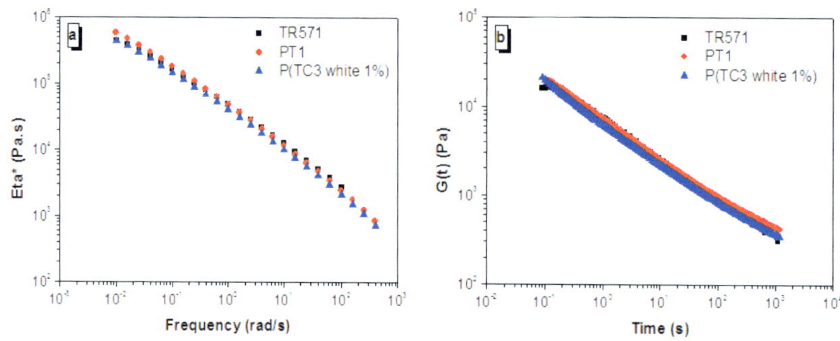

Figure 14. (a) Dynamic frequency sweep and (b) stress relaxation in nonlinear region with strain = 300.0 % of HMMPE [TR571], HMMPE/TLCP 1.0 wt% [PT1] and HMMPE/(TC3 white 1.0 wt %) [P(TC3 white 1%)] at 190 °C.

Elongation deformation

The effects of purified TLCP and TC3 white on the HMMPE matrix were also characterized using a pressure-driven rheometer. Here, the controlled piston speed mode with the round hole capillary dies at 190 °C and 230 °C was used. Capillary flows are usually considered as simple shear flows. The shear stress is highest near the capillary die wall, where the polymer chains are most likely to be stretched to an extended configuration. This is valid only if the melt shows negligible entrance effects. Entrance effects are caused by the elongational flow due to the converging melt flowing from the reservoir into the capillary die with large contraction ratios. The polymer melt along the centerline will experience the highest stretching rate. It has been observed that such effects are extremely important when anisotropic melts such as TLCPs are studied.

Rheological behavior at 190 °C.

Apparent shear viscosities as a function of shear stress at wall for HMMPE [TR571], HMMPE/(purified TLCP 1.0 wt%) [PT1] and HMMPE/(TC3 white 1.0 wt %) [P(TC3 white 1%)] with the same L/D ratio (30) and different diameters (0.7 mm and 1.0 m) at 190 °C are shown in Figure 15. At this temperature, TLCP shows the nematic structures [20]. Significant viscosity reductions are initially observed in different die diameter tests. Based on an equivalent wall stress of $10^5\,Pa$, the viscosity reductions at 190 ºC with L = 21 mm and die radius equal to 0.271 mm for the different blends are: for PT1, a similar viscosity to that of HMMPE, because no yielding occurred; for P(TC3 white 1%), 98.5 % viscosity reduction compared to HMMPE (corresponding apparent shear rate 317.6 1/s). For the maximum processing rate, HMMPE is ~ 39 1/s, PT1 is ~ 318 1/s, whereas P(TC3 white 1%) is up to ~ 700 1/s. P(TC3 white 1%) can achieve a shear rate almost 20 times higher than that of HMMPE and twice as high as that of PT1.

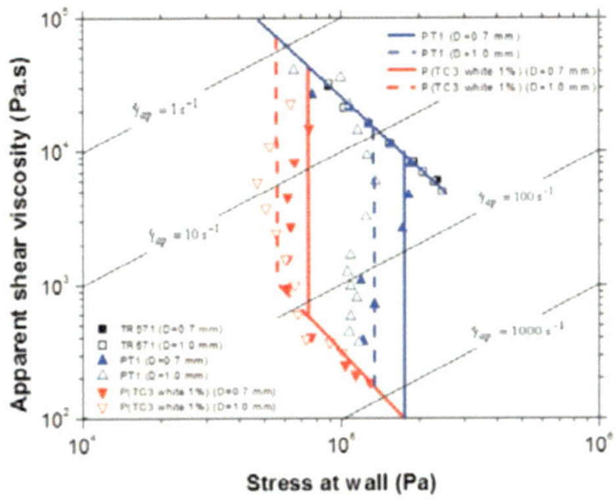

Figure 15. Apparent shear viscosities as a function of shear stress at wall for HMMPE [TR571], HMMPE/TLCP 1.0 wt% [PT1] and HMMPE/(TC3 white 1.0 wt %) [P(TC3 white 1%)] with the same L/D ratio (30) and different diameters (0.7 mm and 1.0 m) at 190 ºC with capillary rheometer. (The points were measured data and the lines were simulated data).

A yielding-like behavior is shown by all blends when the wall stress is almost constant over a region of rapidly decreasing viscosity. However, yielding stresses and the corresponding beginning and ending apparent shear rates are different for each blend. Table 1 details the yielding behaviors of the blends. From the table, it is clear that with the organoclay modified, much lower values of yielding stress were needed. Moreover, the apparent shear rate of all blends at the beginning of transition is much lower than that of PT1. This explains why it is difficult to obtain the first power-law region in capillary rheometer tests for P(TC3 white 1%) in simulation, as we describe later in this study. Due to the low yield stress and the initial transition shear rate, the blend can easily move through the transition zone and reach the zone with lower viscosity. The low energy input needed to process the blend holds promising potential for industrial application.

Rheological behavior at 230 ºC

The rheological behaviors of three materials at 230 ºC are presented with a plot of apparent shear viscosities as a function of shear stress at wall in Figure 16. At this temperature, TLCP shows nematic/isotropic biphasic structures. Based on an equivalent wall stress of $10^5\,Pa$, for PT1, the yielding stress is higher than $10^5\,Pa$, and has similar viscosity to that of

HMMPE; for P(TC3 white 1%), a viscosity reduction of > 93% was achieved at 230 °C. The maximum processing shear rates are ~ 66 1/s for HMMPE, ~ 315 1/s for PT1 and ~ 904 1/s for P(TC3 white 1%). Table 2 lists the experimental data for yielding stress and transition shear rates for the blends at 230 °C. Similar with the parameters at 190 °C, lower yielding stresses are obtained in PT1 and P(TC3 white 1%). A more obvious phenomenon concerns the transition zone, which is narrow in the range of 8 1/s to 23 1/s for P(TC3 white 1%) and still cannot obtain transition ending shear rate for PT1. A small force can be used to pass through the narrow transition zone to arrive at the low viscosity region at this temperature for P(TC3 white 1%).

Yielding behaviors	Stress (Pa)		Beginning $\dot{\gamma}_{ap}$ (1/s)		Ending $\dot{\gamma}_{ap}$ (1/s)	
Dia. (mm)	0.7	1.0	0.7	1.0	0.7	1.0
PT1	**1.63×10^5**	**1.34×10^5**	**41.0**	**24.1**	**855.8**	**476.9**
	1.61×10^5	1.36×10^5	39.8	22.9	---	---
P(TC3 white 1%)	**7.22×10^5**	**5.78×10^5**	**5.6**	**3.3**	**215.7**	**120.1**
	7.08×10^5	5.90×10^5	6.0	3.5	207.4	110.8

Table 1. Typical parameters with experimental and simulated tests for HMMPE blends at 190 °C (Bold data are predicted results).

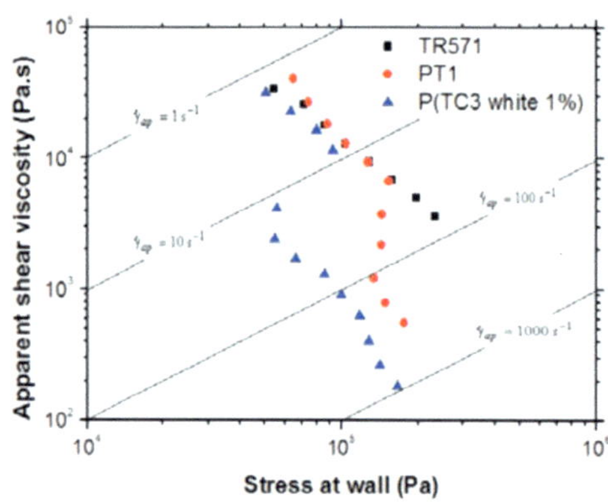

Figure 16. Apparent shear viscosities as a function of shear stress at wall for HMMPE [TR571], HMMPE/TLCP 1.0 wt% [PT1] and HMMPE/(TC3 white 1.0 wt %) [P(TC3 white 1%)] with L/D ratio = 30 and diameter = 1.0 mm at 230 °C with capillary rheometer.

The yielding-like behavior in P(TC3 white 1%) presents an obvious negative gradient. We suggest that the negative gradient is due to TLCP phase transition in the TLCP/organoclay hybrid from isotropic to nematic, or maintaining the isotropic phase at that temperature. TLCPs are known and theoretically understood to undergo shear-induced phase transitions when the domain orientation is sufficient high [33, 34]. Chan et al. [2] have presented the results of optical microscopy/shearing experiments demonstrating a phase transition from isotropic to nematic for this type of TLCP. A pre-transitional order in the isotropic phase of a homologous series of liquid crystals close to the isotropic-to-nematic transition has also been experimentally observed. De Schrijver et al. [35] used a transmission ellipsometric technique to observe this surface-induced isotropic ordering. For the P(TC3 white 1%) blend, since fully exfoliated organoclay structures were formed in the TLCP, no confinement existed to hold an ordered structure and cause phase transition, but there were interactions, such as long-range non-bond forces, which also gave structures more order and retained the orientation even during the relaxation period [22].

Yielding behaviors	Stress (Pa)	Beginning $\dot{\gamma}_{ap}$ (1/s)	Ending $\dot{\gamma}_{ap}$ (1/s)
PT1	1.27×10^5	13.6	---
P (TC3 white 1%)	0.56×10^5	8.0	22.9

Table 2. Typical parameters with experimental tests for HMMPE blends at 230 °C.

3.6. Morphological studies

The SEM diagrams of etched extrudates are shown in Figure 17 with magnification 20,000. For the HMMPE, as shown in Figure 17(a), the etched stand gives a rough and highly topological contrast. Moreover, a fine line texture is disclosed after NaOH etching. This indicates that some surface materials or even layers are removed during the etching process. PE is a material that strongly resists attack by NaOH. Therefore, the detached material is thought not to be pure PE. It would be too difficult for NaOH to diffuse into the PE lattice and remove it from the surface. As has already been illustrated by Chan et al. [36] the material removed is an anti-oxidant enrich polyethylene layer, which is caused by migration of anti-oxidant during shear. The surfaces of the blends are smooth at those apparent shear rates. At higher magnification, some interesting features are revealed. Long and thin cavities are seen, with the long dimension paralleling to the flow direction. These cavities are due to the removal of TLCP filaments. In these well defined morphologies: a PT1 strand (Figure 17(b)) shows only fibrillar structures aligned along the flow direction; also within a P(TC3 white 1%) strand (Figure 17(c)) only longitudinal fibrillar striations exist. Both images show that in situ fibril formation occurs during elongation in both blends.

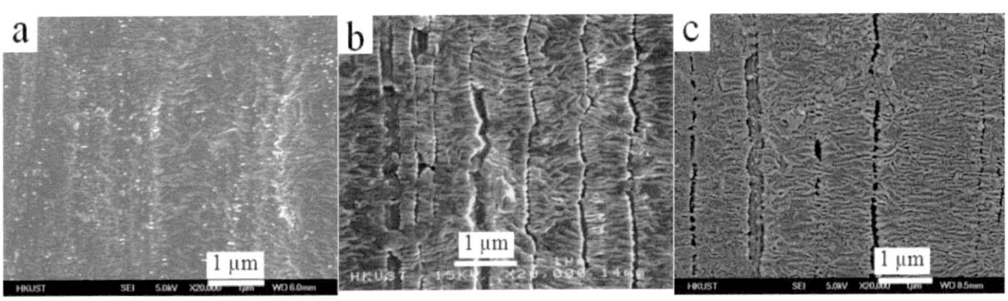

Figure 17. SEM images of (a) HMMPE [TR571], (b) HMMPE/TLCP 1.0 wt% [PT1] and (c) HMMPE/(TC3 white 1.0 wt %) [P(TC3 white 1%)] with magnification 20,000X.

TEM images of P(TC3 white 1%) are exhibited in Figure 18 with different magnifications. Global alignment of PE lamellae can be clearly seen in Figure 18(a), with typical row-nucleated shish-kebab structures (a schematic drawing in the insert of Figure 18(a)), a structure which is usually observed when PE crystallization takes place under stress [37, 38]. The long fiber crystals formed by the extended high molecular mass fractions act as nucleation sites for the growth of folded PE crystals. Detail micrographs (Figure 18(b) and 5(c)) clearly show the strong interfacial compatibilities between the aligned TC3 white filament and the adjacent PE matrix. Also the embedded TC3 white fiber exhibits a regular banded structure, all the bands being perpendicular to the direction of chain alignment. The above observations are similar to those in our earlier studies of PT1 systems [39], indicating that they have a similar viscosity reduction mechanism [40].

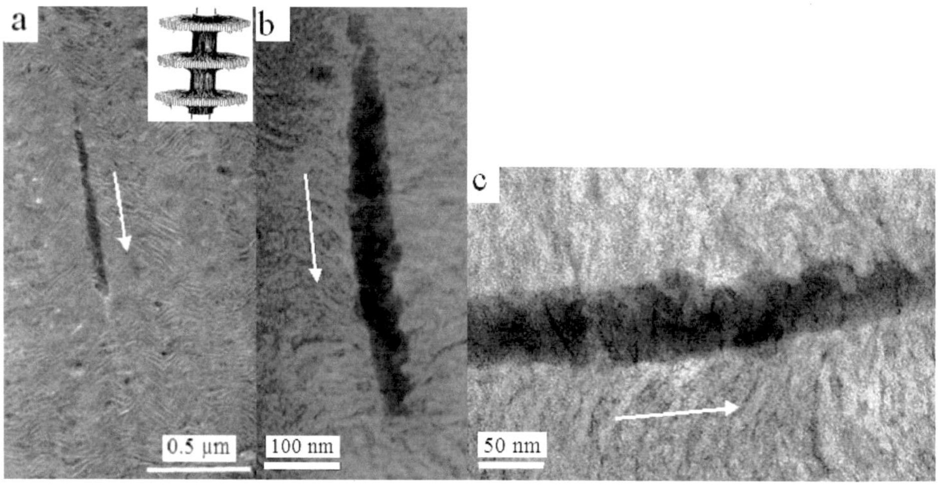

Figure 18. TEM micrographs of the lateral section of the HMMPE/ (TC3 white 1.0 wt %) [P(TC3 white 1%)] blend extrudate surface prepared parallel to the flow direction at different magnifications.

3.7. Texture studies

The textures of purified TLCP and TC3 white at 230 °C and 250 °C are presented in Figure 19 with different magnifications. These samples all underwent the same thermal history with the following steps: (1) sheared at shear rate 0.5 1/s for 3600 seconds at 185 °C; (2) maintained at this temperature to obtain stable texture; (3) temperature ramped to 230 °C at 5.0 °C/min; (4) obtained texture structure after structure evolution for a specified period. From previous results [22] the purified TLCP and TC3 white displayed a similar texture after steps (1) and (2) at 185 °C. The fully exfoliated organoclay did not affect the liquid crystallinity and mesophase structure at the nematic state at 185 °C. In Figure 19(a) and (b), an isotropic phase is clearly presented alongside the nematic phase after steady shear at 0.5 1/s for 600 seconds at 230 °C and relaxation for 600 seconds for purified TLCP. The nematic phase exists in dispersed and discrete regions containing defect lines, which are highly birefringent and contain domains of anisotropy. The isotropic phase is continuous. There is a distinctly biphasic, nematic/isotropic, texture in purified TLCP at 230 °C. With the elapse of time, a spherical shaped nematic region with a more relaxed state (indicated by the presence of fewer line defects inside the sphere) occurs. Figure 19(c) presents an image of purified TLCP after shearing at 5.0 1/s for 60 seconds followed by relaxation for 600 seconds at 230 °C. The area of continuous isotropic phase has become larger than the nematic phase. With increased of temperature to 250 °C, as Figure 19(d) shows, the nematic phase gradually diminishes in size and population within the isotropic phase matrix. For TC3 white, as shown in Figure 19(e) and (f), after steady shearing at 0.5 1/s for 600 seconds followed by relaxation for 600 seconds, a dominant nematic texture occurs with few dispersed isotropic regions at 230 °C. The defects which were stable at 185 °C become unstable at 230 °C, due to the effect of the high temperature. As time elapses, the isotropic phase occurs in a minority of discrete regions, and the nematic phase is still dominant. Even after steady shearing at a high shear rate 5.0 1/s for 60 seconds followed by relaxation for 600 seconds at 230 °C, the nematic phase still exists as a continuous structure with a few discrete isotropic structures, as shown in Figure 19(g). The exfoliated organoclays enhance the rigidity of the TLCP molecules and keep them in ordered structures at the high temperature. The competition between the high thermal energy and the internal molecular interactions of the organoclay and TLCP molecules causes the nematic phase to be dominated by biphasic structures for TC3 white at 230 °C. Even at a higher temperature, i.e. 250 °C, after 5.0 1/s shearing for 60 seconds followed by relaxation for 300 seconds, the nematic phase is still in continuous mode, as Figure 19(h) shows.

For the organoclay-modified TLCP, i. e. TC3 white, the combination of the above mentioned nematic dominated structure and shear-induced isotropic-nematic transition had the effect that its rheological behavior at 230 °C was similar to that at 190 °C, with even a higher processing window and a lower transition region.

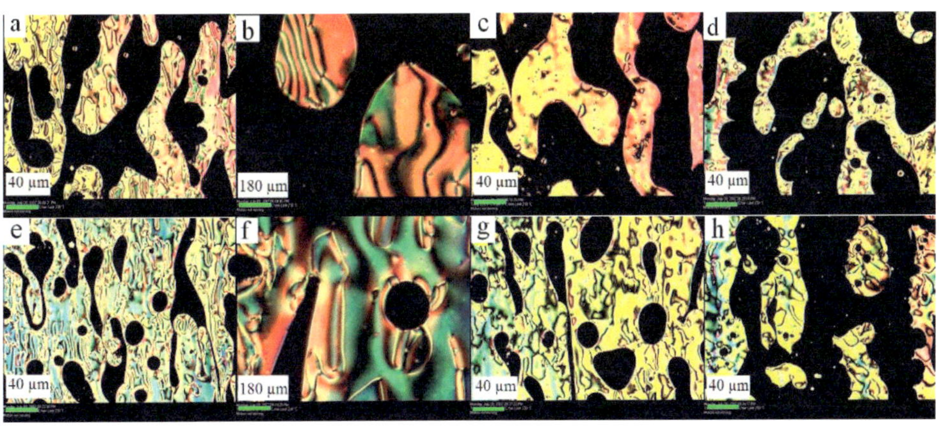

Figure 19. Polarized optical microscope images of (a)(b) purified TLCP and (e)(f) TC3 white relaxed for 600 seconds at 230 °C after steady shear with 0.5 1/s for 600 seconds; (c)purified TLCP and (g) TC3 white relaxed for 600 seconds at 230 °C after steady shear with 5.0 1/s for 60 seconds; (d)purified TLCP and (h) TC3 white relaxed for 300 seconds at 250 °C after steady shear with 5.0 1/s for 60 seconds (all samples have been sheared with shear rate 0.5 1/s for 3600 seconds and relaxed to steady state at 185 °C).

4. Predictions based on a phenomenological model

A binary flow pattern model, previously used for prediction of the effects of a small amount of TLCP in HMMPE (TR570) [40] was used to simulate the rheological responses of the blends in this study.

4.1. Model

To account for the structural effects due to elongational flow along the centerline region in the capillary die, the overall melt flow characteristics are divided into three regions depending on the magnitude of the maximum fluid velocity in the capillary that is usually along the centerline in converging flows. This critical fluid velocity corresponds to the maximum stretching rate of the TLCP (or TC3 white) domains at which irreversible TLCP (or TC3 white) domain elongation into slender filament occurs. The three regimes are:

Region I: The fluid velocity is below the critical velocity for irreversible TLCP (or TC3 white) domain elongation, and the flow of the blends is dominated by the melt flow behavior of the matrix polymer HMMPE melt, independent of TLCP (or TC3 white). A schematic illustration of the melt structure during flow in Region I is shown in Figure 20(a). PE chains formed random coiled conformation and TLCP (or TC3 white) has ellipse shapes with uniformed dispersion in PE matrix. The insert in Figure 20(a) shows the organoclay and TLCP chain conformation at this region. Organoclays of uniform size were well and irregularly dispersed in the nematic phase TLCP.

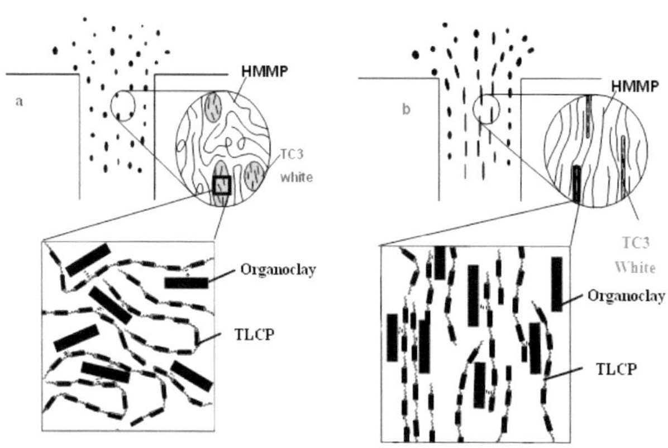

Figure 20. Schematic drawing of HMMPE chains, TLCP chains and organoclay morphological change in capillary die at 190 °C (a) before and (b) after critical shear rates.

Region II: The maximum fluid velocity within the capillary reaches the critical velocity at the entrance of the capillary and irreversible elongational deformation of the TLCP (or TC3 white) domains into long slender fibrous forms begin to occur. This causes a rapid chain elongation and disengagement in the PE melt adjacent to the TLCP (or TC3 white) domains. Consequently, a region of low viscosity melt is formed in the center core of the capillary die.

This center region expands as the flow rate increases until all fluid within the capillary is filled with such melt. The simulated velocity profile developments of fluid flowing through a capillary die to describe the above phenomenon will be presented later in this study. A schematic illustration of the melt structure during flow in Region II is shown in Figure 20(b). The insert in Figure 20(b) shows the chain conformations of TLCP molecules with help of organoclay. Shear-induced molecular alignment occurs with TLCP molecules and organoclay oriented along the elongation direction.

Region III: After all low viscosity fluid is formed across the entire capillary die diameter, a homogeneous melt flow corresponding to the low viscosity melt may be assumed again.

4.2. Velocity profiles

The velocity profile developments of fluid flowing through capillary dies are shown in Figure 21.

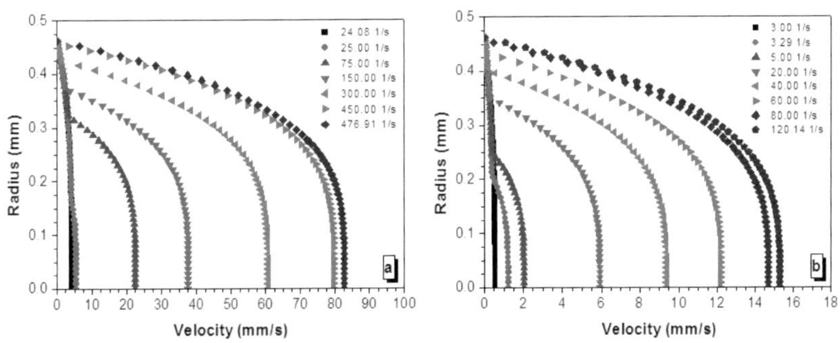

Figure 21. Velocity profile development in region II of flow at R = 0.462 mm for (a) HMMPE/TLCP 1.0 wt% and (b)HMMPE/TC3 white 1.0 wt% at 190 °C by simulation.

As shear rate increases, the center core region characterized by low viscosity melt flow characteristics expands from the center core towards the die wall. Close to the wall, the velocity profiles are independent of apparent shear rates. This implies that the shear rates at the wall are independent of flow rates of fluid during the melt structure transition period. Consequently, the wall shear stresses will remain constant throughout this transition period. In the velocity profiles for the different blends, the real die diameters were used instead of the nominal diameters. For nominal diameters 1.0 mm and 0.7 mm, the real calibrated diameters were 0.924 mm and 0.542 mm. Table 1 shows the predicted yielding stress and transition shear rates with the experimental data. The predicted data coincide well with the experimental results. The prediction results also give the end transition shear rates for PT1 at 190 °C with different die diameters, which cannot be obtained experimentally due to the flow oscillation.

4.3. Flow curves simulation

The flow curves are divided into three regimes as described above. In Regions (I) and (III), simple power-law constitutive relations are assumed. In Region (I), because of the relatively lower yielding shear rates for these blends, dynamic frequency sweep data performed on ARES were used combined with data from CR for simulation. The flow curves in Region (II) were obtained using the velocity profiles together with the power-law parameters in Regions (I) and (III). Figure 22 shows the schematic drawing of apparent shear viscosity as a function of shear stress at wall at the different stages in capillary die. Precise flow curves with Matlab program simulation for the blends are plotted in Figure 15 together with experimental data. Excellent agreement between the model prediction and the experimentally measured flow curves are obtained.

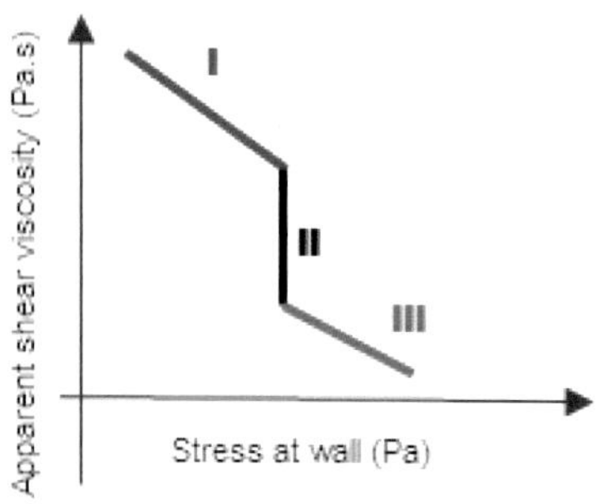

Figure 22. Schematic drawing of apparent shear viscosity evolution as a function of stress at wall at different stages in capillary die at 190 °C.

5. Conclusions

An organoclay-modified TLCP nanocomposite (TC3 white) with the organoclays of uniform size 15-25 nm in length well dispersed in thermotropic liquid crystalline polymer (TLCP) with fully exfoliated structures was designed and prepared by a combination method. Polarized optical microscope images showed that the organoclay did not affect the liquid crystallinity and mesophase structures of the TLCP matrix. However, thermal stability and thermal properties were affected by the organoclay, enhancing the thermal stability of TLCP and shifting the transition temperatures to the high ends. The presence of organoclays caused the nanocomposite to present different rheological behaviours with TLCP at the nematic temperature, i.e. 185 °C. Dynamic experiments demonstrated that TC3 white displayed higher pseudo-solidlike behaviour than TLCP alone in the low frequency region. TC3 white had a similar but even lower viscosity and the first normal stress difference (N1) than TLCP, but the rate of N1 increase in TC3 white was greater than that in TLCP. When enhanced with organoclays, TLCP became more rigid, and with a slight deformation in the TC3 white melt, organoclay helped the TLCP molecules to align in the shear direction and to retain the orientation.

The rheological behaviours of purified TLCP and TC3 white in high molecular mass polyethylene (HMMPE) were characterized by capillary rheometer (CR) with nominal dies of L/D = 30 and diameters 0.7 mm and 1.0 mm at 190 °C, where purified TLCP and TC3 white showed similar nematic phase structures. At 230 °C, purified TLCP presented as a continuous isotropic phase with a minority of discrete nematic phase, whereas TC3 white displayed a continuous nematic phase with a few isotropic phases. The interactions between organoclays with TLCP molecules at the molecular level enhanced the rigidity of the TLCP molecules, displaying the nematic order structure even at higher temperature. The rheological experiments using CR with a nominal die of L/D = 30 and diameter 1.0 mm showed even higher viscosity reduction ability with a wider processing window for TC3 white than for the purified TLCP in the PE matrix. In addition, a much lower yielding stress with a narrower transition window was obtained in the TC3 white/PE blend than in the purified TLCP/PE blend. These findings have promising potential for industrial application to save energy and increase processing efficiency when used in processing such thermoplastics. Mechanism study confirmed that the binary flow model can be applied to describe the rheological behaviours of both blends and shear induced phase transitions and alignment of in-situ formation of fibrils are the primary reasons for viscosity reduction.

Author details

Youhong Tang
Flinders University, Adelaide, South Australia, Australia

Ping Gao
The Hong Kong University of Science and Technology, Kowloon, Hong Kong, China

Acknowledgement

This project was funded by Research Grant Council (RGC) of Hong Kong and Australian Research Council – Discovery Early Career Researcher Award (ARC-DECRA).

6. References

[1] Romo-Uribe, A.; Lemmon, T. J. & Windle, A. H. (1997). Structure and linear viscoelastic behaivor of main-chain thermotropic liquid crystalline polymers. *Journal of Rheology*, Vol. 41, No. 5, pp. 1117-1145, ISSN 0148-6055

[2] Chan, C. K.; Whitehouse, C. & Gao, P. (1999). The effect of TLCP melt structure on the bulk viscosity of high molecular mass polyethylene. *Polymer Engineering and Science*, Vol. 39, No.8, pp.1353-1364, ISSN 0032-3888

[3] Whitehouse, C.; Lu, X. H.; Gao, P. & Chai, C. K. (1997). The viscosity reducing effects of very low concentrations of a thermotropic copolyester in a matrix of HDPE. *Polymer Engineering and Science*, Vol 37, No. 12, pp. 1944-1958, ISSN 0032-3888

[4] Baird, D. G. & Wilkes, G. L. (1983). Sandwich injection-molding of thermotropic copolyesters and filled polyester. *Polymer Engineering and Science*, Vol. 23, No. 11, pp. 632-636, ISSN 0032-3888

[5] Bafna, S. S.; de Souza, J. P.; Sun, T. & Baird, D. G. (1993). Mechanical-properties of in-situ composites based on partially miscible blendd of glass-filled polyetherimide and liquid-crystalline polymers. *Polymer Engineering and Science*, Vol. 33, No. 13, pp. 808-818, ISSN 0032-3888

[6] Kulichikhin, V. G.; Parsamyan, I. L.; Lipatov, Y. S.; Shumskii, V. F.; Getmanchuk, I. P.; Babich, V. F. & Postema, A. R. (1997). Rheological, mechanical, and adhesive properties of thermoplastic-LCP blends filled by glass fibers. *Polymer Engineering and Science*, Vol. 37, No. 8, pp. 1314-1321, ISSN 0032-3888
[7] Pisharath, S. & Wong, S. C. (2003). Processability of LCP-nylon-glass hybrid composites. *Polymer Composites*, Vol. 24, No. 1, pp. 109-118, ISSN 0272-8397
[8] Tchoudakov, R.; Narkis, M. & Siegmann, A. (2004). Electrical conductivity of polymer blends containing liquid crystalline polymer and carbon black. *Polymer Engineering and Science*, Vol. 44, No. 3, pp. 528-540, ISSN 0032-3888
[9] Tjong, S. C. & Meng, Y. Z. (1999). Microstructural and mechanical characteristics of compatibilized polypropylene hybrid composites containing potassium titanate whisker and liquid crystalline copolyester. *Polymer*, Vol. 40, No. 26, pp. 7275-7283, ISSN 0032-3861
[10] Lee, M. W.; Hu, X.; Yue, C. Y.; Li, L. & Tam, K. C. (2003). Effect of fillers on the structure and mechanical properties of LCP/PP/SiO2 in-situ hybrid nanocomposites. *Composites Science and Technology*, Vol. 63, No. 3-4, pp. 339-346, ISSN 0266-3538
[11] Clarke, S. M.; Rennie, A. R. & Convert, P. (1996). A diffraction technique to investigate the orientational alignment of anisotropic particles: Studies of clay under flow. *Europhysics Letters*, Vol. 35, No. 3, pp. 233-238, ISSN 0295-5075
[12] Mourchid, A.; Lecolier, E.; van Damme, H. & Levitz, P. (1998). On viscoelastic, birefringent, and swelling properties of laponite clay suspensions: Revisited phase diagram. *Langmuir*, Vol. 14, No. 17, pp. 4718-4723, ISSN 0743-7463
[13] Schmidt, G.; Nakatani, A. I. & Han, C. C. (2002). Rheology and flow-birefringence from viscoelastic polymer-clay solutions. *Rheologica Acta*, Vol. 41, No. 1-2, pp. 45-54, ISSN 0035-4511
[14] Pignon, F.; Magnin, A.; Piau, J. M. & Fuller, G. G. (2001). Fibrous clay gels: Extensional flow properties. Dichroism and SALS measurements, *The 2nd International Conference on Self-Assembled Fibrillar Networks*, Autrans, France, Nov 24-28, 2001.
[15] Zhang, G. L.; Jiang, C. L., Su, C. Y. & Zhang, H. Z. (2003). Liquid-crystalline copolyester/clay nanocomposites. *Journal of Applied Polymer Science*, Vol. 89, No. 12, pp. 3155-3159, ISSN 0021-8995
[16] Chang, J. H.; Ju, C. H. & Kim, S. H. (2006). Synthesis and characterization of a series of thermotropic liquid crystalline copolyester nanocomposites. *Journal of Polymer Science Part B: Polymer Physics*, Vol. 44, No. 2, pp. 387 -397, ISSN 0887-6266
[17] Vaia, R. A. & Giannelis, E. P. (2001). Liquid crystal nanocomposites: direct intercalation of thermotropic liquid crystalline polymers into layered silicates. *Polymer*, Vol. 42, No. 3, pp. 1281-1285, ISSN 0032-3861
[18] Huang, W. Y. & Han, C. D. (2006). Dispersion characteristics and rheology of organoclay nanocomposites based on a segmented main-chain liquid-crystalline polymer having pendent pyridyl group. *Macromolecules*, Vol. 39, No. 1, pp. 257-267, ISSN 0024-9297
[19] Huang, W. Y. & Han, C. D. (2006). Dispersion characteristics and rheology of organoclay nanocomposites based on a segmented main-chain liquid-crystalline

polymer having side-chain azopyridine with flexible spacer. *Polymer*, Vol. 47, No. 12, pp. 4400-4410, ISSN 0032-3861

[20] Gao, P., Lu, X. L. & Chai, C. K. (1996). Rheology of low nematic transition temperature thermotropic liquid crystalline copolyester HBA/HQ/SA. *Polymer Engineering and Science*, Vol. 36, No. 22, pp. 2771-2780, ISSN 0032-3888

[21] Tang, Y. H.; Gao, P.; Ye, L.; Zhao, C. B. & Lin, W. (2010). Organoclay/thermotropic liquid crystalline polymer nanocomposites. Part II: shear-induced phase separation. *Journal of Materials Science*, Vol. 45, No. 16, pp. 4422-4430, ISSN 0022-2461

[22] Tang, Y. H., Gao, P.; Ye, L. & Zhao, C. B. (2010). Experimental measurement and numerical simulation of viscosity reduction effects in HMMPE containing a small amount of exfoliated organoclay-modified TLCP composite. *Polymer*, Vol. 51, No. 2, pp. 514-521, ISSN 0032-3861

[23] Chiu, F. C.; Lai, S. M.; Chen, J. W. & Chu P. H. (2004). Combined effects of clay modifications and compatibilizers on the formation and physical properties of melt-mixed polypropylene/clay nanocomposites. *Journal of Polymer Science Part B: Polymer Physics*, Vol. 42, No. 22, pp. 4139-4150, ISSN 87-6266

[24] Pazzagli, F.; Paci, M.; Magagnini, P.; Pedretti, U.; Corno, C.; Bertolini, G. & Veracini, C. (2000). Effect of polymerization conditions on the microstructure of a liquid crystalline copolyester. *Journal of Appllied Polymer Science*, Vol. 77, No. 1, pp. 141-150, ISSN 0021-8995

[25] Onogi, S. & Asada, T. (1980), *Rheology*, vol. I, Astarita, G., Marrucci, G. & Nicolais, L. (eds.), 127-128, Plenum Press, ISBN 030-640-4656 , NY, USA

[26] Fornes, T. D.; Yoon, P. J.; Keskkula, H. & Paul, D. R. (2001). Nylon 6 nanocomposites: the effect of matrix molecular weight. *Polymer*, Vol. 42, No. 25, pp. 9929-9940, ISSN 0032-3861

[27] Magda, J. J.; Lee, C. S.; Muller, S. J. & Larson, R. G. (1993). Rheology, Flow instabilities, and shear-induced diffusion in polystyrene solutions. *Macromolecules*, Vol. 26, No. 7, pp. 1696-1706, ISSN 0024-9297

[28] Macossko, C. W. (1994). *Rheology principles, measurements and applications*, Wiley-VCH, ISBN 1-56081-579-5, NY, USA

[29] Doi, M. & Edwards, S. F. (1986). *The theory of polymer dynamics*, Oxford Press, ISBN 978-019-8520-33-7, London, UK

[30] Larson, R. G. (1990). Arrested tumbling in shearing flows of liquid-crystal polymers. *Macromolecules*, Vol. 23, No. 17, PP. 3983-3992, ISSN 0024-9297

[31] Baek, S. G.; Magda, J. J. & Larson, R. G. (1993). Rheological differences among liquid-crystalline polymers. 1. The 1st and 2nd normal stress differences of PBG solutions. *Journal of Rheology*, Vol. 37, No. 6, pp. 1201-1224, ISSN 0148-6055

[32] Baek, S. G.; Magda, J. J. & Larson, R. G. (1994). Rheological differences among liquid-crystalline polymers. 2. Disappearance of negative N(1) in densely packed lyotropes and thermotropes. *Journal of Rheology*, Vol. 38, No. 5, pp. 1473-1503, ISSN 0148-6055

[33] Mather, P. T.; Romo-Uribe, A.; Han, C. D. & Kim, S. S. (1997). Rheo-optical evidence of a flow-induced isotropic-nematic transition in a thermotropic liquid-crystalline polymer. *Macromolecules*, Vol. 30, No. 25, pp. 7977-7989, ISSN 0024-9297

[34] Han, C. D. & Kim, S. S. (1995). Shear-induced isotropic-to-nematic transition in a thermotropic liquid-crystalline polymer. *Macromolecules*, Vol. 28, No. 6, pp. 2089-2092, ISSN 0024-9297

[35] de Schrijver, P,; van Dael, W. & Thoen, J. (1996). Surface-induced pretransitional order in the isotropic phase near the isotropic-nematic phase transition. *Liquid Crystals*, Vol. 21, No. 5, pp. 745-749, ISSN 0267-8292

[36] Chan, C. K.; Whitehouse, C.; Gao, P. & Chai, C. K. (2001). Flow induced chain alignment and disentanglement as the viscosity reduction mechanism within TLCP/HDPE blends. *Polymer*, Vol. 42, No. 18, pp. 7847-7856, ISSN 0032-3861

[37] Yu, T. H. & Wilkes, G. L. (1996). Influence of molecular weight distribution on the melt extrusion of high density polyethylene (HDPE): Effects of melt relaxation behavior on morphology and orientation in HDPE extruded tubular films. *Journal of Rheology*, Vol. 40, No. 6, pp. 1079-1093, ISSN 0148-6055

[38] Zhou, H. & Wilkes, G. L. (1998). Orientation-dependent mechanical properties and deformation morphologies for uniaxially melt-extruded high-density polyethylene films having an initial stacked lamellar texture. *Journal of Material Science*, Vol. 33, No. 2, pp. 287-303, ISSN 0022-2461

[39] Chan, C. K. & Gao, P. (2005) Shear-induced interactions in blends of HMMPE containing a small amount of thermotropic copolyester HBA/HQ/SA. *Polymer*, Vol. 46, No. 24, pp. 10890-10896, ISSN 0032-3861

[40] Chan, C. K. & Gao, P. (2005). A phenomenological model for the viscosity reductions in blends of HMMPE containing a small quantity of thermotropic liquid crystalline copolyester HBA/HQ/SA. *Polymer*, Vol. 46, No. 19, pp. 8151-8156, ISSN 0032-3861

Chapter 14

Application of Thermo-Viscoelastic Laminated Plate Theory to Predict Warpage of Printed Circuit Boards

Takaya Kobayashi, Masami Sato and Yasuko Mihara

Additional information is available at the end of the chapter

http://dx.doi.org/10.5772/49978

1. Introduction

Along with rapid assembly for the purpose of creating thinner printed circuit boards, the side effect of warping during the reflow process is inevitable. As a result, the assembly process encounters serious challenges, such as difficulty in implementing a highly integrated assembly, as well as degradation in the reliability of connectivity. Aside from attempts to simulate these issues employing the FE analysis method, it is also necessary to conduct an estimation of the warpage at the early stage of design, for which development of more simplified estimation tools is strongly desired. From a material behavior point of view, if any glass transition points exist within the temperature range of the reflow process, the relaxation effect (due to viscoelastic characteristics of the material of the boards) appears as a deformation, which acts as a barrier to achieving a distinct estimation of the amount of warpage. Generally speaking, the viscoelastic behavior of resin materials exhibits very sensitive temperature dependency. This makes it difficult to accurately capture the characteristics of resin materials in actual experiments, and accordingly, to build numerical models based on such an experimental measurement. There have been many analysis cases published using sophisticated FE approaches (Shrotriya et al., 2005 and Valdevitet al., 2008), but these approaches have not extended beyond applications as a handy design tool.

To easily estimate the thermal deformation in the laminated structure, some approaches employ the multilayered beam theory (Lim, 2008 and Yuju, 2003). The method proposed in this paper enhances these approaches to incorporate the layered plate theory, and includes the effect from the temperature-dependent viscoelasticity and the temperature-dependent coefficient of thermal expansion of resin materials. The program, which is equipped with the developed method, can give an arbitrary temperature history to a multilayered plate consisting of an arbitrary number of layers. As well, the practical approach for measurement

INTECH
open science | open minds

© 2012 The Author(s). Licensee InTech. This chapter is distributed under the terms of the Creative Commons Attribution License http://creativecommons.org/licenses/by/3.0), which permits unrestricted use, distribution, and reproduction in any medium, provided the original work is properly cited.

of viscoelasticity characteristics has been specifically developed by the author's group, with the aim of performing measurements to obtain highly accurate data (Kobayashi, 2008). The method proposed herein was verified using the achievements from this development. Verification results are also shown in this paper. This method instantly calculates the amount of warpage and stress in a multilayered plate by giving the values of the thickness and material constants of the plate. Advantages of the proposed method may cover a wide range of real world applications, such as design optimization problems.

2. Viscoelasticity theory and its incremental form solution

In order to incorporate viscoelasticity into the multilayered plate theory, the one-dimensional linear viscoelasticity theory can be expanded into the plane stress field, and then treated as an incremental form of solution. The generalized Maxwell model is applied to exhibit linear viscoelastic behavior. This model is composed of a parallel series of multiple Maxwell models, each of which is assembled with a serial connection of a dashpot and a spring defined by:

$$E_r(t') = E_e + \sum_{n=1}^{N} E_n \exp(-t' / \tau_n) \tag{1}$$

where $E_r(t')$ is called the relaxation modulus of longitudinal elasticity defining the stress relaxation keeping the strain constant. E_n, τ_n, and N are the material constants denoting the coefficient to the Prony series, the relaxation time, and the number of terms in the Maxwell model, respectively; t' is the reduced time. Where the reduction rule of time-temperature is applicable, the time-temperature conversion factor, $a_T(T)$, is obtainable from the following formula:

$$t' = \int_0^t \frac{1}{a_T(T)} dt \tag{2}$$

where t is the real time and T is the temperature.

Using Eq. (1) and Eq. (2), the one-dimensional behavior of the viscoelastic material with temperature dependency can be represented. Considering the plane stress state, where uniform in-plane deformation is assumed to take place, the stress-strain (σ-ε) relation can be represented with the relaxation form as follows:

$$\sigma(t) = \frac{1}{1-\nu} \int_0^t E_r(t' - \tau') \frac{\partial \varepsilon(\tau)}{\partial \tau} d\tau \tag{3}$$

where ν is the Poisson's ratio. This study is conducted in accordance with the plate theory of shin shells assuming the thicknesses of a circuit board are not relatively large. Based on this pre-condition, the effect from the out-of-plane shear deformation of the plate is not taken into consideration. Therefore, the bending behavior of the plate is mainly governed by only the Young's modulus. Note that Poisson's ratio is incorporated into Eq. (3) for the purpose of expanding the use of the beam theory into the plate theory. Accordingly, the Poisson's ratio in this study is treated as constant value. Further, the Poisson's ratio of viscoelastic materials in its strict sense is allowed to have its own time-dependency independent of the Young's modulus, such that the Poisson's ratio can be experimentally measured and the measured Poisson's ratio becomes available for the associated analysis.

Equation (3) should be converted into an incremental form so that it can be treated numerically. Taking t_m as arbitrary time, and assuming that the strain varies with a constant gradient of $\Delta\varepsilon(t_m)/\Delta t_m$ during a time increment of $\Delta t_m = t_m - t_{m-1}$, the relationship between the stress increment, $\Delta\sigma(t_m)$, and the strain increment, $\Delta\varepsilon(t_m)$, can be represented using Eq. (1) and Eq. (3) by following the expression below (Eq. (4)). Equation (4) takes a form with respect to the strain increment; in the multilayered plate theory described in the section below, necessary equations are derived with respect to the continuity of the strain in each layer, where σ_n is the stress in the n-th term of the Maxwell model.

$$\Delta\varepsilon(t_m) = \frac{1}{A}\left\{\Delta\sigma(t_m) + \sum_{n=1}^{N} a_n(\Delta t'_m)\sigma_n(t_m - \Delta t_m)\right\} \tag{4}$$

where

$$\begin{cases} \Delta t'_m = t'_m - t'_{m-1} \\ a_n(\Delta t'_m) = 1 - \exp(-\Delta t'_m / \tau_n) \\ b_n(\Delta t'_m) = a_n(\Delta t'_m) \cdot \tau_n / \Delta t'_m \\ A = \dfrac{1}{1-\nu}\left\{E_e + \displaystyle\sum_{n=1}^{N} b_n(\Delta t'_m) \cdot E_n\right\} \\ \sigma_n(t_m - \Delta t_m) \\ = \dfrac{1}{1-\nu}\displaystyle\int_0^{t_m - \Delta t_m} E_n \exp\left\{\frac{-t'_m + \Delta t'_m + \tau'}{\tau_n}\right\}\frac{\partial\varepsilon}{\partial\tau}d\tau \end{cases} \tag{5}$$

3. Multilayered plate theory including viscoelasticity

3.1. Assumptions

For modeling a printed circuit board with a multilayered plate as illustrated in Fig. 1, the following assumptions were made:

- The in-plane property of the plate is homogeneous and isotropic.
- The thickness of the plate is sufficiently thin to generate a no stress component in the direction normal to the surface.
- The plate is not subjected to any constraint.
- The plate will warp under the uniformly distributed temperature.

The bending moment to deform the plate convexly is defined as positive; the deflection along this direction is also defined as positive. H_i and ν_i in the figure denote the plate thickness and the Poisson's ratio, respectively. The curvature induced under temperature variation becomes identical in all in-plane directions of the xy plane, since isotropicity and non-constraint are assumed. Therefore, the warpage of the multilayered plate can be calculated only on the deformation within the xz plane, as shown in Fig. 2.

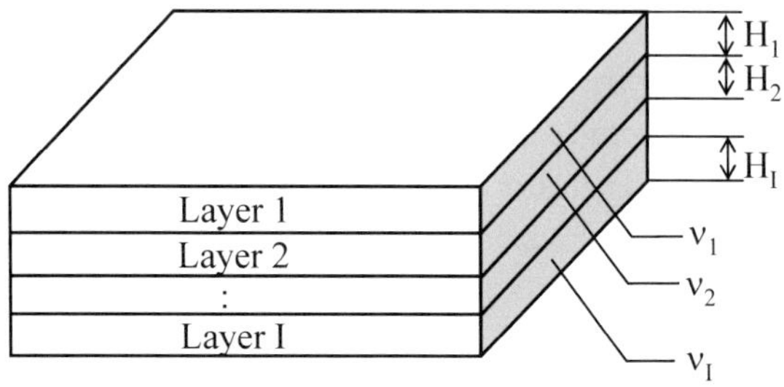

Figure 1. Multilayered plate

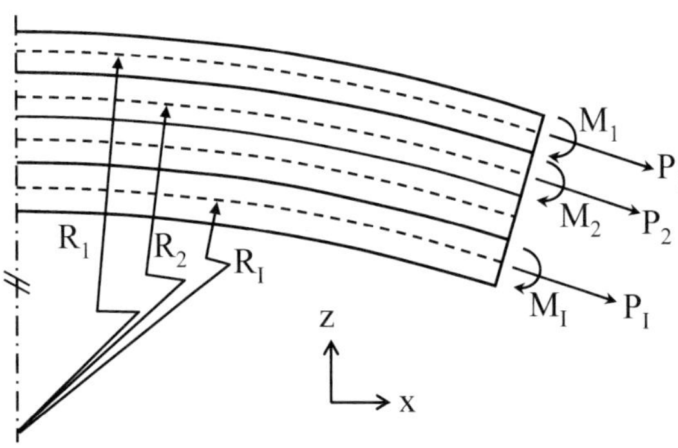

Figure 2. Deformation in xz plane

In multilayered plate theory, global warpage is calculated based on the assumption that the strain in each layer is independent and the respective interface is continuous; the strain generated in each of the layers is composed of three components:

a. Thermal strain
b. Strain due to in-plane force in the plate
c. Strain due to bending moment on the plate

Each strain component is classified as follows:

a. Thermal strain

When the temperature-dependent thermal expansion coefficient of the i-th layer at an arbitrary temperature, T, is expressed as $\alpha_i(T)$, the thermal strain is expressed as the following equation:

$$\varepsilon_i' = \alpha_i(T)\cdot(T-T_0)-\alpha_i(T_1)\cdot(T_1-T_0) \tag{6}$$

where T_1 represents the initial temperature in the analysis. The temperature-dependent thermal expansion coefficient is the average thermal expansion coefficient based on a reference temperature, T_0. The second term in Eq. (6) is necessary in order to prevent strain at the initial temperature from being generated. In order to express it as an incremental form, Eq. (6) is differentiated by T.

$$\frac{d\varepsilon_i'}{dT} = \frac{d\alpha_i(T)}{dT} \cdot (T - T_0) + \alpha_i(T) \tag{7}$$

Therefore, the thermal-strain increment in a given temperature change, ΔT (t_m), can be obtained from the following equation:

$$\Delta\varepsilon_i'(t_m) = \left\{ \frac{d\alpha_i(T)}{dT} \cdot (T - T_0) + \alpha_i(T) \right\} \Delta T(t_m) \tag{8}$$

b. Strain due to in-plane force in the plate

On each layer of the multilayered plate illustrated in Fig. 2, generation of in-plane force and bending moment are considered. Taking the in-plane force in the i-th layer as $P_i(t_m)$, the incremental force can be expressed with Eq. (9). H_i and B are the thickness and width of each layer, respectively; $\sigma''_i(t_m)$ indicates the stress due to the in-plane force.

$$\Delta P_i(t_m) = H_i \cdot B \cdot \Delta\sigma_i''(t_m) \tag{9}$$

Introducing the effect from viscoelasticity using Eq. (4), as previously derived, Eq. (8) can be written in the incremental form as follows;

$$\Delta\varepsilon_i''(t_m) = \frac{1}{H_i \cdot B \cdot A_i} \left\{ \Delta P_i(t_m) + \sum_{n=1}^{N} a(\Delta t_m') \cdot P_{i,n}(t_m - \Delta t_m) \right\} \tag{10}$$

where

$$P_{i,n}(t_m - \Delta t_m) = H_i \cdot B \cdot \sigma_{i,n}''(t_m - \Delta t_m) \tag{11}$$

c. Strain due to bending moment

The incremental strain due to the bending moment in each layer can be similarly expressed by Eq. (13); $\overline{z}_i$ is the distance measured from the neutral plane of the i-th layer.

$$\Delta M_i(t_m) = \frac{H_i^3 \cdot B}{12\overline{z}_i} \cdot \Delta\sigma_i'''(t_m) \tag{12}$$

$$\Delta\varepsilon_i'''(t_m) = \frac{12\overline{z}_i}{H_i^3 \cdot B \cdot A_i} \left\{ \Delta M_i(t_m) + \sum_{n=1}^{N} a(\Delta t_m') \cdot M_{i,n}(t_m - \Delta t_m) \right\} \tag{13}$$

where

$$M_{i,n}\left(t_m-\Delta t_m\right)=\frac{H^3{}_i\cdot B}{12\overline{z}_i}\cdot\sigma'''_{i,n}\left(t_m-\Delta t_m\right) \tag{14}$$

The strain increment, $\Delta\varepsilon'''_i(t_m)$, can be represented as Eq. (15).

$$\Delta\varepsilon'''_i\left(t_m\right)=\overline{z}_i\cdot\Delta C_i\left(t_m\right) \tag{15}$$

where $\Delta C_i(t_m)$ is the increment of the curvature,

$$C_i\left(t_m\right)=\frac{1}{R_i\left(t_m\right)} \tag{16}$$

3.2. Strain continuity and equilibrium equation

The global in-plane strain generated at each layer is expressed as the sum of the above-mentioned strain components a. through c. As this global strain must be kept continuous across the interface on each layer, the following expression can be written:

$$\begin{aligned}&\alpha_i\Delta T\left(t_m\right)+\frac{1}{H_i\cdot B\cdot A_i}\left\{\Delta P_i\left(t_m\right)+\sum_{n=1}^{N}a\left(\Delta t'_m\right)\cdot P_{i,n}\left(t_m-\Delta t_m\right)\right\}-\frac{H_i}{2}\cdot\Delta C_i\left(t_m\right)\\&=\alpha_{i+1}\Delta T\left(t_m\right)+\frac{1}{H_{i+1}\cdot B\cdot A_{i+1}}\left\{\Delta P_{i+1}\left(t_m\right)+\sum_{n=1}^{N}a\left(\Delta t'_m\right)\cdot P_{i+1,n}\left(t_m-\Delta t_m\right)\right\}\\&+\frac{H_{i+1}}{2}\cdot\Delta C_{i+1}\left(t_m\right)\qquad i=1\ to\ I-1\end{aligned} \tag{17}$$

The applied load must also satisfy the following equilibrium equations:

$$\sum_{i=1}^{I}\Delta P_i\left(t_m\right)=0 \tag{18}$$

$$\begin{aligned}&\sum_{i=1}^{I}\Delta M_i\left(t_m\right)+\Delta P_1\left(t_m\right)\cdot\left(\overline{z}'-\frac{H_1}{2}\right)+\Delta P_2\left(t_m\right)\cdot\left(\overline{z}'-H_1-\frac{H_2}{2}\right)\\&+\cdots+\Delta P_I\left(t_m\right)\cdot\left(\overline{z}'-H_1-H_2-\cdots-\frac{H_I}{2}\right)=0\end{aligned} \tag{19}$$

Assuming the thickness of each layer is negligibly small compared with its resultant curvature, the respective curvature on each layer is identical as Eq. (20). Therefore $\Delta P_i\ (t_m)$, $\Delta M_i(t_m)$ and $\Delta C_i(t_m)$ are obtainable by simultaneously resolving Eq. (17), Eq. (18), and Eq. (19).

$$\Delta C_1(t_m) \approx \Delta C_2(t_m) \approx \cdots \approx \Delta C_I(t_m) = \Delta C(t_m) \tag{20}$$

3.3. Derivation of stress and deflection

By way of the above-mentioned procedure, in-plane forces and curvatures are obtained. Using these results, stress and deflection can be derived following the steps below. Firstly, the stress generated in each layer is calculated as the sum of the in-plane stress and the bending stress, as follows:

$$\sigma_i(t_m) = \frac{P_i(t_m)}{H_i B} + \frac{12z}{H_i^3 B} M_i(t_m) \tag{21}$$

As this model is assumed to be isotropic and unconstrained, the curvature must be identical in all directions within the xy plane. As per Fig. 3, L is taken as the distance from the center to the corner of each layer; the cross-section of each layer is assumed to have the shape shown in Fig. 4, and θ is taken as the slope at the tip of each layer. Accordingly, the maximum tip deflection, $\delta(t_m)$, is obtained as follows:

$$\delta(t_m) = (1 - \cos\theta) / C(t_m) \tag{22}$$

Assuming θ is relatively small and in-plane elongation is also small, resulting in L = L', the following expression can be obtained:

$$\delta(t_m) = \frac{L^2 \cdot C(t_m)}{2} \tag{23}$$

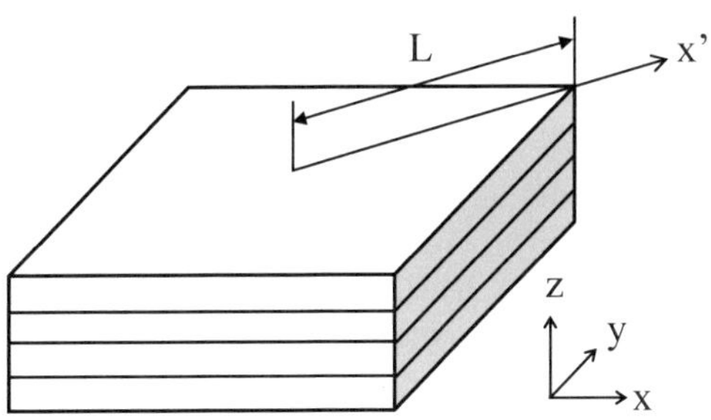

Figure 3. Definition of L

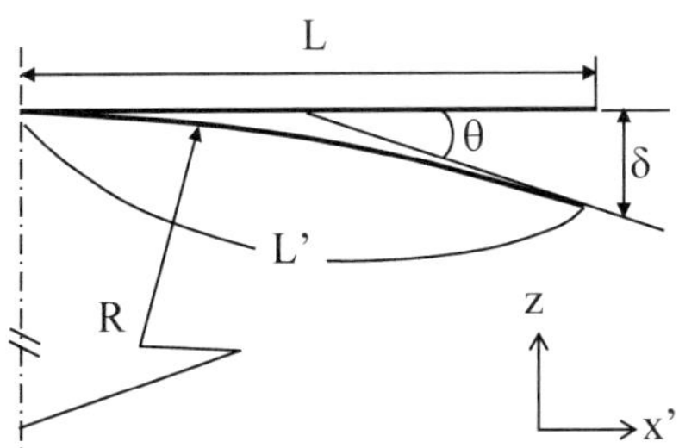

Figure 4. Deformation of the cross section

4. Measurement of viscoelastic material properties

Aside from the fact that thermal stress analysis taking viscoelasticity into account has not been adequately carried out to date, one must recognize that it is difficult, in practice, to obtain sufficient accuracy in such experimental measurements. Accordingly, any data measured in such experiments is not accurate enough to relate to the viscoelastic numerical model. Viscoelastic materials exhibit very sensitive temperature dependency, particularly in the vicinity of the glass transition point. Thus, intricate temperature control is required throughout the entire duration of the measurement operation. In addition, the time-domain constants obtained usually span a wide range of digits, in the range of 20 to 30. This leads to the additional task of determining desirable factors, which may prove difficult unless advanced optimization techniques are applied.

This paper also presents a test case for obtaining the characteristics of epoxy resin material. A device for measuring dynamic viscoelasticity, RSA III (TA Instruments), was used. Dynamic viscoelasticity characteristics were measured for angular velocities 3.16, 10, 31.6, and 100 rad/sec with an ascending temperature rate of 2 °C/min, in the temperature range of -40 to 60 °C. As shown in Fig. 5 and Fig. 6, the storage modules, E′, and the loss modulus, E′′, of the epoxy resin material were measured. This measurement device is equipped with a temperature-controlled oven with a solid structure and a large volume flow rate, providing excellent performance in temperature control.

Figure 7 shows the master curve, representing the relationship between the storage/loss modulus and reduced angular velocity, for the reference temperature, T_R. To create the master curve, the WLF formula as shown below was applied for the temperature-time reduction.

$$\log\left(a(T)\right) = -\frac{C_1\left(T-T_R\right)}{C_2+\left(T-T_R\right)} \tag{24}$$

$$T_R = T_g + 50 \qquad C_1 = 8.86 \qquad C_2 = 101.6 \quad T_g = -16^{\circ}\mathrm{C}$$

In this study, a simple optimization technique was employed in which generally recommended values for C_1 and C_2 were used, and T_g was estimated using the quasi-Newton method (Kobayashi, 2008). As a result, a single smooth master curve could be used to cover the wide range of angular velocities. The coefficients for the Maxwell model were obtained from the master curve by the optimization approach so that the relaxation curve from Eq. (1) could be calculated numerically. The result is shown in Fig. 8. The relationship between time-domain constants and frequency-domain constants is expressed by Eq. (25). Using this formula, the numerical model that gives a close approximation of the actual measurement data can be obtained, as seen in Fig. 8.

$$E' = E_e + \sum_{n=1}^{N} \frac{\tau_n^2 \omega'^2}{1+\tau_n^2\omega'^2} E_n \quad E'' = \sum_{n=1}^{N} \frac{\tau_n \omega'}{1+\tau_n^2\omega'^2} E_n \tag{25}$$

These procedures for identification have been organized in Excel; please refer to the Appendix.

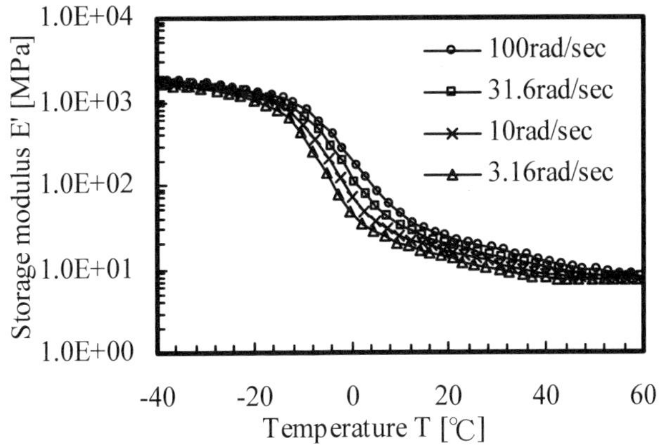

Figure 5. Storage modulus of epoxy resin

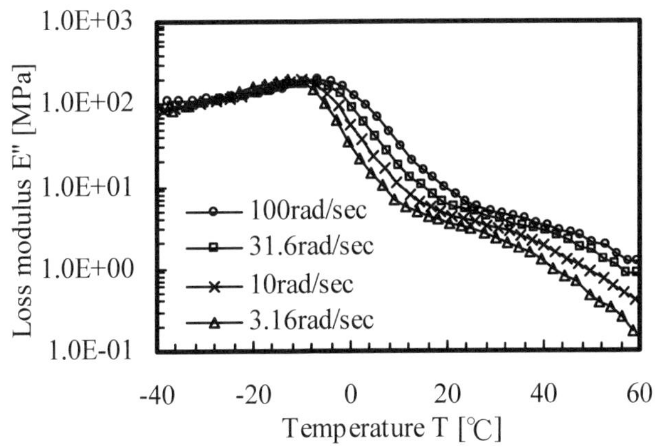

Figure 6. Loss modulus of epoxy resin

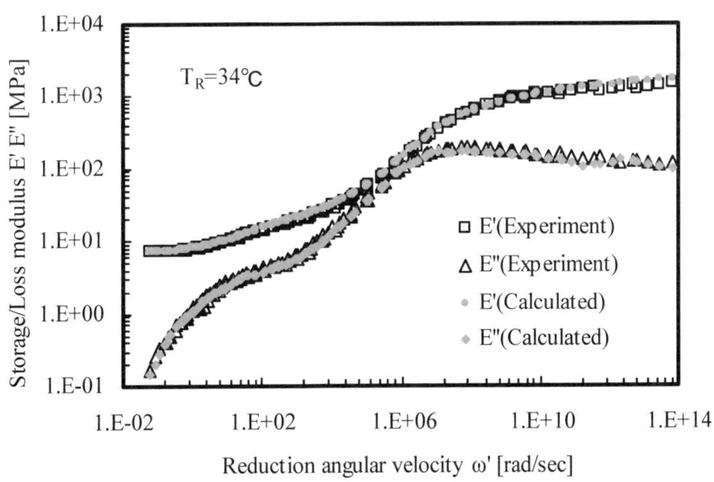

Figure 7. Master curve for Storage/Loss modulus

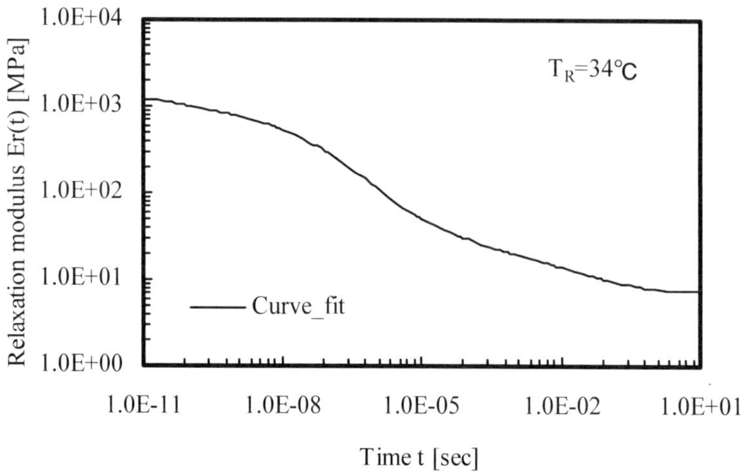

Figure 8. Master curve for relaxation modulus

5. Numerical verification by FEM

To verify the calculation method proposed in this study, a three-layered plate—40 mm in length and 20 mm in width (Fig. 9), consisting of epoxy resin with 0.5 mm thickness (viscoelastic, T_g = 105 °C), FR-4 substrate with 0.5 mm thickness (viscoelastic, T_g = 120 °C), and aluminum with 0.1 mm thickness (elastic)—was analyzed. The initial temperature of the laminated plate was set at 180 °C, and it was cooled to 25 °C within 2,000 sec, as shown in the temperature history (Fig. 10). Because the laminated plate was gradually cooled, its inside temperature was assumed to be uniform.

Figure 11 shows the master curves of the relaxation moduli of the epoxy and FR-4 substrates. The temperature-dependent viscoelasticity was assumed to be in accordance with the Arrhenius formula, as expressed by Eq. (26).

$$\log\left(a(T)\right) = -\frac{\Delta H}{2.303R}\left(\frac{1}{T} - \frac{1}{T_0}\right) \tag{26}$$

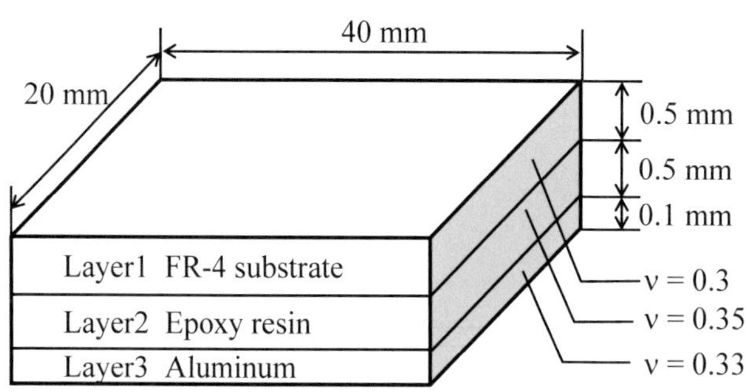

Figure 9. Three-layered plate model

The values of the coefficients for Epoxy resin shown in Eq. (26) were as follows: $\Delta H_1 = 2.4420 \times 10^5$ (before T_0); $\Delta H_2 = 3.0480 \times 10^5$ (after T_0); and $T_0 = 97.7$ °C. Those for the FR-4 substrate were as follows: $\Delta H_1 = 8.8300 \times 10^4$ (before T_0); $\Delta H_2 = 4.3220 \times 10^5$ (after T_0); and $T_0 = 113.2$ °C.

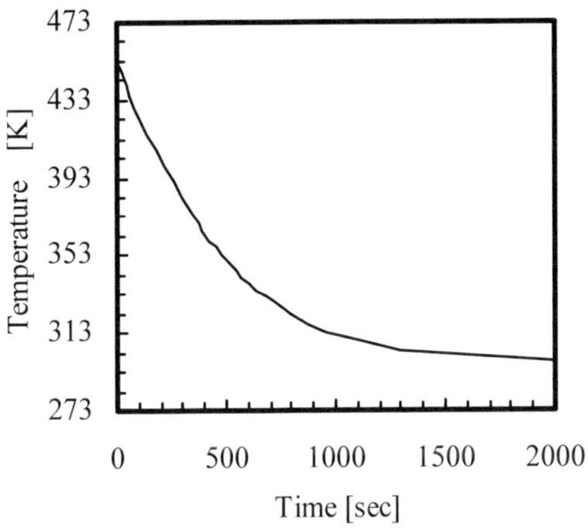

Figure 10. Applied temperature history

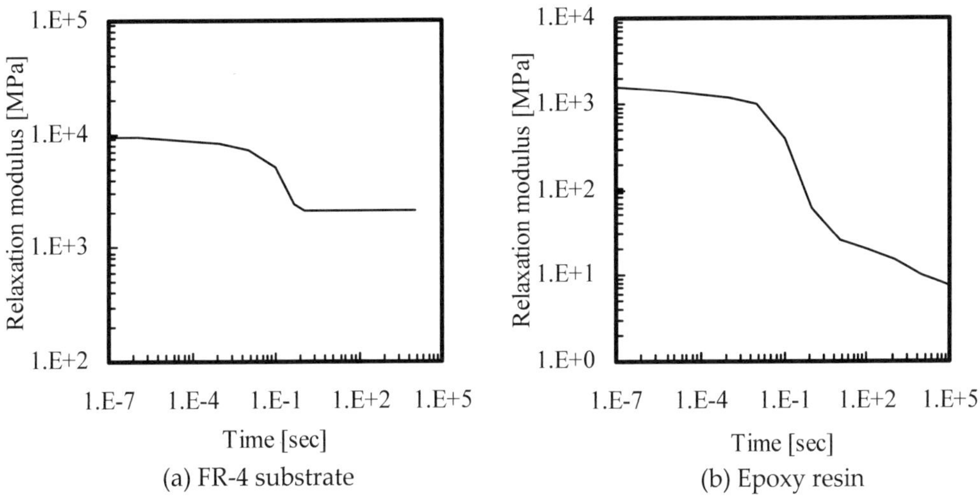

Figure 11. Relaxation modulus

Please note that the material constants of the epoxy resin used for this verification analysis differ from those expressed in the measurement example in the preceding section. Figure 12 shows the thermal expansion coefficients of the epoxy and FR-4 substrates. The aluminum that composed Layer 3 was an elastic body with the Young's modulus E = 70,000 MPa and the thermal expansion coefficient $\alpha_3 = 23.2 \times 10^{-6}$/°C.

For comparison with the multilayered plate theory developed in this study, an FE analysis of the same model was performed using the shell elements of Abaqus ver. 6.8. Figure 13 shows the deformation of the plate obtained by the FE analysis. Figure 14 shows the relationship between the deformation and temperature. Figure 15 shows the relationship between stress and temperature for each layer. As shown in Figs. 14 and 15, the deformation and stress of the laminated plate obtained by the multilayered plate theory were confirmed to be identical to those obtained by the FE analysis.

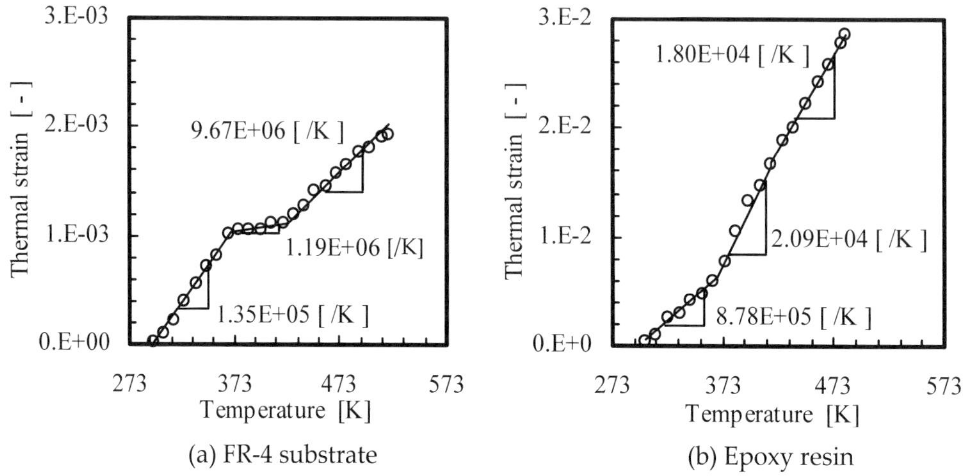

Figure 12. Thermal Expansion curves

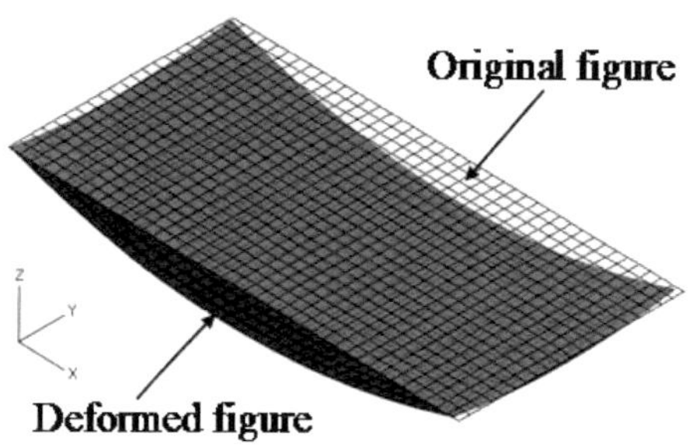

Figure 13. Deformation of plate (FEM)

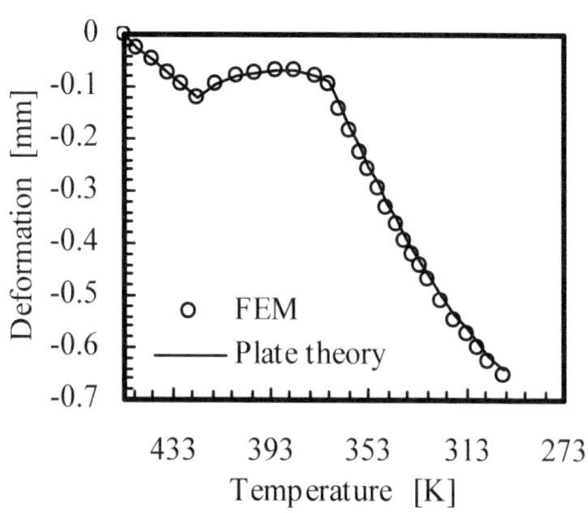

Figure 14. History of plate deflection

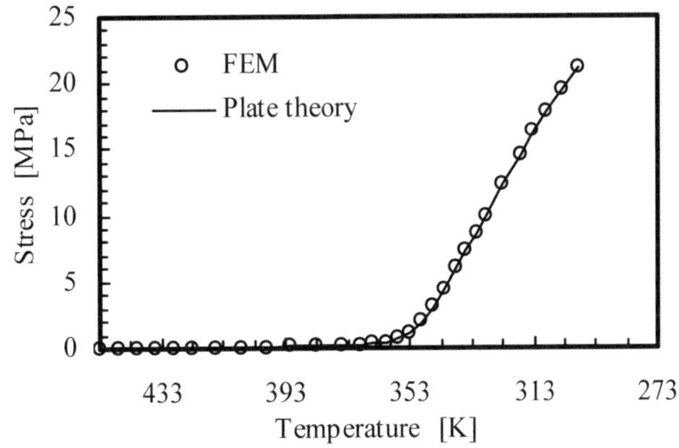

(a) FR-4 substrate (Upper surface of Layer 1)

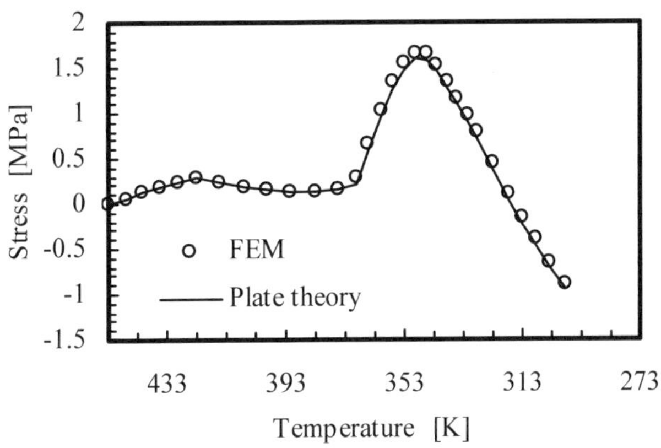

(b) Epoxy resin (Interface between Layer 1 and Layer 2)

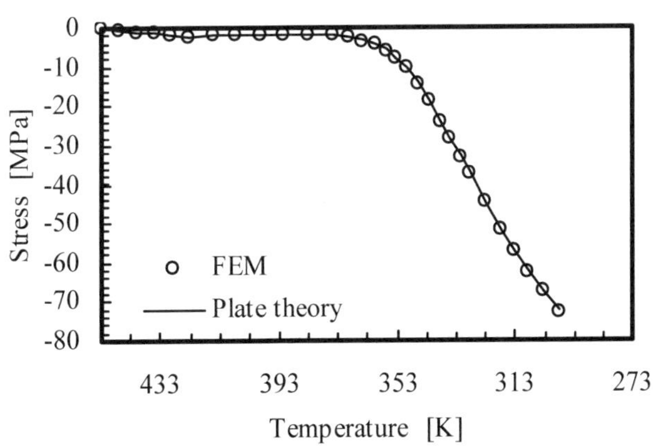

(c)Aluminum (Bottom surface of Layer 3)

Figure 15. History of stress

6. Conclusions

We were able to establish a simplified and easy-to-use tool for estimating the warpage in printed circuit boards based on the multilayered plate theory combined with the effects from the temperature-dependent thermal expansion coefficient and the temperature-dependent viscoelastic characteristics of the resin material. The results derived from this method are confirmed to be in agreement with the FE analysis results.

Appendix (Identification of the generalized Maxwell model and development of a curve fit program)

When the generalized Maxwell model for time domain is identified based on the master curve in frequency domain using Eq. (25), the following three points should be noted.

E_e, E_n, τ_n of positive definite

Since the generalized Maxwell model is regarded as a mechanical model, it is preferable that all the values of these coefficients are always positive. However, the input rule for them is different among general purpose FEMs. For example, some codes allow negative value input, while others strictly prohibit negative value input. Hence, there is no unified rule among all the codes. Since it is empirically observed that master curves may oscillate due to the affect of terms with negative values, it is considered reasonable to control the input data to make them positive definite.

Number of terms in the generalized Maxwell model

It is common practice to give the abscissas of a master curve, i.e., frequencies using a logarithmic scale covering a range of 20 to 30 digits. To make this master curve approximate a smooth curve, it is said that the number of terms (number of two-element Maxwell models) should be selected so they are equal to or above the number of digits of the frequencies. In order to confirm this, a simple calculation was carried out using a single two-element Maxwell model. The calculation was performed under the following conditions.

Elastic model: E = 100 Pa
Viscoelastic model: E_1 = 100 Pa, τ_1 = 1 sec

The relaxation modulus calculated using this model is shown in Fig. 16. In the figure, the curve with the solid line in the relaxation module is noted to decay one digit over time. It shows that a single Maxwell model is capable of representing relaxation behavior over a time domain of about one digit. Accordingly, when each dashpot is provided with the sequence of τ_n such that the next one has an order of time greater by one digit than the former, the relaxation behavior over the full range of a time domain can be expressed without a break. If the abscissa of a master curve, for example, is represented with a time domain of 20 digits, the number of terms in the generalized Maxwell model may be selected to include 20 or more.

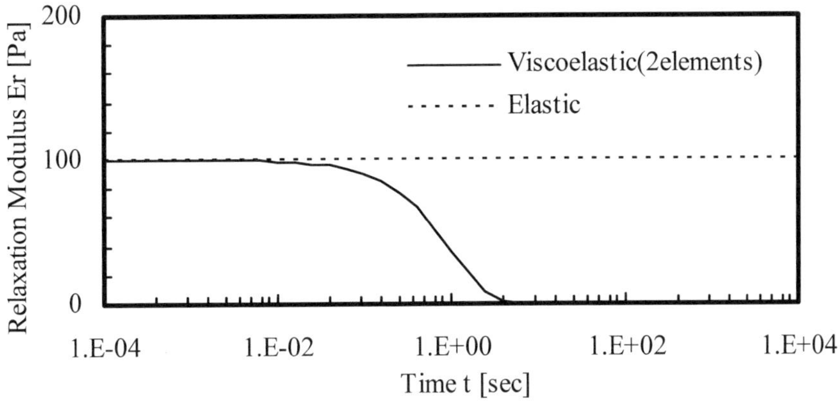

Figure 16. Relaxation behavior of single two-element Maxwell model

Smoothness of relaxation spectra

The next task is to organize the model so that the contribution from each term is approximately smoothed. In accordance with the knowledge derived by Emri et al. (1993), keeping the smoothness of discrete relaxation spectra is effective in securing the desirable accuracy of approximation results. The Kronecker's delta in the expression is denoted by δ.

$$H_n(\tau) = \sum_{n=1}^{N} E_n \tau_n \delta(\tau_n - \tau) \tag{27}$$

An example of these relaxation spectra is shown in Fig. 17. An attempt was made in this example such that the envelope for these discrete spectra is approximated to be piecewise quadratic so that the smoothness can be maintained subject to the curvature change along this envelope being not too large. Through the testing of such provisions, followed by an approximate calculation, it becomes possible to perform a curve fit operation for a master curve even though data is missing. For viscoelastic materials with sharp temperature dependency, it becomes very difficult for the temperature control in the measurement device to catch up with the actual material response, and as a result, critical defects are bound to occur (Fig. 18(b)); therefore, smoothing manipulation for those relaxation spectra is a highly effective measure.

In the curve fit program developed by the author's company, the generalized Maxwell model is identified based on the master curve shown in Fig. 7. This program is designed to completely fulfil the constraint conditions discussed in the preceding section. A sample output from this program is shown in Fig. 19. The user is only required to enter "Input data," "Number of Prony terms," and "Poisson's Ratio" in the specified input field, and then press the "Optimization" button. The program automatically performs an approximate calculation. The optimization operation uses the quasi-Newton method. For the quasi-Newton method, it is necessary to set up the initial condition in the vicinity of the optimized value. However, this program incorporates an algorithm that can automatically estimate, from the test results, an initial condition that easily converges.

The master curve for epoxy resin material shown in Fig. 8 covers a range of about 12 digits in terms of angular frequency, so the number of terms in the Maxwell model is set to 12. As previously mentioned, positive values for all of E_e, E_n, and τ_n are maintained during the calculation. The coefficients of the identified Maxwell model are automatically written into the respective format of the input file for Abaqus, Marc, or LS-DYNA.

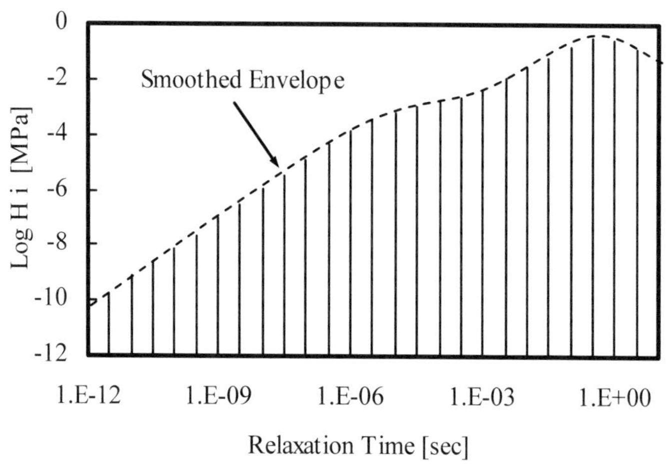

Figure 17. Smoothing manipulation for relaxation spectra

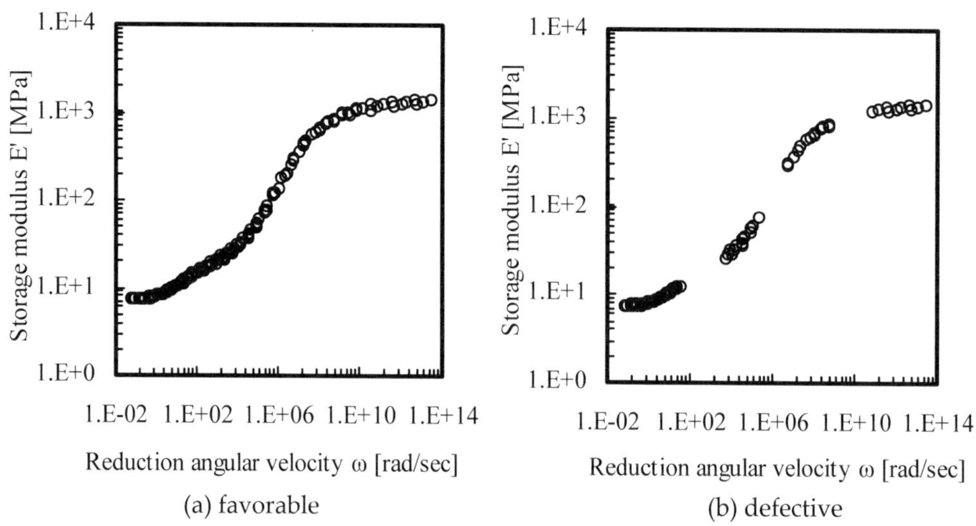

Figure 18. Examples of measured master curve

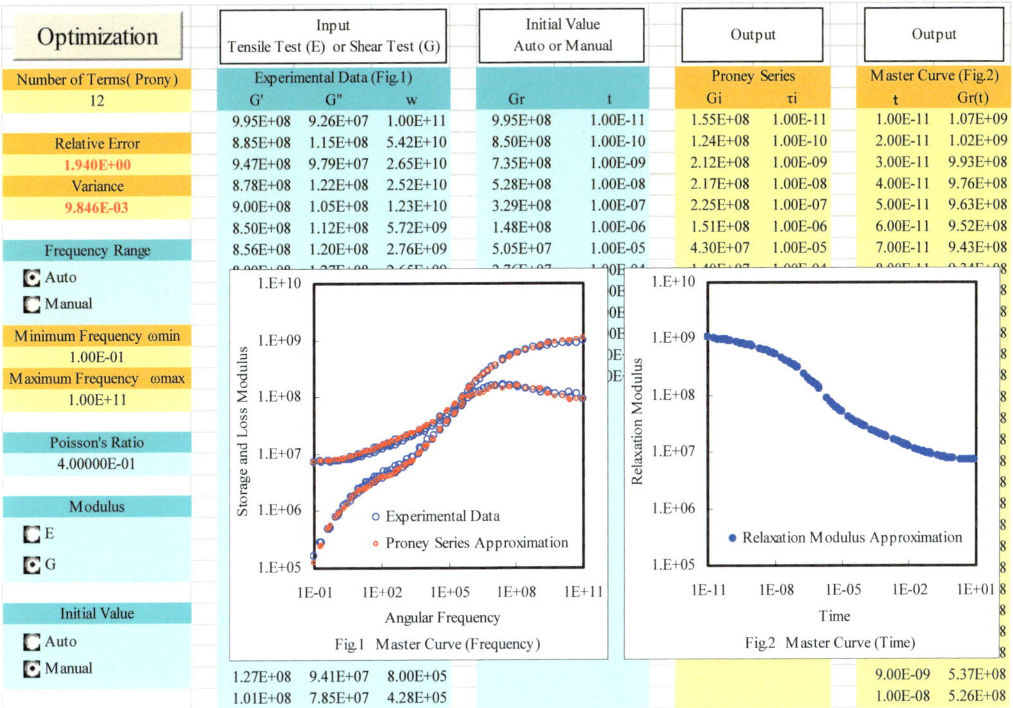

Figure 19. Viscoelastic curve fit program using Excel

Author details

Takaya Kobayashi, Masami Sato and Yasuko Mihara
Mechanical Design & Analysis Corporation, Japan

7. References

Emri, I. & Tschoegl, N. W. (1993), Generating line spectra from experimental responses. Part I: Relaxation modulus and creep compliance, *Rheologica Acta*, Vol. 32, pp. 311-322.

Kobayashi, T.; Mikami, T. & Fujikawa, M. (2008), Application of Abaqus for Advanced Inelastic Analysis (I: Linear Viscoelastic Materials), *2008 Abaqus Users' Conference Proceedings*, Dassault Systems Simulia Corp., pp. 360-373.

Lim, J. H.; Han, M.; Lee, J.; Earmme, Y. Y.; Lee, S. & Im, S. (2008), A study on the thermomechanical behavior of semiconductor chips on thin silicon substrate, *Journal of Mechanical Science and Technology*, Vol. 22, pp. 1483-1489.

Shrotriya, P. & Sottos, N. R. (2005), Viscoelastic response of woven composite substrates, *Composites Science and Technology*, Vol.65, pp.621-634.

Valdevit, L.; Khanna, V.; Sharma, A.; Sri-Jayantha, S.; Questad, D. & and Sikka, K. (2008), Organic substrates for flip-chip design: A thermo-mechanical model that accounts for heterogeneity and anisotropy, *Microelectronics Reliability*, Vol.48, pp.245-260.

Yuju, W. & Basaran, C. (2003), Thermomechanical Stress Analysis of Multi-Layered Electronic Packaging, *Journal of Electronic Packaging*, Vol. 125, pp. 134-138.

Chapter 15

An Approach for Dynamic Characterisation of Passive Viscoelasticity and Estimation of Anthropometric Inertia Parameters of Paraplegic's Knee Joint

B.S. K. K. Ibrahim, M.S. Huq, M.O. Tokhi and S.C. Gharooni

Additional information is available at the end of the chapter

http://dx.doi.org/10.5772/45994

1. Introduction

Functional electrical stimulation (FES) is a promising way to restore mobility to individuals paralyzed due to spinal cord injury (SCI). Modelling and parameter identification of both the passive and active joint properties are needed to improve control of this nonlinear time varying system. In order to develop a suitable control strategy for the FES to move of the leg correctly, a proper model of stimulated muscle has to be used. The muscle is assumed to consist of two components: an active force generator and parallel passive properties. Riener and Edrich (1999) suggested passive muscle properties should be identified separately from active muscle properties because it is easier to consider the passive elastic forces as contributions to the total joint moment. Other researchers such as Zajac (1989) and Pandy et al. (1990) used a musculo-tendon model, in which the passive and the active forces are generated by single muscles. However, such models have too many parameters that cannot be identified non-invasively due to the muscle-joint redundancy of the musculoskeletal system (Riener and Edrich, 1999).

Conventionally the joint passive resistance is modelled as an elastic element like a torsion spring and a viscous element like a rotary damper (Lamb et al., 1991). These two resistances are non-linear, but the viscous resistance is often approached as a linear function of the joint angular velocity (Chizeck et. al., 1999; Mansour and Audu, 1986). These characteristics are important to estimate muscle load or fatigue during motion, especially in the field of biomechanics. Some researchers have shown further that these resistances influence the basis of motion effectively. For example, the resistance imposed by passive joint properties can impede the functionality of FES systems during limb movements (Amankwah, 2001).

INTECH
open science | open minds

© 2012 The Author(s). Licensee InTech. This chapter is distributed under the terms of the Creative Commons Attribution License http://creativecommons.org/licenses/by/3.0), which permits unrestricted use, distribution, and reproduction in any medium, provided the original work is properly cited.

Segment mass of human body is an elementary inertial parameter for kinetic analyses of human motion. Many methods exist to estimate body segment properties. In the past, the most popular approach used to estimate segment parameters has been based on data obtained from elderly male cadavers. This database is quite limited in that a small number of cadavers have been studied. Thus, the database for making inertial estimates is not representative of the subjects often under investigation in many exercise and sport biomechanics studies. Dempster (1956) addressed the problem of mass model using cadaveric studies to establish segmental masses expressed as a percentage of total body mass (Winter, 1990). Other techniques have been developed in which inertial properties are directly measured for an individual. Zatsiorsky and Seluyanov (1985) used gamma mass scanning as a means of quantifying mass distribution in analyzing human body segment inertial characteristics. Both of these methods were promising and widely used but gave different measurement of the segment mass since calculated properties can vary drastically depending on the method used. Segment properties can significantly affect such variables, especially during swing phase (Doane and Quesada, 2006).

The pendulum test of Wartenberg is a technique commonly applied to evaluate passive properties in which the leg is allowed to drop from an initially extended position under the influence of gravity and then allowed to oscillate freely (Wartenberg, 1951). The test is very attractive in that it requires no special equipment and is very simple and the oscillatory movements of the lower leg recorded at the knee joint, are captured with electro-goniometers (Bajd and Vodovnik, 1984; Le Cavorzin et al., 2001), uniplanar video-based methods (Jamshidi and Smith, 1998) and 3D motion analysis systems (White et al., 2007). Most of the analyses of the pendulum motion depend on a second-order linear model to extract the elastic and viscous moments from the recorded leg oscillations. However a second order linear model does not provide an adequate description of the motion for either spastic or normal legs (Lin and Rymer, 1991).

In this paper a new approach for estimating passive properties of the paraplegic's knee joint based on pendulum test is described. On the basis of these experimental and optimization results, a non-linear fuzzy model is proposed which can be used to estimate the passive viscoelastic knee joint moment as a function of knee angle and knee velocity. The model of a dynamic system of the lower limb is derived using Kane's equations (Josephs and Huston, 2002) with accessibility to estimate of the foot mass, shank mass, moment of inertia about COM and position of COM along the segmental length of the subject.

2. Material and methods

The passive behaviour of the knee joint depends on the knee joint elastic moment (M_s) and the viscous moment (M_d). Considering the inertial (M_i) and gravitational (M_g) moments, a moment balance equation can be written as (Ferrarin and Pedotti, 2000):

$$M_i = M_g + M_s + M_d \tag{1}$$

or

$$M_i - M_g = M_s + M_d \tag{2}$$

In this research the $M_i - Mg$ represented by equation of motion for dynamic model of the lower limb and $M_s + M_d$ represented by fuzzy model as viscoelasticity. The subject was a 48 year-old T2&T3 incomplete paraplegic male with 20 years post-injury with height = 173cm and weight = 80kg. Informed consent was obtained from the subject.

In this section, first the procedure to perform the pendulum test is presented to get the experimental data. Second, the equations of motion for dynamic model of the lower limb are introduced. Third, the estimations of anthropometric inertia parameters lower limb model are described briefly. Forth, the optimization of fuzzy model as passive viscoelasticity is outlined. Lastly new method for estimation and optimization of passive properties using GA by comparing with experimental data is elaborated. The procedure for estimation of the anthropometric inertia parameters and optimisation of FIS as passive viscoelasticity of the knee joint model are shown in Figure 1.

2.1. Pendulum test

Pendulum test can be used to evaluate passive properties such as viscosity and elasticity moments of the knee. Viscoelasticity is combination of elasticity and viscosity and represents passive resistances to joint motion associated with the structural properties of the joint tissue and of the muscular-tendon complex. Elasticity can be considered as an intrinsic property of the tissue to resist deformation, while viscosity is related to cohesive forces between adjacent layers of tissues. Both parameters may influence the joint range of motion affecting knee angle (Valle et. al, 2006). A genetic optimization algorithm is used to identify the unknown viscoelasticity by minimizing the error between the data obtained experimentally and from the simulation model. The pendulum test was conducted to measure the passive knee motion of an SCI patient. The subject sat on a chair, which allowed the lower leg to swing freely, while ankle joint was monitored to be at 0°. Reflexive or voluntary activation of muscles acting on the knee occurred during the pendulum test has been monitored to avoid the influence of pendulum movements.

In the pendulum test the knee was slowly extended, by having the experimenter lift it with minimal acceleration at the starting position (1) and then it was released as shown in Figure 3. The knee angle was recorded using electro-goniometer until the final position (2). A Biometric electro-goniometer was used to measure knee movements with sampling time of 0.05s. The electro-goniometer arrangement is shown in Figure 4.

2.2. Equations of motion for dynamic model of the lower limb

The inertial (M_i) and gravitational (M_g) moments are represented by mathematical model of a dynamic system of the lower limb based on Kane's equations as follows:-

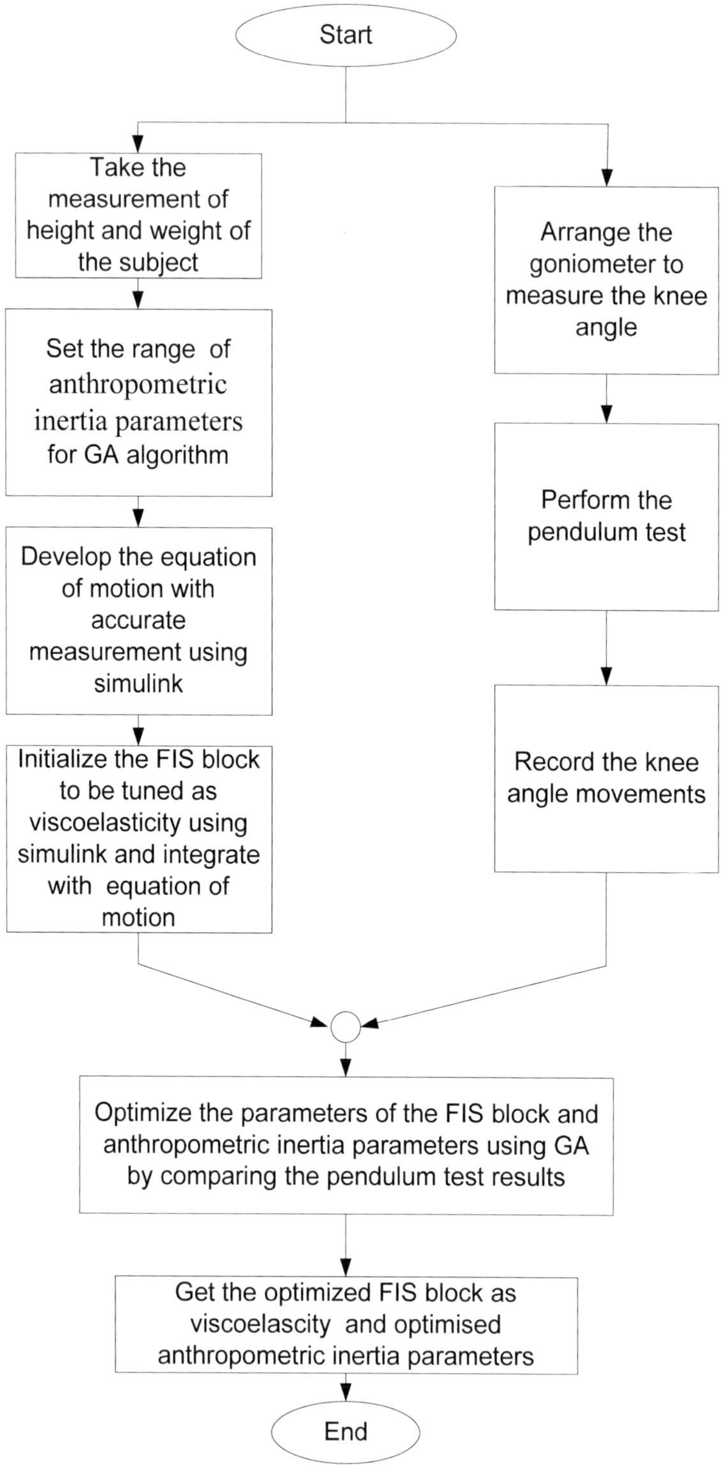

Figure 1. Optimization and Estimation Procedure

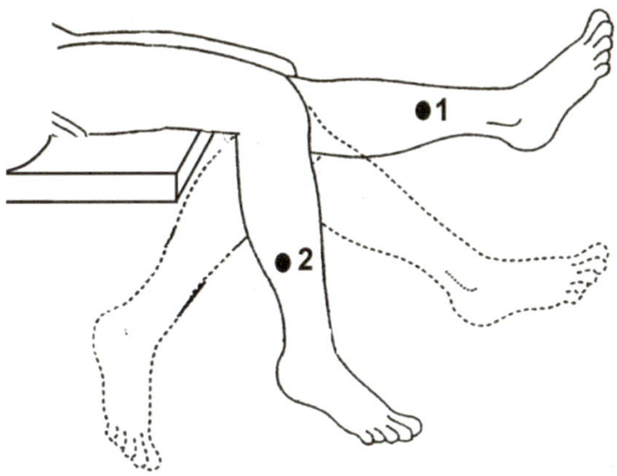

Figure 2. Limb oscillation during pendulum test (Valle et. al, 2008)

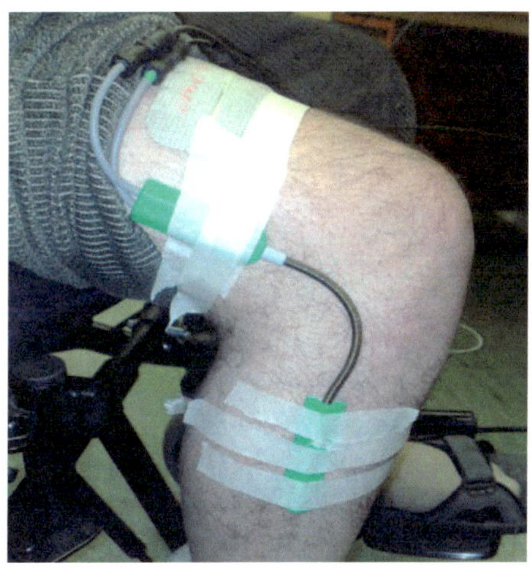

Figure 3. Electro-goniometer arrangement for pendulum test

$$M_i - M_g = -m_2 q_2 \dot{\theta}_1^{\,2} r_2 + m_1 g \cos\theta_1 r_1 - I_1 \ddot{\theta}_1 + m_2 g \cos\theta_1 q_2 - m_1 r_1^{\,2} \ddot{\theta}_1 - m_2 q_2^2 \ddot{\theta}_1 \quad (3)$$

where,

m_1 = shank mass, m_2 =foot mass, I_1 = moment of inertia about COM, q_2 = shank length r_1 = position of COM along the shank, r_2 = position of COM along the foot, θ_1 =knee angle, $\dot{\theta}_1$ =knee velocity, $\ddot{\theta}_1$ =knee acceleration, θ_2 =ankle angle g =gravity.

The lower limb model with the angles sign convention is shown in Figure 4. Anthropometric measurements of length of the lower limb were made as shown in Table 1.

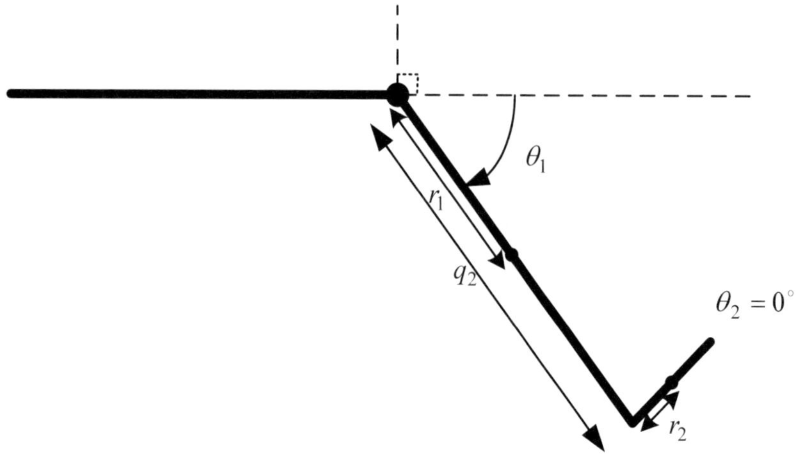

Figure 4. Lower limb model

Segment	Length (m)
Shank	0.4256
Foot	0.0675
Approximated position of COM of shank	0.2128
Approximated position of COM of soot	0.03375

Table 1. Anthropometric data of subject

2.3. Estimation of anthropometric inertia parameters

The lower limb dynamics is complex and less well defined because the form is not composed of simple geometrical shapes. Indeed, even the location of the mass center of elemental parts is imprecise, and comprehensive analyses of joint kinematics are extremely difficult (Josephs and Huston, 2002). To quantify limb dynamics, accurate estimates are needed of anthropometric inertia parameters (mass, COM location, and moments of inertia). Therefore these equations have accessed to optimise mass, moment of inertia about COM and position of COM along the segmental length of the subject using GA tuning method to fit the experimental data.

2.3.1. Moments of Inertia

Moments of inertia are fundamental parameters describing the mass distribution of body segments which enter into all computations involving segmental rotations. Methods based on the geometrical segment body models and appropriate anthropometric measures were used in measuring the moment of inertia of various segments in cadavers and in living subject (Dempster, 1955; Hatze, 1980; Jensen, 1986; Schneider and Zernicke, 1992). Different procedures were applied to identify the segment moment of inertia from the dynamic equation of motion. There is no unique way of calculating the moment of inertia To obtain more accurate moment of inertia of the knee, the minimum (0.32 Nm^2) and maximum (0.58 Nm^2) range as reported by M. K. Lebiedowska (2003) is optimized using GA.

2.3.2. *Mass and position of COM of shank and foot*

Vatnsdal et al (2008) addressed the masses and positions of the COM of the lower limb are possible error source in the lower limb model. Dempster (1956) estimated the mass from the total height and weight of the person using standard human dimensions (Winter, 1990). Zatsiorski and Seluyanov (1983) estimated the mass on the basis of regression equations based on statistics obtained from measuring cadavers. However, both of the methods give different measurements for a selected subject as shown in Table 2. Therefore they can potentially introduce large errors, especially while modelling the passive properties, as the pendulum test results may be affected by those parameters. Thus, fine tuning these parameters around the predicted values with an efficient stochastic search algorithm such as GA could lead to good solution with immense potential of extracting accurate subject specific results. GA is used to optimise the foot and shank mass between the two measurements with ±10% tolerance. While the approximated positions of COM along the segment length are assumed as in the centre of the segment length. To get the accurate positions, the position of the COM of shank and foot are optimised between ±10% tolerances.

Segment Mass	Zatsiorski and Seluyanov's Method	Dempster's Method
Shank	1.0499kg	1.16kg
Foot	3.3973kg	3.72kg

Table 2. Mass of patient using different methods

2.4. Optimization of fuzzy model as passive viscoelasticity

Modelling based on conventional mathematical tools is not well suited for dealing with complex and nonlinear nature systems. By contrast, a fuzzy inference system employing fuzzy 'if–then' rules can model the qualitative aspects of human knowledge and reasoning processes without employing precise quantitative analyses (Dinakaran, 2009). Besides that, the purpose of modeling passive viscoelascity is to be used as a part of the simulation platform and need to integrate with the active properties of the knee joint model for the controller application. Controller such as fuzzy control, neural network does not require the mathematical model of the plant. Therefore fuzzy model are well suited for modeling non-linear models such as passive viscoelasticity. The passive viscoelastic joint moment is represented as a non-linear function of knee angle and knee angular velocity in a fuzzy model. An automatic Mamdani fuzzy system design method integrates three stages; determines membership functions, the rule-consequent parameters, and scaling factor at the same time. Gaussian membership functions are preferred, because of their continuously differentiable curves and smooth transitions. There are 58 parameters to optimise including

30 parameters that determine the centre and width of the 15 Gaussian membership functions, 25 weights associated with the fuzzy rules, and 3 scaling factors for the normalization and denormalization of 2 inputs and 1 output of fuzzy model respectively. The weights of the 25 fuzzy rules have been optimised between 0 and 1. The rules are identified based on expert knowledge that refers to relationship between knee angle and velocity and passive torque. Table 3 summarizes the rule base for the fuzzy model encompassing possible AND combinations of the input fuzzy values.

		Knee angular velocity				
		NB	NS	ZO	PS	PB
Knee angle	NB	PB	PB	PB	PS	ZO
	NS	PB	PB	PS	ZO	NS
	ZO	PB	PS	ZO	NS	NB
	PS	PS	ZO	NS	NB	NB
	PB	ZO	NS	NB	NB	NB

Table 3. Rule base of the fuzzy model
NB=Negative big NS=Negative small
Z=Zero PS=Positive small PB=Positive big

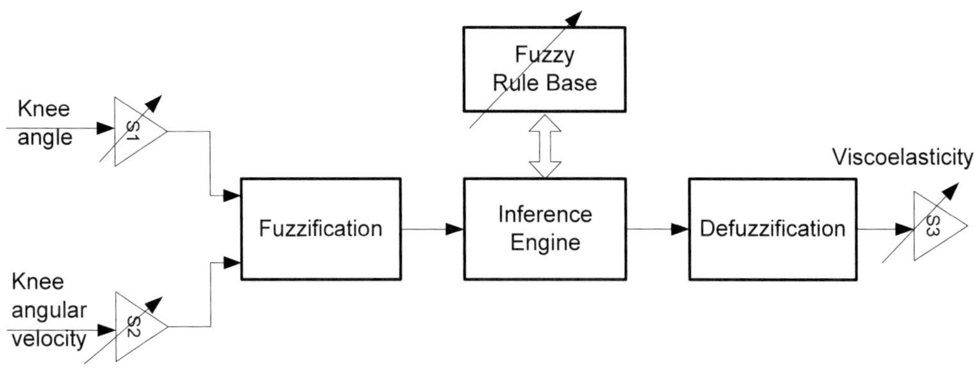

Figure 5. A fuzzy expert system model

The configuration of the fuzzy expert system model is shown in Figure 5. In the fuzzification step, crisp inputs are fuzzified into linguistic values to be associated to the input linguistic variables. After fuzzification, the inference engine refers to the fuzzy rule base containing fuzzy IF-THEN rules to derive the linguistic values for the intermediate and output linguistic variables. Once the output linguistic values are available, the defuzzifier produces the final crisp values from the output linguistic values. The defuzzification method was based on calculating the centre of gravity of the fuzzy output. Scaling factors is applied to ensure that the domain of discourse covers the whole range (Reznik, 1997). Therefore, two input scaling factors are used to transform the crisp inputs into the normalised inputs so as to keep their value within -1 and +1. The scaling factors are S1 for knee angle, θ and S2 for knee angular velocity, $\dot{\theta}$. An output scaling factor S3 provides a transformation of the defuzzified crisp output from the normalised universe of the model output into an actual physical output (passive torque, τ_{pas}). These scaling factors influence the performance of the model and optimise simultaneously using GA. The ranges of the scaling factors are set as follows:-

a. S1 is between $1/\theta_{\max}$ and $1/\theta_{\min}$ of the experimental data.

b. S2 is between $1/\dot{\theta}_{\max}$ and $1/\dot{\theta}_{\min}$ of the experimental data.

c. S3 is between 1 and 50.

2.5. Estimation and optimization of passive properties using GA

The GA is based on natural selection and population genetics theory (Goldberg, 1989). This evolutionary algorithm is chosen to estimate passive knee joint properties of paraplegic because the search space is large and complex. The advantage of the GA approach is robust, searches many points simultaneously, and is able to avoid local optima that the traditional algorithms might get stuck in (Lee, 1993).

GA is used to estimate a dynamic characterization of passive viscoelasticity of the knee joint using fuzzy model and estimation of the anthropometric inertia parameters such as foot mass, shank mass, moment of inertia about COM and positions of COM along the segmentals length of the lower limb as shown in Figure 6. The goal of GA optimization process is to minimize the error between the knee angle obtained experimentally and and from the model. The output prediction error is defined as:-

$$e(t) = y(t) - \hat{y}(t) \tag{4}$$

where $y(t)$ is the experimental data and $\hat{y}(t)$ is the estimated current output of knee angle. The 'goodness of fit' of the identified model is determined using the objective function by minimizing the mean-squared error (MSE);

$$f = \min\left\{\frac{\sum_{i=1}^{N}\left(y(t) - \hat{y}(t)\right)^2}{N}\right\} \tag{5}$$

The GA optimization procedure is shown in Figure 7 and implemented in MATLAB with GA Toolbox. First, an initial population of individuals is generated. Each individual corresponds to a chromosome, which is a set of specific genes from the biological point of view. Then, the performance of each member of the population is assessed through an objective function imposed by the problem. This fires the process of selecting pairs of individuals which will be mated together during reproduction. Selection is based probabilistically on a gene's fitness value; the higher the fitness of a gene, the more likely it can reproduce. After selecting two parents, crossover is performed according to a crossover probability. If crossover is to be performed, offspring are constructed by copying portions of

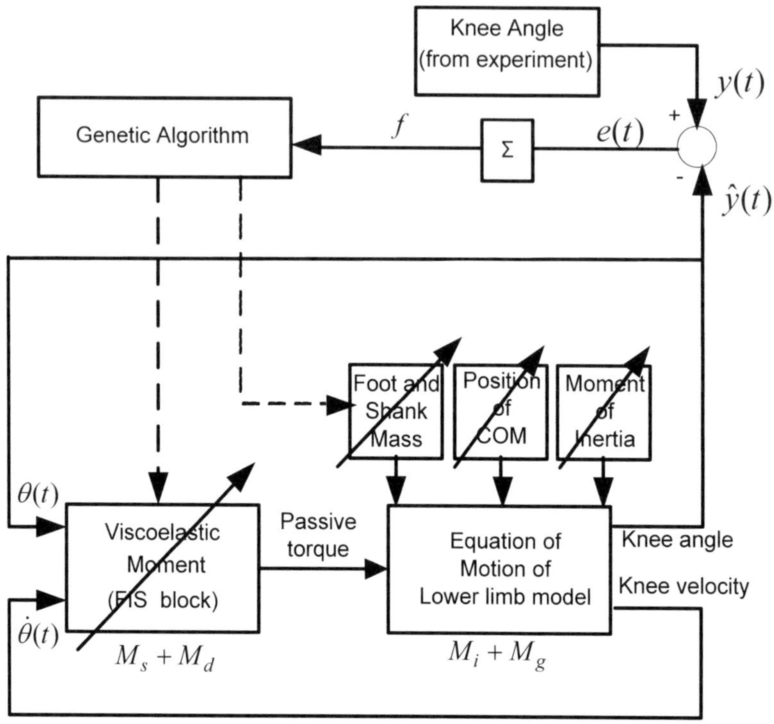

Figure 6. Optimization of passive properties

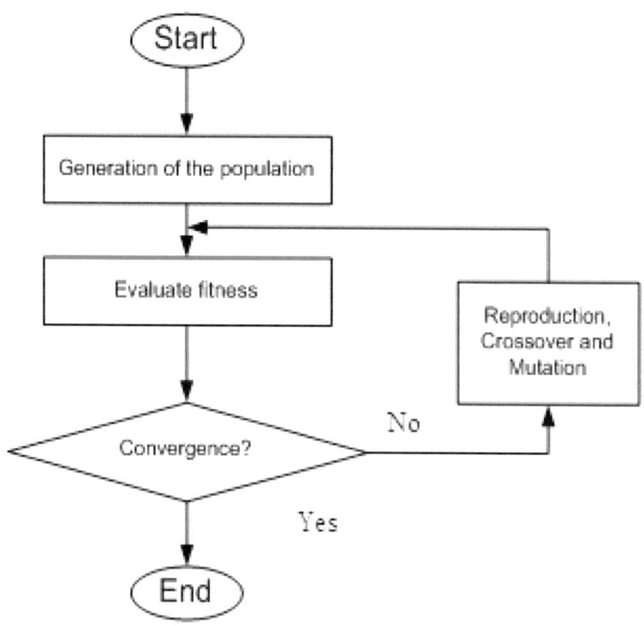

Figure 7. The GA optimization procedure

parent genes designated by random crossover points. Otherwise, an offspring copies its entire gene from one of the parents. As each bit is copied from parent to offspring, the bit has the probability of mutating. Mutation is believed to help in reinjection of any information that may have been lost in previous generations (Goldberg, 1989).

3. Results

In this section, the results of the new method to model the passive viscoleasticity and estimate the anthropometric inertia parameters by GA optimization process are presented. Population size of GA was set to 50 and crossover and mutation probabilities were 0.8 and 0.001 respectively. The automatic GA optimization process was generated up to 200 generations of solutions. The best solution was kept and the rest were discarded until there is no significant change in the mean square error (MSE) was observed after the 165th generation. The minimum MSE achieved was 1.87. The response of the model was tested and the result is shown in Figure 8. The results showed that the model parameters were estimated well and that the fit between the model and the experimentally data was good.

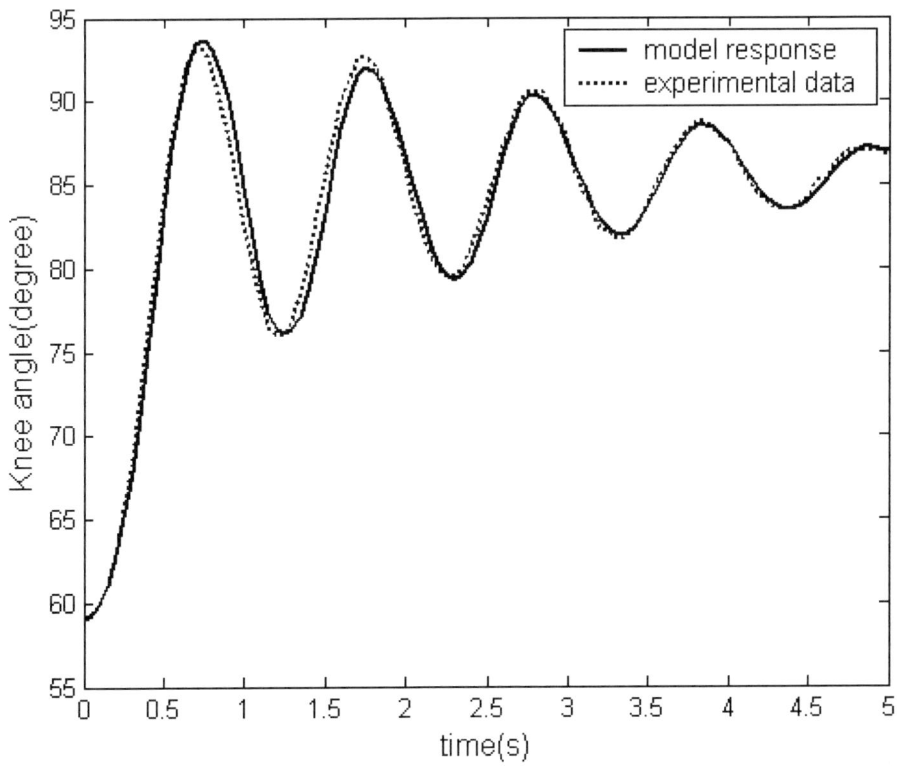

Figure 8. Responses of the pendulum test and the model response

3.1. Optimised value of anthropometric inertia parameters

The optimized value of the anthropometric inertia parameters such masses, moment of inertia about COM and positions of COM along the segmentals length of the subject's lower limb are tabulated in Table 4. The accuracy of these optimised anthropometric inertia parameters been validated through repeated process of optimisation.

Parameter	Optimised Value
Foot Mass	0.95 kg
Shank Mass	3.5 kg
Moment of Inertia	0.35823Nm^2
Position of COM of Foot	0.035 m
Position of COM of Shank	0.22 m

Table 4. Optimised Value ofAnthropometric Inertia Parameters

3.2. Optimised Fuzzy Model as Viscoelasticity

The vicoelasticity is represented by a fuzzy model. The GA optimization is used to optimise 48 parameters including 30 associated with the membership function, 15 weights associated with the minimized fuzzy rules and 3 scaling factors of fuzzy model. The scaling factor for the normalization and denormalization of two inputs and output were 0.0091247, 0.0053982 and 35 respectively. The fuzzy model takes into account the nonlinear component of passive viscoelasticity. A three-dimensional plot that represents the mapping from knee angle and knee angular velocity to viscoelastic moment is shown in Figure 9. This surface plot shows the normalised viscoelasticity (unit less) changes as a function of the normalised knee angle (unit less) and normalised velocity (unit less). The presence of the non-linearities in the viscoelasticity can be noted on this uneven surface shape from both angles. The valley for knee angle between 0.5 and 1 gives high viscoelasticity could be due to the high stiffness near extended knee joint.

3.3. Model validation

Model validation is possibly the most important step in the modelling process. The model and the optimised parameters obtained from the optimization process were validated in terms of consistency and the prediction error. Two different approaches of model validity are conducted. Different sets of experimental conditions are usually required to define the domain of a model's intended applicability. Therefore, in the first validation process, the model is validated with a different set of data than the training data. To avoid any change in the actual plant, the validation data is obtained in the same experimental arrangement but with a different initial knee joint angle. The responses of the optimised model and new experimental data are shown in Figure 10. It is noted that the two agree closely with one another.

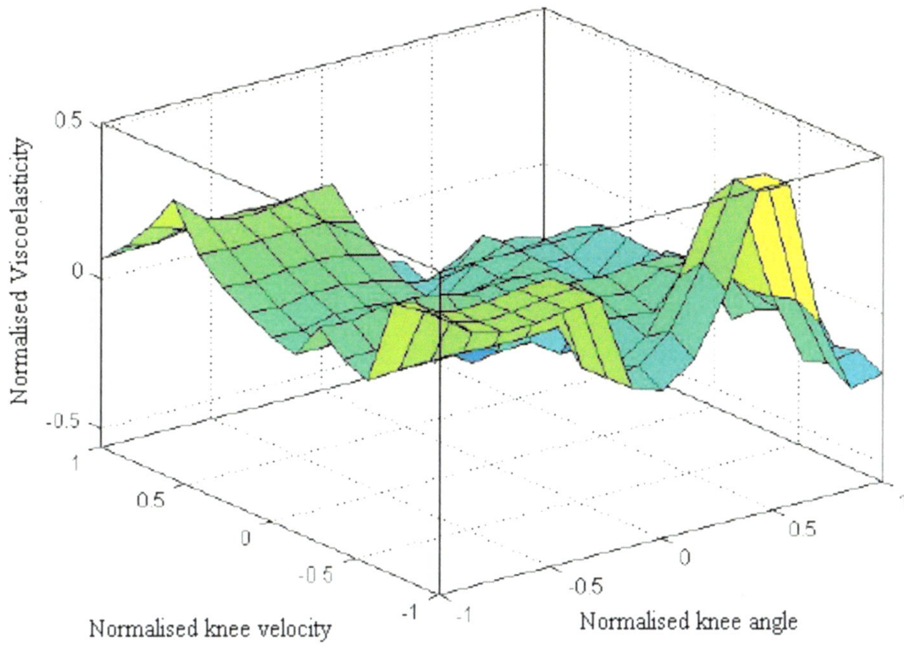

Figure 9. 3D Fuzzy surface of normalised viscoelastic moment

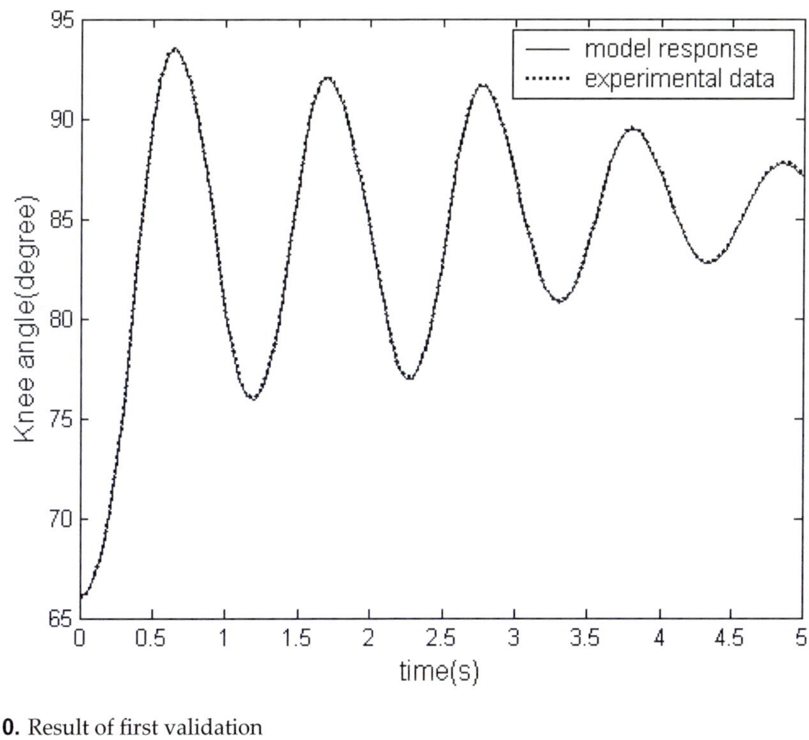

Figure 10. Result of first validation

Second, the optimised segmental masses are validated by repeating the same optimization process for further four times. The five sets of the optimised parameters emanating from five different runs of the GA routines are shown in Table 5. The results of the optimised anthropometric inertia parameters from different simulation runs exhibit acceptable repeatability with only a slight difference between each other with small standard deviation. Therefore, it can be concluded that the optimised masses obtained are valid.

Iteration	Foot Mass (kg)	Shank Mass (kg)	Moment of inertia (Nm^2)	Position of COM of Foot (m)	Position of COM of Shank (m)
1	0.95	3.5	0.35823	0.0350	0.220
2	0.948	3.51	0.35830	0.0352	0.221
3	0.951	3.49	0.35828	0.0355	0.220
4	0.949	3.51	0.35815	0.0351	0.222
5	0.952	3.49	0.35820	0.0345	0.221
Mean	**0.95**	**3.5**	**0.3582**	**0.0351**	**0.2208**
Standard deviation	**0.0016**	**0.01**	**0.00006**	**0.00036**	**0.00083**

Table 5. Five sets of the optimised parameters

4. Discussion

In this approach the passive properties have been divided into two parts; First, mathematical model is used to represent the combination of knee joint inertial (M_i) and gravitational (M_g) moments. Second, fuzzy model is applied to represent the combination of the elastic moment (M_s) and the viscous moment (M_d) as viscoelastic moment of the knee joint. Mathematical model has been used because of the availability of this model and accessibility to estimate anthropometric inertia parameters. This could be an easy method to estimate these parameters without go through complicated clinical experiment since these parameters vary for each subject. The system such as viscoelasticity could be difficult to model due to complexity and nonlinerity. Therefore, fuzzy model is used to eliminate the development of complex mathematical model and helps to simplify the modelling task. These passive properties model then need to be integrated with active properties of the knee joint model to have a complete model of the knee joint. Finally, these models can be utilized as platform for the simulation purpose of the control system development.

5. Conclusion

A new approach of estimation of the anthropometric inertia parameters and model of the passive viscoelasticity of the knee joint has been presented. Fuzzy model has been used to model the passive viscoelasticity and could be an effective tool for the modeling of uncertain nonlinear systems without represents using complicated mathematical model. The anthropometric inertia parameters of the lower limb such as foot mass, shank mass, moment of inertia about COM and positions of COM along the segmental length of the lower limb have been optimized between the given ranges to obtain the accurate value of the equations of motion. These optimizations have been performed simultaneously using GA with the objective to minimize the error between prediction model and the experimental data. Each person has a unique composition of muscle tissue in their body therefore this passive properties are not generally applicable.

Author details

B.S. K. K. Ibrahim, M.S. Huq, M.O. Tokhi and S.C. Gharooni
Department of Automatic Control and System Engineering, University of Sheffield, United Kingdom

6. References

B. Vatnsdal, P. E. Ingvarson, D. Rafolt, V.Gudmudsdottir, S.Knutsdottir,S. Yngvason and T.Helgason, *Monitoring of knee joint mechanics during electrical treatment of denervated degenerated muscle*, Proceeding of the 13th annual conference of the IFESS, 2008.

C. Fonseca, and P. Fleming, "Genetic algorithms for multiobjective optimization: formulation, discussion and generalization," Genetic Algoritms: Proceeding of the Fifth International Conference, San Mateo, CA, pp. 416-423, 1993.

D.A.Winter. *Biomechanics and motor control of human movement.* 2nd edition, Wiley-Interscience, New York, 1990.

D.C. Lin, W.Z. Rymer. *A quantitative analysis of pendular motion of the lower leg in spastic human subjects.* IEEE Trans Biomed Eng. 1991 Sep;38(9):906-18.

D. E. Goldberg, *Genetic algorithms in search, optimization and machine learning*. Reading, MA: Addison-Wesley, 1989.

F.E. Zajac, *Muscle and tendon properties: models, scaling, and application to biomechanics and motor control.* CRC Critical Reviews in Biomedical Engineering 17, pp. 359-411, 1989.

H. Hatze, "*A three-dimensional multivariate model of passive human torques adn articular boundaries,*" Clinical Biomechanics, vol. 12, pp. 128-135, 1997.

H.J. Chizeck, S. Chang, R.B. Stein, A. Scheiner, D.C. Ferencz, *Identification of electrically stimulated quadriceps muscles in paraplegic subjects,* IEEE Transactions Biomedical Engineering on Vol 46, Issue 1,pp:51 – 61, 1999.

H. M. Franken, P. H. Veltink, R. Tijsmans, H. Nijmeijer, and H. B. K.Boom, "*Identification of passive knee joint and shank dynamics in paraplegics using quadriceps stimulation,*" Transactions on Rehabilitation Engineering, vol. 1,pp. 154–164, 1993.

H. White, T.L. Uhl, S. Augsburger, C.Tylkowski. *Reliability of the three-dimensional pendulum test for able-bodied children and children diagnosed with cerebral palsy.* Gait Posture 2007;26:97–105.

J. E. Doane and P. M. Quesada, *Subject specific body segment parameter estimation using 2D work and energy principles*, Gait & Posture, Volume 24, Supplement 2, pp.123-125,2006.

J. M. Mansour and M. L. Audu, *The passive elastic moment at the knee and its influence on human gait*, Journal Biomechical., vol. 19, no. 5, pp. 369–373, 1986.

J., F. Lamb, C. G.Ingram, I. A. Johnston, R. M. Pitman, *Essentials of Physiology*, Blackwell Scientific Publications, UK, 1991.

K. Amankwah, R.J. Triolo, R.F. Kirsch. *The effects of spinal cord injury on lower limb passive joint moments revealed through a nonlinear viscoelastic model.* Journal of Rehabilitation Research and Development 41 (1), 15–32, 2004.

L. Reznik, *Fuzzy Controllers*, Newnes-Butterworth-Heinemann, Oxford, 1997.

M. Ferrarin, and A. Pedotti, *The relationship between electrical stimulus and joint torque: a dynamic model.* IEEE Transactions on Rehabilitation Engineering, 8 (3), pp 342-352, 2000.

M. Jamshidi, A.W. Smith. *Clinical measurement of spasticity using the pendulum test: comparison of electrogoniometric and videotape analyses.* Arch Phys Medical Rehabilitation 1998;77:1129–32.

M. S. Valle, A. Casabona, R. Sgarlata, R. Garozzo, *The pendulum test as a tool to evaluate passive knee stiffness and viscosity of patients with rheumatoid arthritis.* BMC Musculoskeletal Disorder, 2006.

M.G. Pandy, F.E. Zajac, E.Sim, W.S. Levine, *An optimal control model for maximum-height human jumping*, Journal of Biomechanics 23, pp.1185-1198, 1990.

P.H. Veltink et al. *Modeling the optimal control of cyclical leg movements induced by functional electrical stimulation*, Int. J. Art. Org., vol. 15, pp. 746-755, 1992.

P. L. Cavorzin, X Hernot, O Bartier, G Carrault, P Rochcongar, F Chagneau. *A computed model of the pendulum test of the leg for routine assessment of spasticity in man*. ITBM-RBM 2001;22:170–7.

R. Massoud. *Intelligent control techniques for spring assisted FES-cycling.* PhD Thesis. The University of Sheffield, Sheffield, UK, 2007.

R. Riener and T. Edrich, "*Significance of passive elastic joint moments in FES*," Proceedings of the 2nd Conference of IFESS, pp. 103-106, 1997.

T. Bajd, L.Vodovnik , *Pendulum testing of spasticity.* Journal of Biomedical Engineering 1984;6:9–16.

V. Zatsiorsky and V. Seluyanov, "*The mass and inertia characteristics of the human body*", Biomechanics VIII-B, University Park Press, pp. 1152-1159, 1983.

W. T. Dempster. *Space requirements of the seated operator.* WADC Technical Report (TR-59-159), Wright-Patterson Air Force Base, OH, 1955.

Wartenberg, R. (1951). *Pendulousness of the leg as a diagnostic test.* Neurology, 1, 18-24.

Chapter 16

Non Linear Viscoelastic Model Applied on Compressed Plastic Films for Light-Weight Embankment

Hayssam El Ghoche

Additional information is available at the end of the chapter

http://dx.doi.org/10.5772/50290

1. Introduction

This present study aims to identify the behavior of a new cellular material, including a biphasic structure (air + solid plastic films) which consists of a flexible plastic film. The finished product is in the form of compressed and attached cubic blocks whose volume is about $1m^3$ and whose density is ranging from 0.3 to 0.6.

This material has diverse applications in Civil Engineering and Public Works. It can be used as a light weight embankment on poor mechanical characteristics soils, to minimize the loads on buried pipelines (Marston effect) and to make soundproof wall or dams to protect from falling stones.

This research was initiated by a thorough understanding of the physical reality of the material. Then, the behavior of this material has been identified by targeted experiments, and finally a theoretical model is developed.

The first form of this model is nonlinear which was modified to be able to take into account the hardening of material while loading. Finally, this model was linearized to be applied in special cases in a field of stresses which can reach 200 kPa and covers the main domain of real applications.

2. Manufacturing mode

The raw materials used in the manufacture of flexible films are mainly made out of polyethylene.

INTECH
open science | open minds

© 2012 The Author(s). Licensee InTech. This chapter is distributed under the terms of the Creative Commons Attribution License http://creativecommons.org/licenses/by/3.0), which permits unrestricted use, distribution, and reproduction in any medium, provided the original work is properly cited.

As the volume and the density are fixed, we put the necessary quantity of plastic in a parallelepiped shape press. After being compressed under high pressure (about 7 bar), and legated the resulting block stayed in shape after it is released. [1].(Fig.1)

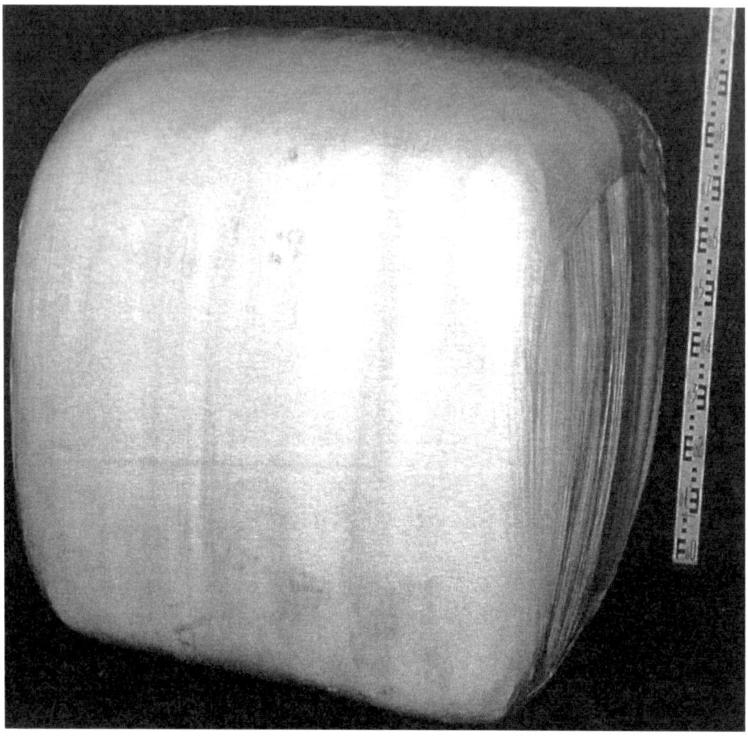

Figure 1. Photo of a block

3. Physical reality of material

After the manufacturing process which is at unidirectional pressing, the material is in the form of stacked almost identical layers. A certain volume of air is trapped inside the plastics sheets and forms more or less closed cells. This action gives lightness and makes the material alveolar.

In reality, the material is anisotropic and discontinuous, but macroscopically, it may be considered an orthotropic, continuous material with a revolution case whose axis coincides with the direction of pressing during manufacturing. (fig.2). In this case, we will be able to apply the principles of the Continum Mechanics

On the other hand, the internal structure of the material is formed by a network of cells which are filled with air and have irregular geometric shapes. The flexible Walls of cells are not airtight.

The global deformation of loaded material is the superposition of the proper deformation of the plastic sheets and the deformation due to the changing of the volume of air including into the cells, by compressibility or and by fleeing.

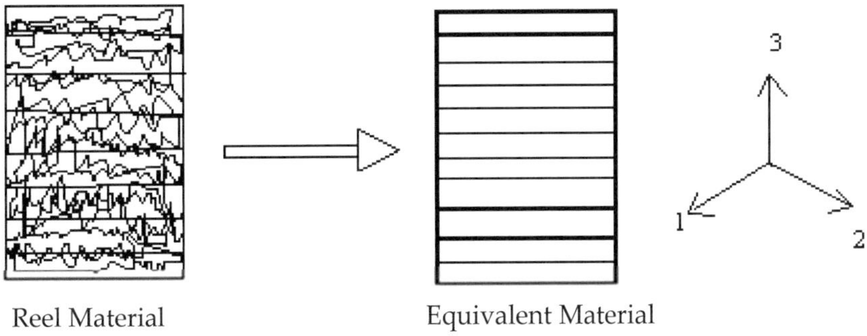

Figure 2. The principle of the real material and the equivalent one.

4. Experimental studies

The experimental studies of the behavior of the material began by compression and creep tests on full size blocks and completed by several kinds of tests on reduced size samples.

In the purpose to obtain a reduced size sample, tow apparatus were conceived and built. The first one enabled us to make cylindrical samples and the second one is conceived to produce cubic samples.

The dimensions of real blocks or samples are carefully chosen in order to neutralize the microscopic details and to be able to apply the Continum Mechanics principles.

In addition, two experimental machines are designed and constructed specifically for this type of material.

A first apparatus is intended to perform simple compression tests (Fig. 3). The load applied by a bar, was transmitted by a rigid plate that moves while remaining horizontal. This system that prevents the differential deformation is in the form of three rod lubricated to facilitate sliding.[6].

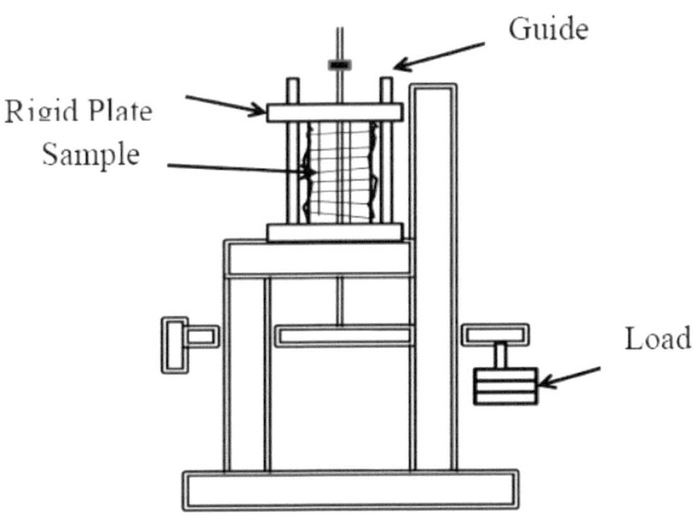

Figure 3. Apparatus for simple compression test

Secondly, a biaxial apparatus is developed (Fig. 4). In a direction, the applied load using the weight is transmitted through a rigid plate. In the other direction, the stress is measured by a dynamometer attached to another plate of the same nature. Its movement has approximately the same deformation of the ring. Other strains are measured by conventional means (comparator). The guidance system consists of four lubricated rigid rods.

The choice of biaxial testing was justified by the inability to perform simple compression tests, at the same time in the directions (1) and (2). In fact, the material is in the form of a stack of sheets clamped by metal wires. The stress in the direction, parallel to the layers causes an increase in tension in the wires and the behavior of material will be managed by their rigidity and their number. From a mechanical point of view, the sample will be charged on certain places on time (in contact with the wire) and the deformation will not be uniform on both open sides [6].

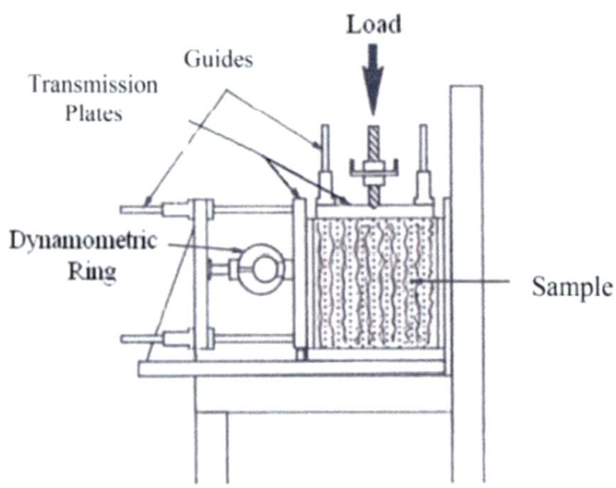

Figure 4. Biaxial apparatus

4.1. Testing

After testing and verifying the proper functioning of both apparatus, wide open test campaigns are made.

The carried out tests are :

- Compression and creep tests corresponding to the revolution axis of material.
- Cyclical and monodirectionnal tests with constante deformation velocity.
- Biaxial tests.

For all kind of test, real blocks and reduced samples of the same or different densities are tested according to a path of applied stresses step by step. This approach helps us to understand more the behavior of this discontinuous and anisotropic material.

4.2. Simple compression test (creep)

Simple compression tests along the axis of revolution of the material are carried out on real size blocks as on samples of small size with different densities (0.4, 0.5 or 0.6). The stress is applied through stages. Each strain is maintained long enough to highlight the phenomenon of creep (Fig. 5, 6, 7) [2].

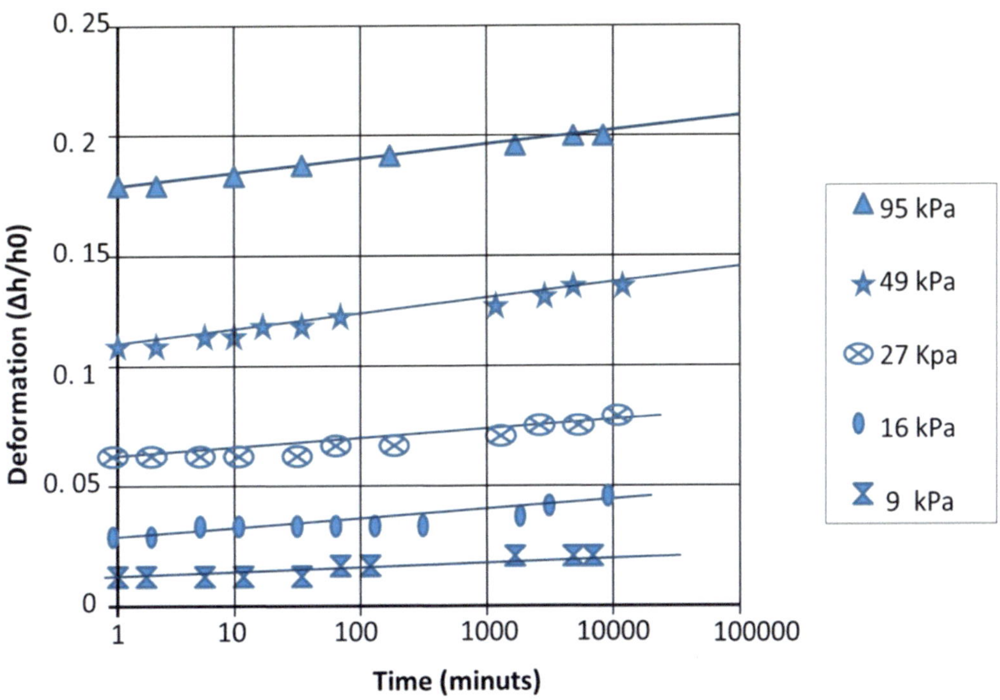

Figure 5. Deformation- time curve in logarithmic scale, under constant stress, initial density 0.4, loading by steps. Real blocks size.

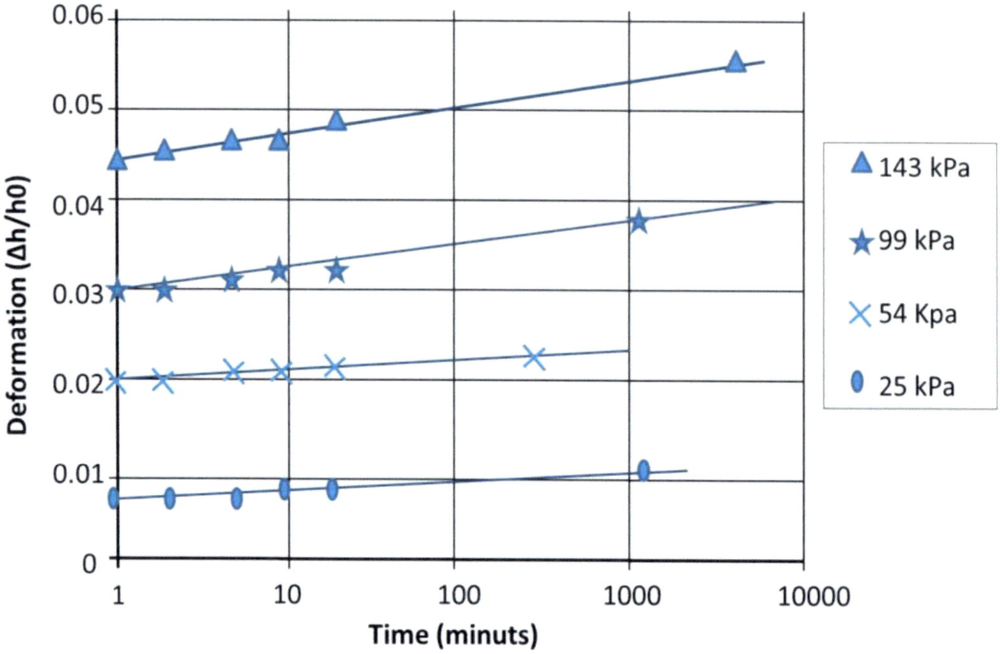

Figure 6. Deformation- time curve in logarithmic scale, under constant stress, initial density 0.5, loading by steps. Reduced size sample.

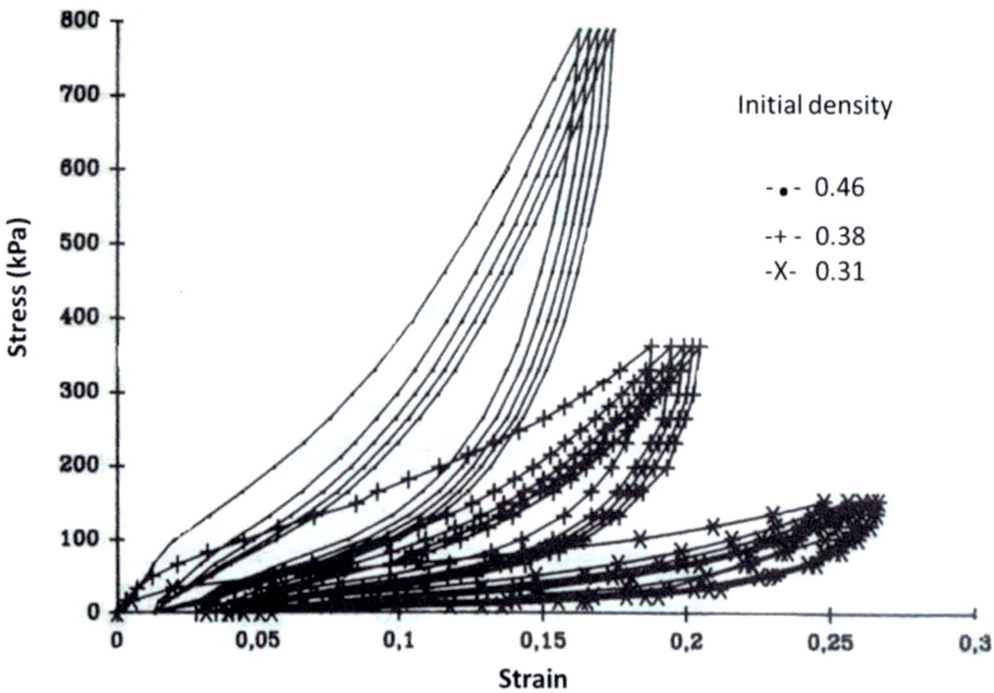

Figure 7. Stress – strain curves under constant deformation velocity (0.0003/s)

The examination and analysis of the experimental results enabled us to identify the behavior of the material and to find the necessary characteristics and parameters in order to begin the theoretical modeling:

The material behavior is a viscoelastic and nonlinear one. The deformation variation according to compression and creep tests under constant stress can be considered linear according to time in logarithmic scale (Fig. 6, 7):

$$\varepsilon_3 = \varepsilon_{0,3} + A_3 log(\frac{t}{t_0})$$

ε_3 : $unidirectional\ deformation$, according to direction 3

$\varepsilon_{0,3}$: $instantaneous\ deformation$, according to direction 3

A_3: $slope\ of\ the\ deformation\ straight\ line\ according\ to\ time\ logarithm$

t_o : $time\ unit$

In this scale, the slope and the ordinate at origin (instantaneous deformation) are variable according to the initial density and applied stress (fig. 8, 9).

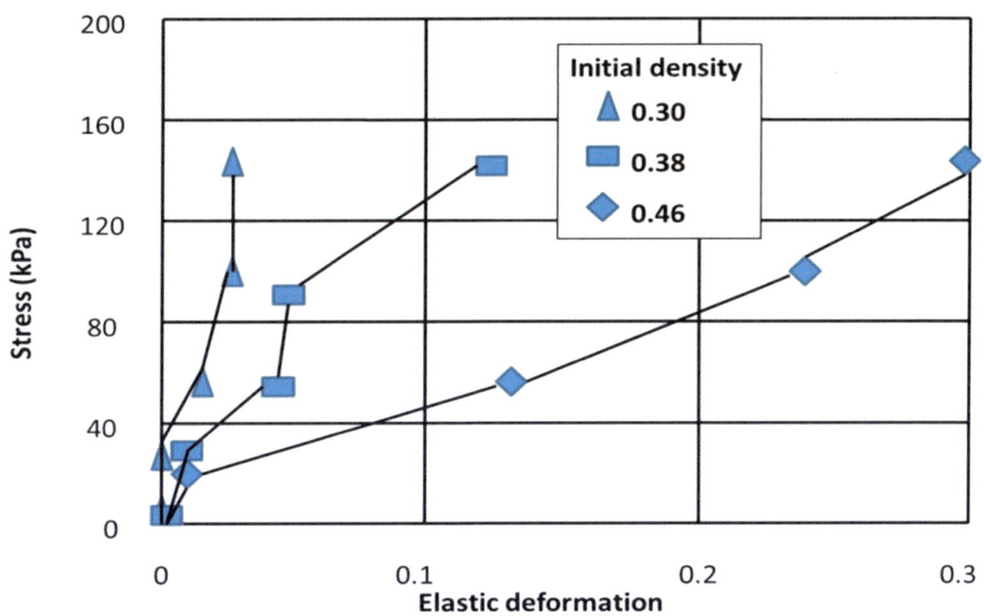

Figure 8. Stress – elastic strain curves for different initial densities

The behavior depends too much on the initial density (fig.7).

Many complex rheological phenomena are observed (hardening, aging, recovery....) (fig.7).

The proper plastic deformation can be neglected in comparison with the total one (fig.10). This result was confirmed by a test realized on solid block of plastic (polyethylene whose density is 0.94). Then, the deformation is the direct result of air volume variation by compressibility or expulsion or by both.

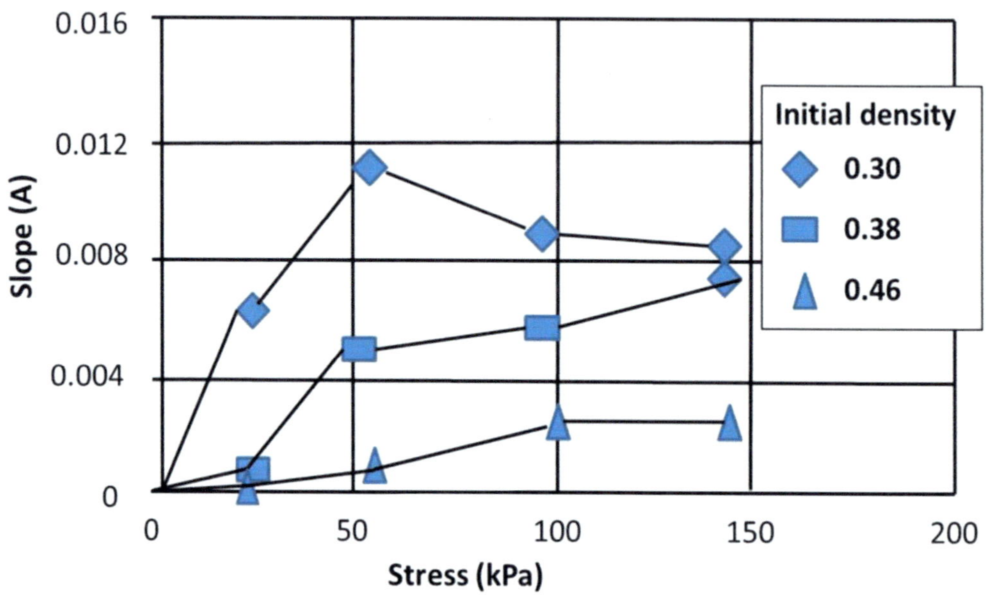

Figure 9. Slope (A) - Stress curves for different initial densities

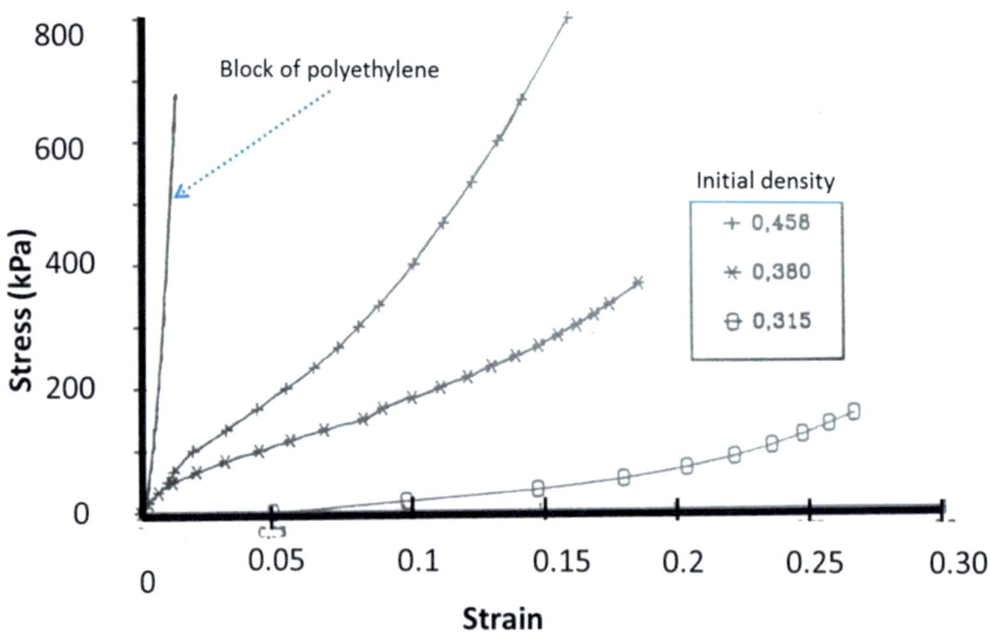

Figure 10. Stress – strain curves with constant velocity deformation for the material) in comparison with the curve of solid block of polyethylene.

4.3. Biaxial tests

This test campaign is initiated by long test duration. Then, eighteen samples of four different initial densities are tested. The same stress path of the unidirectional tests is adopted. The residual stress is measured at the time of manufacture. The variation of the stress perpendicular to the layers according to the stress in the parallel direction and the time is not significant (Fig. 11, 12).

The variation of the deformation parallel to the layers can be considered as linear with the logarithm of time (Fig.13):

$$\varepsilon_2 = \varepsilon_{0,2} + A_2 log(\frac{t}{t_0})$$

ε_2 : *unidirectional deformation*, according to direction 2

$\varepsilon_{0,2}$: *instantaneous deformation*, according to direction 2

A_2: *slope of the deformation straight line according to time logarithm*

t_o : *time unit*

In this scale, the slope A_2 and the ordinate at origin $\varepsilon_{0,2}$ (instantaneous deformation) are variable according to the initial density and applied stress (fig. 14, 15,16,17).

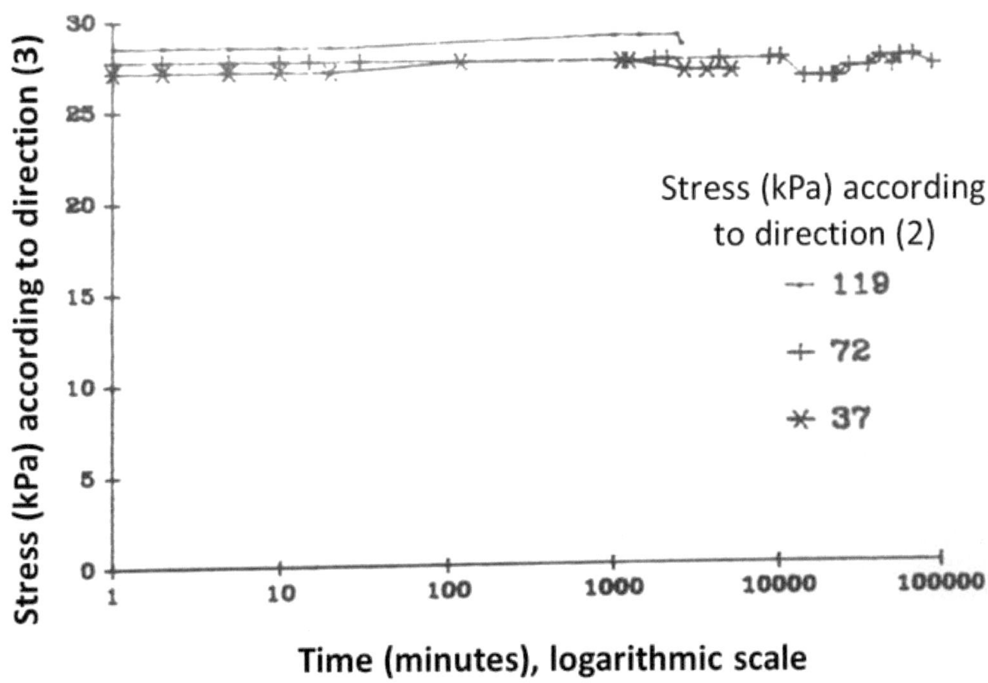

Figure 11. Stress (σ3) – time curve in logarithmic scale for initial density 0.41

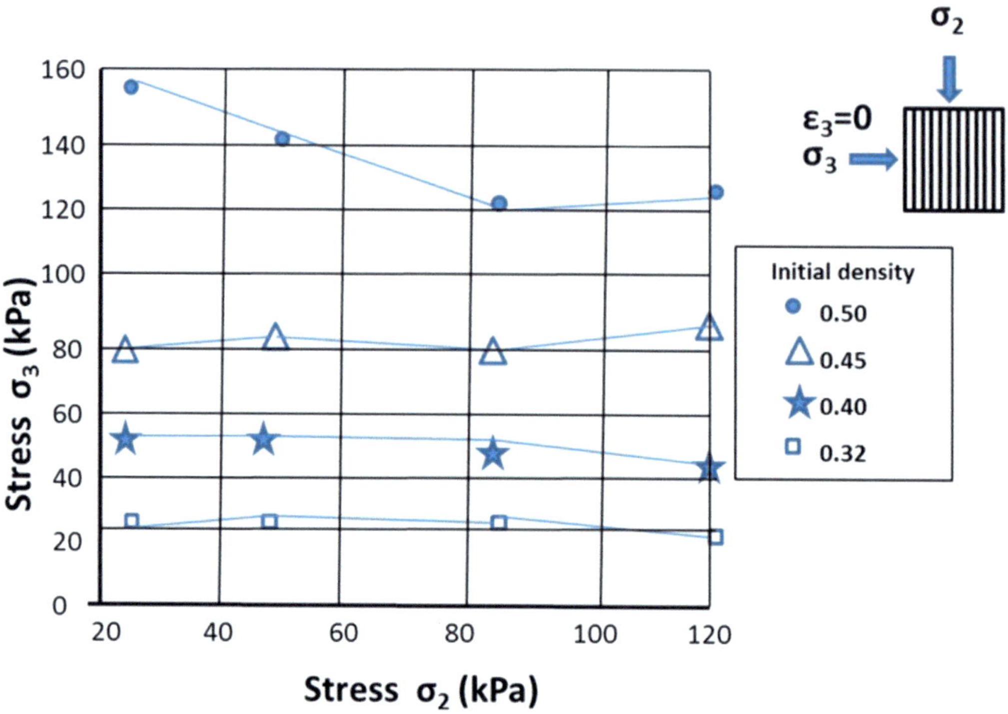

Figure 12. Stress (σ_3) – Stress (σ_2) for different densities

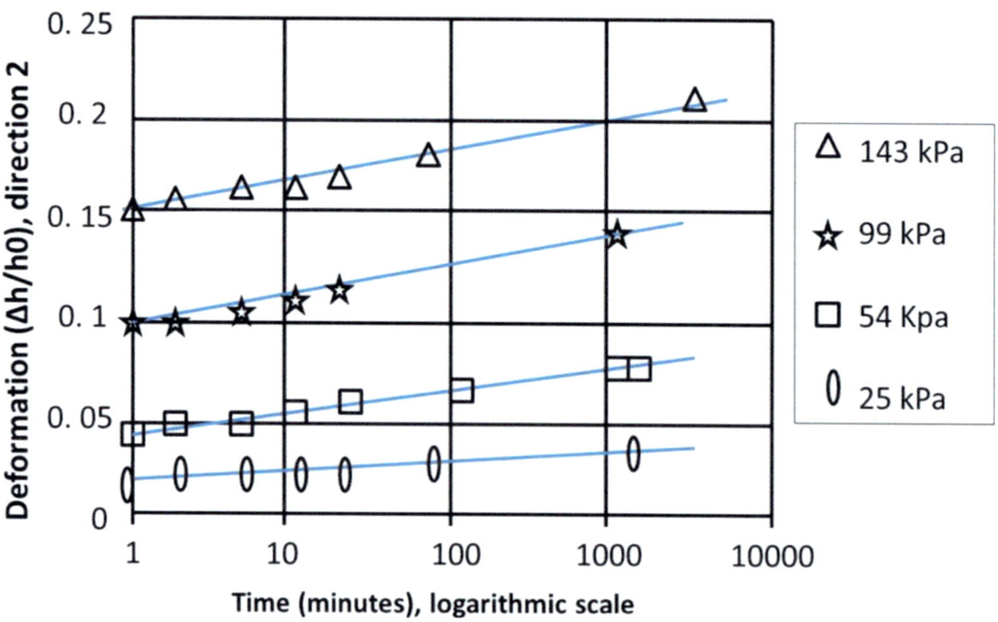

Figure 13. Deformation- time curve in logarithmic scale, under constant stress according to direction 2, initial density 0.32

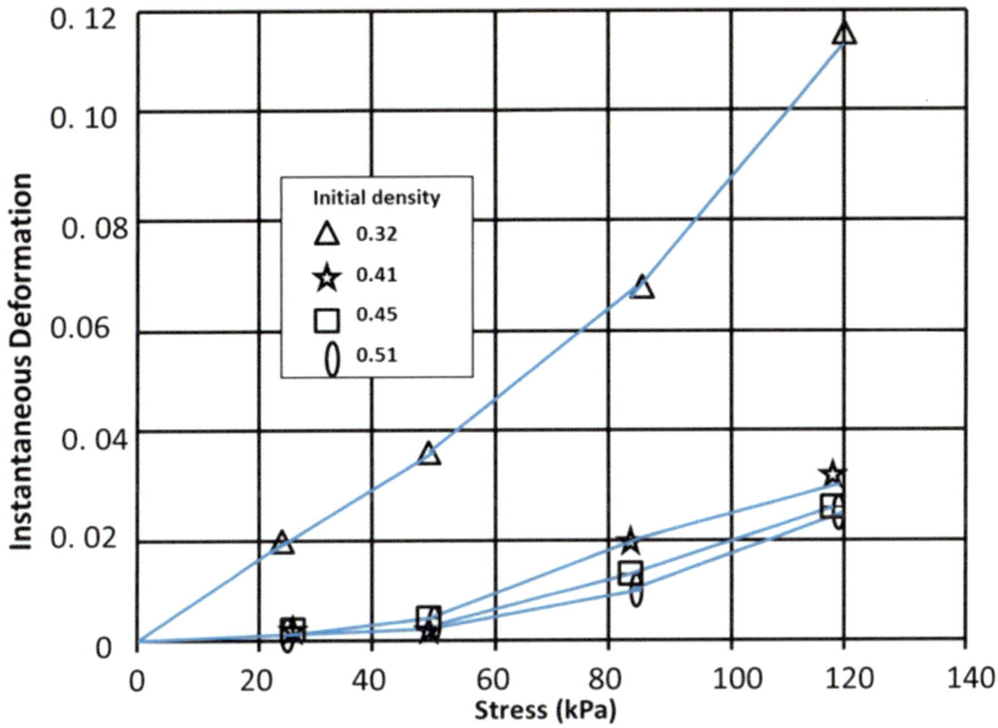

Figure 14. Instantaneous deformation (direction 2) – stress curves at constant initial density

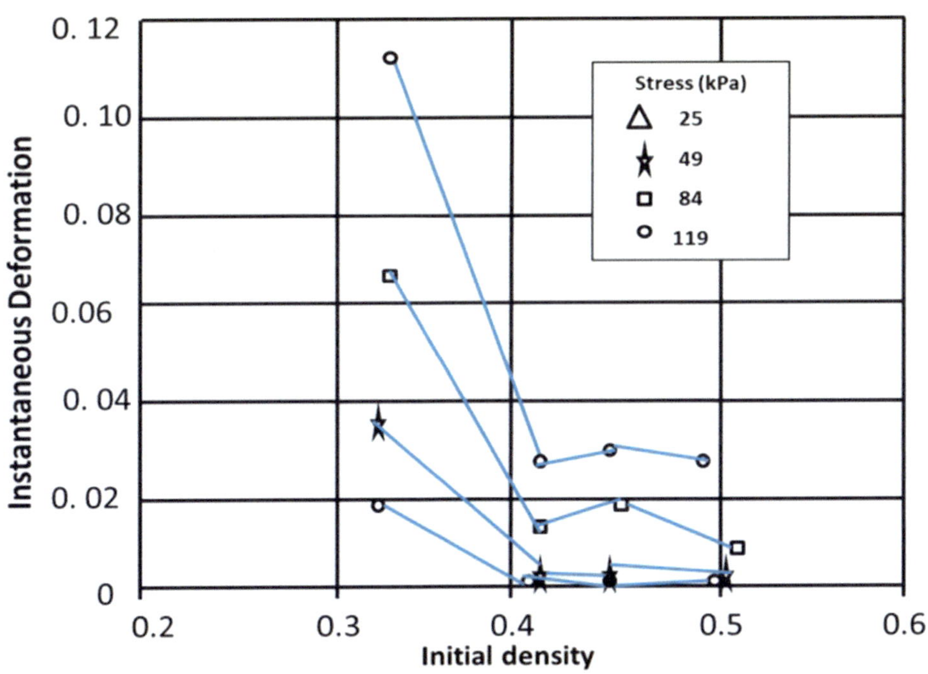

Figure 15. Instantaneous deformation (direction 2) – initial density curves at constant stress

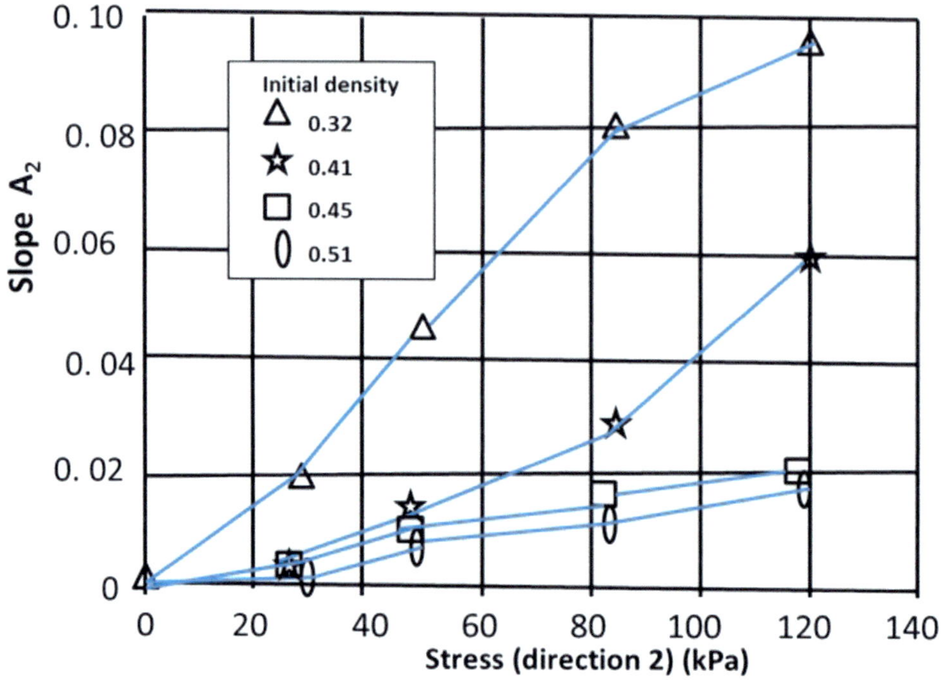

Figure 16. Slope A2 (direction 2) – stress curves at constant initial density

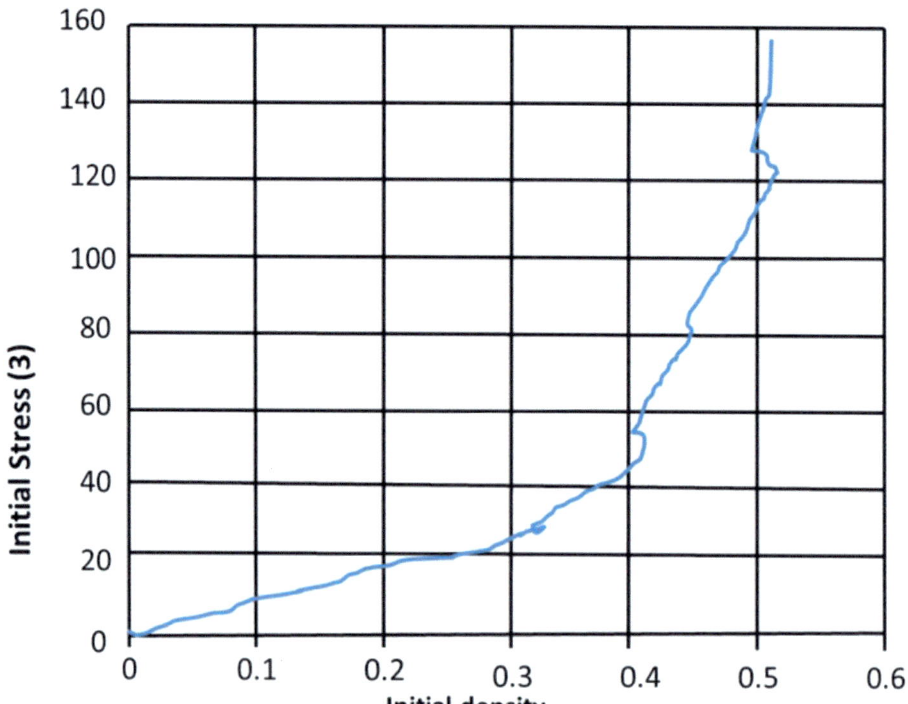

Figure 17. Initial residual stress (3) - initial density curve

5. Rheological Behavior and Modeling in the case of mono dimensional loading according to direction (3), perpendicular to the layers

Test campaigns on real blocks and small size samples [2] are able to show that the viscoelastic behavior is nonlinear (Fig. 5). Moreover, it is the subject of several rheological complex phenomena (hardening, aging, accommodation...) (Fig. 7) [3].

Initial density is an important parameter which influences behavior. In fact, the more great it is the more the behavior improves. The same density changes during the solicitation resulting hardening.

At first glance, the complex reality of the material makes it difficult to apply principles of mechanics of continuous materials. This is made possible by the adoption of some simplifying assumptions.

The assumptions are:

First, we assume that there is no slippage or detachment between the elementary leaves and the volume used is large enough to be able to erase the influence of microscopic details. Then, it is considered that the material is continuous, homogeneous and orthotropic revolution.

Finally, the aging process can be ignored.

On the other hand, specific plastic deformation is negligible compared to the overall deformation of material due to the change in volume of air trapped into the cells, or by compressibility and evacuation (fig.10).

Because the material is assimilated to the alveolar one, the behavior is represented by the following model (fig. 18) [8]:

The material is the subject of superposition of tow behaviors:

- Compression of the gas into closed cells
- Deformation of the walls of cells

In the case of alveolar material with open cells, the behavior of gas does not take place.

When the stress is applied quickly (case of variable stress or imposed deformation velocity), the material can be considered as alveolar with closed cells. But, when it is maintained constant for a long time (case of creep), the material can be considered as alveolar with open cells.

In conclusion, the alveolar material is characterized by an elastic limit σ_e.

For the material under study, the elastic limit was identified by simple compression tests with constant deformation velocity (fig. 19).

The variation of elastic limit (σ_e) of this material was identified experimentally and can be assimilated by a straight line whose equation is:

$$\sigma e = Pa\left(4.49\ d0 + 1.06\right)$$

Pa : atmospheric pressure

d0 : initial density

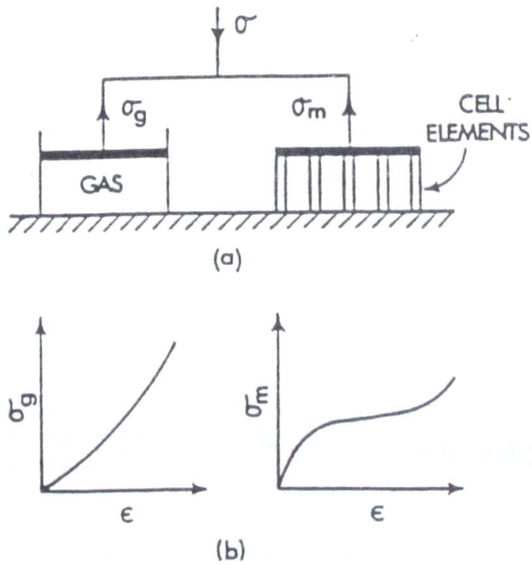

Figure 18. Rheological model of alveolar material with closed cells

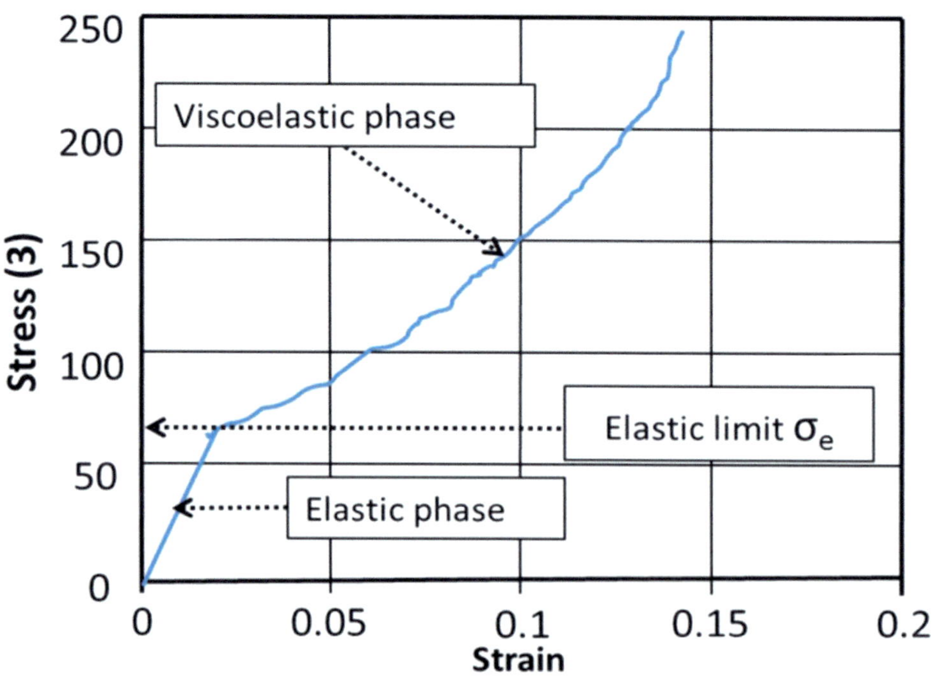

Figure 19. Stress-strain curve of material in case of closed cells

5.1. Nonlinear model with no hardening

Based on a simple analogical model (spring + dash-pot + skate), the behavior may be in the form:

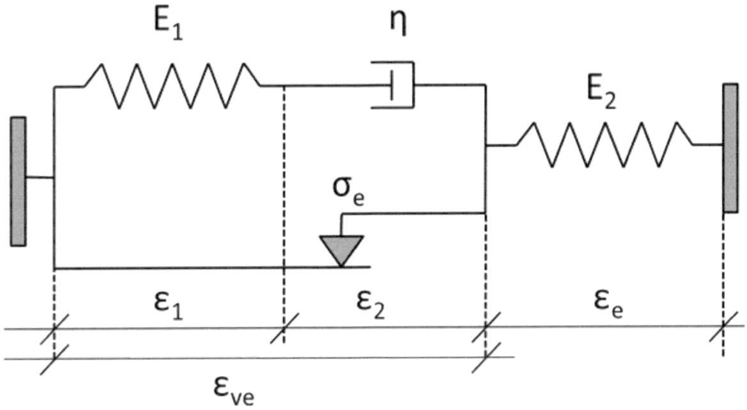

$$\varepsilon = \varepsilon_1 + \varepsilon_2 + \varepsilon_e = \varepsilon_{ve} + \varepsilon_e$$

The skate limits the elastic behavior on maximum value (σ_e). Above this value the viscoelastic phase is activated and the behavior is described by Maxwell analogical model characterized by the Young modulus E1 and the viscosity η[2].

The nonlinear behavior is taken into account by the nonlinear form of E_1 and η [2].

Then

$$\varepsilon = \varepsilon_1 + Alog(\frac{t}{t_0})$$

$$\varepsilon = \varepsilon_1 + \varepsilon_2$$

$$\sigma = E_1\varepsilon_1 = \eta\dot{\varepsilon_2}$$

$$\varepsilon^{\bullet} = \frac{d\varepsilon}{dt} \quad \& \quad \dot{\sigma} = \frac{d\sigma}{dt}$$

All analysis is done, the model takes the form of a differential equation of first degree:

$$\varepsilon^{\bullet} = \left(\frac{d_p - d_0}{d_0^n}\right)\left(\frac{1}{P}\right)\sigma^{\bullet} + A\exp\left[\frac{-(\varepsilon - \varepsilon_0)}{A}\right]$$

With

$$\varepsilon_0 = \left(\frac{d_p - d_0}{d_0^n}\right)\left(\frac{1}{P}\right)\sigma$$

$$A = \left(\frac{d_p - d_0}{d_0^m}\right)\left(\frac{1}{P}\right)\sigma$$

$n = a \times d_0 + b$

$m = a^0 \times d_0 + b^0$

a, b, a^0, b^0 : parameters to be identify by experience

P : atmpspheric pressure

d_P : proper density of plastic

d_0 : initial density

5.2. Nonlinear model with hardening

Until now, the initial density is taken as constant. To be able to take into account the hardening phenomenon, we will introduce the density as a function according to the strain. Necessary analysis and calculation are done. The new version of the model is presented a differential equation as the following [1]:

$$\varepsilon^{\bullet} = \left(\frac{d_p - \omega}{\omega^{a\omega + b}}\right)\left(\frac{1}{P}\right)\sigma^{\bullet} + \left(\frac{d_p - \omega}{\omega^{a_1\omega + b_1}}\right)\exp\left[\frac{-(\varepsilon - \varepsilon_0)}{A}\right] \qquad \varepsilon_0 = \left(\frac{d_p - \omega}{\omega^{a\omega + b}}\right)\left(\frac{1}{P}\right)\sigma$$

$$A = \left(\frac{d_p - \omega}{\omega^{a_1\omega + b_1}}\right)\left(\frac{1}{P}\right)\sigma \quad ; \quad \omega = \frac{d_0}{1 - \varepsilon}$$

dp : proper density of plastic

d_0 : initial density

5.3. Linearized model with no hardening

The simplification of the nonlinear model is carried out in two ways [7]:

Adoption of linear forms of the elastic and the delayed deformations according to the stress.

Adoption of linear forms of the elastic and a constant delayed one according to the stress. In another word, the linear viscoelasticity is applied.

In any case, the initial density is taken as a parameter and the model is presented as the following [5]:

$$\varepsilon^{\bullet} = \left(\frac{d_p - d_0}{d_p^n}\right)\left(\frac{1}{P}\right)\sigma^{\bullet} + A\exp\left[\frac{-(\varepsilon - \varepsilon_0)}{A}\right]$$

$$\varepsilon_0 = \left(\frac{d_p - d_0}{d_p^n}\right)\left(\frac{1}{P}\right)\sigma$$

$\varepsilon 0$: instantaneous deformation

$\varepsilon^{\bullet}$: unidirectional deformation velocity

A, n, m : parameters to be identified

d0 : initial density

The experimental results enabled us to identify all necessary parameters.

5.4. Validation

The comparison of theoretical and experimental results becomes possible by numerical resolution of the differential equation in its three forms. The concordance between reality and theory is good and the simplification gives us reasons of satisfaction.(fig.20, fig.21, fig.22, fig.23) [1],[3],[7].

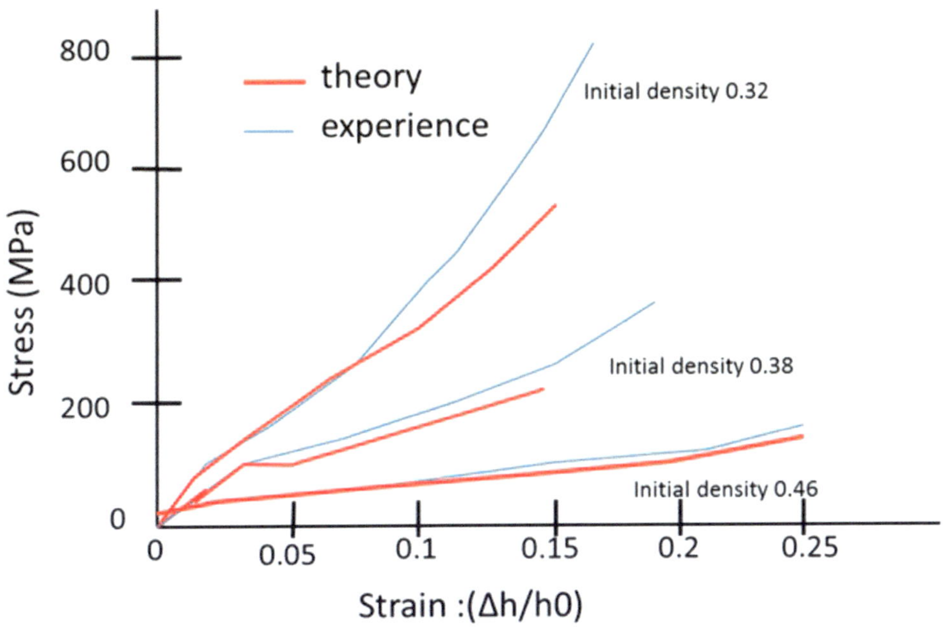

Figure 20. Non linear model, compression with constant deformation velocity

6. Three-dimensional model, constant stress

The hypothesis of non-aging behavior is adopted. The relation "stress-strain" in case of constant stress and small strain is represented by:

$$\vec{\sigma} = F \otimes \vec{\varepsilon}$$

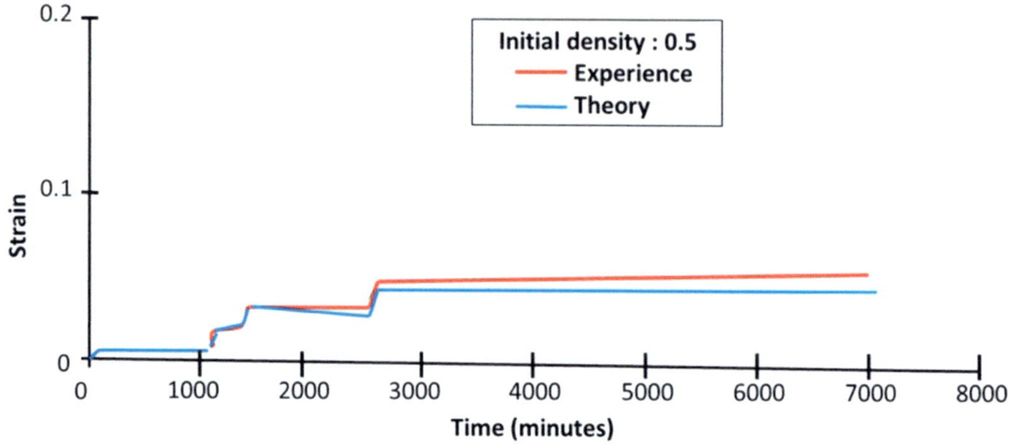

Figure 21. Non linear model, creep tests by stages

Figure 22. Non linear hardening model, settlement of embankment.

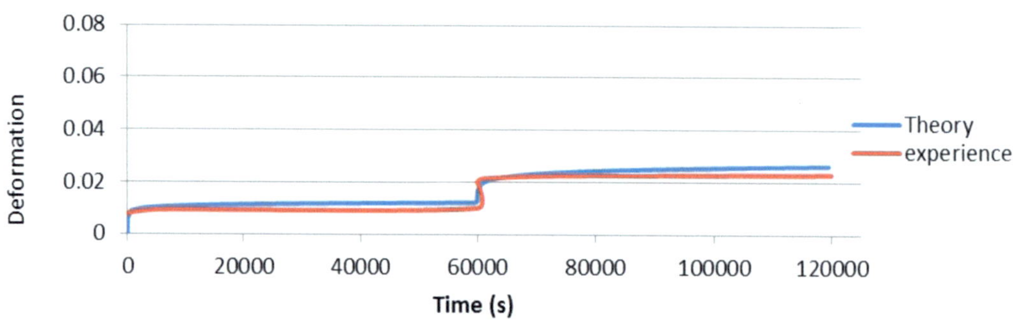

Figure 23. Linearized model. Measured and calculated deformations for an initial density: 0.493, creep tests by stages.

F is a matrix composed by f_1, f_{12}, f_3, f_{13}, f_4, five functions to be identified. They are a function of time, stress and initial density. The role of the stress perpendicular to the layers (σ_3) is paramount. It marks the nonlinearity of the behavior. For the other stresses, the linearization of the behavior is adopted with a quite acceptable accuracy [6].

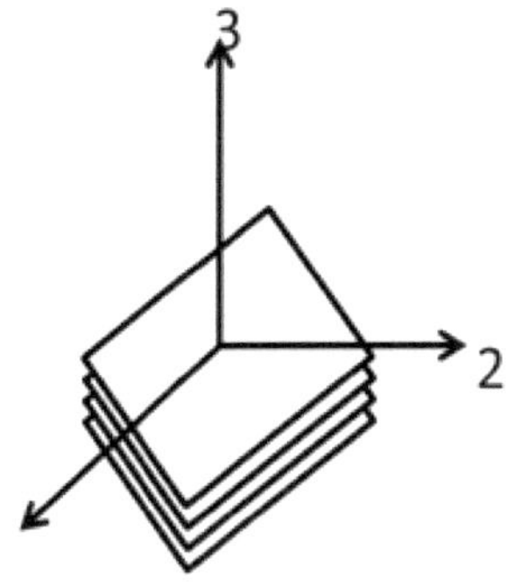

$$\begin{bmatrix} \varepsilon_{11} \\ \varepsilon_{22} \\ \varepsilon_{33} \\ \varepsilon_{12} \\ \varepsilon_{23} \\ \varepsilon_{13} \end{bmatrix} = \begin{bmatrix} f_1 & f_{12} & f_{13} & 0 & 0 & 0 \\ f_{12} & f_1 & f_{13} & 0 & 0 & 0 \\ f_{13} & f_{13} & f_3 & 0 & 0 & 0 \\ 0 & 0 & 0 & f_1 - f_{12} & 0 & 0 \\ 0 & 0 & 0 & 0 & f_4 & 0 \\ 0 & 0 & 0 & 0 & 0 & f_4 \end{bmatrix} \begin{bmatrix} \sigma_{11} \\ \sigma_{22} \\ \sigma_{33} \\ \sigma_{12} \\ \sigma_{23} \\ \sigma_{13} \end{bmatrix}$$

The identification of the functions is carried out from experimental results:

f_1 and f_{13} identified by performing tests of simple compression along the axis of revolution.

f_2 and f_{12} are determined by simple compression tests according to the two other axes or by biaxial testing.

f_4 is equal to $\frac{1}{G13}$ where G_{13} is the shear modulus around the directions (1) and (2). It is identifiable by a distortion test. Without this test, it can be approached by assuming that the material is isotropic respectively with the characteristics specified in the directions (1) and (3)....

After necessary calculation, the characteristic functions take the following forms [6]:

$$f_3 = \frac{1 - \dfrac{d_0\left(\dfrac{\sigma_{33}}{Pa} + B\right)}{d_p\left(\dfrac{\sigma_{33}}{Pa} + B\right) - A} + \dfrac{d_p - d_0}{d_0}\left[\left(\dfrac{\sigma_{33}}{Pa} + 1\right)^m - 1\right]\left[\ln\left(\dfrac{t}{t_0}\right)\right]}{\dfrac{\sigma_{33}}{Pa}}$$

$$f_{13} = \alpha_3 f_3$$

$$f_1 = \left(d_p - d_0\right)\left[\frac{1}{d_0^k} + \frac{1}{d_0^l}\ln\left(\frac{t}{t_0}\right)\right] +$$

$$\alpha_3^2 f_3 .\ f_{12} = \alpha_1\left(d_p - d_0\right)\left[\frac{1}{d_0^k} + \frac{1}{d_0^l}\ln\left(\frac{t}{t_0}\right)\right] +$$

$$\alpha_3^2 f_3 . f_4 = \frac{f_3\left(1-\alpha_3\right) + K\left(1-\alpha_3\right)}{2} ;$$

$$K = \left(d_p - d_0\right)\left[\frac{1}{d_0^k} + \frac{1}{d_0^l}\ln\left(\frac{t}{t_0}\right)\right]$$

Given the approximation of f_4, we can neglect α_1 and α_3 by comparison to (1) and f_4 becomes:

$$f_4 = \frac{\left(f_3 + K\right)}{2},$$

t: time,
t_0: initial time,
d_p: density of plastic
d_0: initial density
A, B, k, l: parameters depending on the initial density

Figure 24. One of realized projects in France

7. Conclusion

The development of mono axial and biaxial testing, the design and implementation of an appropriate apparatus helped us to perform many tests. Therefore, the characteristic functions of the material were determined satisfactorily. This allowed us to identify the essential parameters to the theoretical models developed.

In a particular field of stress up to 200 kPa, the simplified model gives good satisfaction. The results are acceptable and the difference between theoretical and experimental ones does not overstep the bounds of uncertainty both in physical measurement and caracteristics variability due to the material anisotropy.

However, in several cases of application, the model in its many forms helps us to conceive and design structures by using this technique. Indeed, many projects were conceived and realized in France (fig.24).

Author details

Hayssam El Ghoche
Lebanese University, University Institute of Technology (Saïda, Lebanon), Civil Engineering Department, Liban

8. References

[1] El Ghoche H., Boudissa M., Coulet C. - " Validation d'un Modèle Viscoélastique Appliqué à des Chantiers de Remblais Allégés En Blocs de Matières Plastiques Compressibles." ANNALES de l'Institut Technique du Bâtiment et des Travaux Publics. Revue Technique et Scientifique. Septembre 1995.
[2] El Ghoche H., Coulet C., Gielly J. - "Modèle Viscoélastique Non Linéaire Ecrouissable pour des Blocs Compressibles de Matières Plastiques. Huitièmes Rencontres du Centre Jacques Cartier : Elasticité, Viscoélasticité et Contrôle Optimal. Aspects Théoriques et Numériques. Université Claude Bernard. Lyon, France - 6 - 8 Décembre 1995
[3] El Ghoche H..- " Modèle Viscoélastique Non Linéaire Ecrouissable Pour des Blocs Compressible De Matières Plastiques". Première Conférence Syrio-Libanaise de l'Ingénierie (CSLI-1). 13-14 octobre 1999. Université de Damas. Damas, Syrie.
[4] El Ghoche H. - "Modélisation d'un matériau Alvéolaire. Application en Génie Civil." , Troisième Colloque Franco-Libanais Sur Les Sciences des Matériaux (CSM3), 16 - 18 Mai 2002 – Beyrouth Liban.
[5] El Ghoche H, J. Alhajjar, Kh. Mochaorab. - "Géotechnique et Environnement : Substitution des Sols". Deuxième Partie Conseil National de la Recherche Scientifique-Liban, Juin 2004. 166 pages.
[6] El Ghoche H - "Conception and Construction of Biaxial Apparatus for Identification of axial Orthotropic Material". Advanced Materials Research Vol. 324 (2011), pp 368-371.

[7] El Ghoche H - "Linearization of a viscoelastic model for light-weight Embankment ". Advanced Materials Research Vol. 324 (2011), pp 364-367.
[8] Hilyard N.C. - "Mechanics of cellular plastics". Mac Millan ed. 1982, p. 401

Made in the USA
Columbia, SC
04 October 2024

4db1c9ae-4224-441c-8869-0e37a21a3bd7R02